Aquaculture: Production and Engineering

Aquaculture: Production and Engineering

Edited by Roger Creed

SYRAWOOD
PUBLISHING HOUSE

New York

Published by Syrawood Publishing House,
750 Third Avenue, 9th Floor,
New York, NY 10017, USA
www.syrawoodpublishinghouse.com

Aquaculture: Production and Engineering
Edited by Roger Creed

© 2018 Syrawood Publishing House

International Standard Book Number: 978-1-68286-509-5 (Hardback)

Cataloging-in-Publication Data

Aquaculture : production and engineering / edited by Roger Creed.
    p. cm.
Includes bibliographical references and index.
ISBN 978-1-68286-509-5
1. Aquaculture. 2. Agricultural engineering. I. Creed, Roger.
SH135 .A68 2018
639.8--dc23

TABLE OF CONTENTS

PREFACE

Aquaculture studies methods of conserving fishery resources. It promotes the cultivation of freshwater population along with saltwater population. The most popular forms of aquaculture include mariculture, algaculture, shrimp farming, fish farming and ornamental fish cultivation, etc. From theories to researches to practical applications, case studies related to all contemporary topics of relevance to this field have been included herein. This book includes contributions of experts and scientists, which will provide innovative insights into this field. Coherent flow of topics, student-friendly language and extensive use of examples make this book an invaluable source of knowledge.

This book is a comprehensive compilation of works of different researchers from varied parts of the world. It includes valuable experiences of the researchers with the sole objective of providing the readers (learners) with a proper knowledge of the concerned field. This book will be beneficial in evoking inspiration and enhancing the knowledge of the interested readers.

In the end, I would like to extend my heartiest thanks to the authors who worked with great determination on their chapters. I also appreciate the publisher's support in the course of the book. I would also like to deeply acknowledge my family who stood by me as a source of inspiration during the project.

Editor

Effect of Soybean Varieties and Nitrogen Fertilizer Rates on Yield, Yield Components and Productivity of Associated Crops Under Maize/Soybean Intercropping at Mechara, Eastern Ethiopia

Wondimu Bekele[1, *], Ketema Belete[2], Tamado Tana[2]

[1]Oromia Agricultural Research Institute, Mechara Agricultural Research Center, West Hararghe Zone, Mechara, Ethiopia
[2]College of Agriculture and Environmental Science, Department of Plant Science, Haramaya University, Dire Dawa, Ethiopia

Email address:
wondubekele@gmail.com (W. Bekele)

Abstract: Due to decreasing land units and decline in soil fertility integrating soybean in to the maize production system is a viable option for increasing productivity and protein source. In view of this, field experiment was conducted during 2012 at Mechara Agricultural Research Center with theobjectives of identifying best compatible combinations of maize with soybean varieties and N rates for maximum yield and yield components of the associated cropsand productivity of intercropping system. Three varieties of soybean (Awasa-95, Cocker-240 and Crowford) were intercropped with early maturing maize variety Melkasa-2 with three rates of nitrogen (32, 64 and 96 kg N ha[-1]). The experiment waslaid out in factorial arrangement in randomized complete block design in three replications. Highest maize grain yield (2196kg ha[-1]) was obtained from soybean variety Crowford and 32 kg N ha[-1] and lowest yield (1352 kg ha[-1]) was recorded from maize intercropped with soybean variety Awasa-95 at 96 kg N ha[-1]. The grain yield of intercropped soybean was increased from 586 kg ha[-1] to 842kg ha[-1] as the nitrogen rates increased from 32 kg N ha[-1] to 96 kg N ha[-1]. The higheist LER (1.10) was obtained from maize intercropped with soybean variety Crowford and lowest LER (1.08) was from maize intercropped with variety Cocker-240 due to main effects of soybean varieties while due to main effects of N, the highest (1.16) and the lowest (1.1) LER were obtained from higher rate of nitrogen (96 kg N ha[-1]) and lowest rate of nitrogen (32 kg N ha[-1]), respectively. On the other hand, the highest Gross Monetary Value (17315 Birr ha[-1]) was recorded from interaction of Cocker-240 at highest rate of nitrogen (96kg N ha[-1]) which was not significantly different from Awasa-95 at 32 kg N ha[-1] (15304 birr ha[-1]) and Crowford at 32 kg N ha[-1] (15103) while lowest GMV (12362birr ha[-1]) was obtained from variety Cocker-240 at 32 kg N ha[-1]. Therefore, variety Awasa-95 at lower rate of nitrogen (32 kg ha[-1]) could best in intercropping system to reduce cost of fertilizer and maximize total productivity.

Keywords: Soybean,Intercropping, Land Equivalent Ratio,Gross Monetary Value

1. Introduction

Crop intensification is one of the strategies to increase productivity per unit area of land. For example, intercropping provides potential for the subsistence farmers who operate in low resources (inputs) situation. It is the practice of growing two or more crops simultaneously in the same field. Insurance against the vagaries of weather, disease and pests and higher productivity per unit area are the major reasons for the existence of intercropping. By growing more than one crop at a time in the same field, farmers maximize water use efficiency; maintain soil fertility, and minimize soil erosion, which are the serious drawbacks of monocropping (Francis, 1986).

In Ethiopia, as it is also true in most tropical countries, traditional cropping systems are based on resource poor farmers' subsistence requirements and are not necessarily the most efficient ones (Kidane *et al.*, 2010). Because of this, crop production per unit land area is usually below National average. Therefore, in diversified crop production systems having production constraints, diversified options need to be assessed.

In western Hararghe zone intercropping maize with sorghum, cereal with pulse, maize and Kchat is common. As most of people in Hararghe are based on cereal consumption, protein from pulse is very low. Evaluating the performance of soybean varieties for increasing of soil fertility under intercropping systems could help to maximize yield in the area.

Intercropping cereals and soybean is not anew practice and has been tried in a number of countries for example in Nigeria (Mueneke *et al.*, 2007), Canada (Carruther, *et al.*, 2000) and United States of America. In Africa, soybean is one of the leguminous crops selected for active research for instance, in Zimbabwe the maize yield was enhanced with soybean intercropping through nitrogen transferred from nitrogen fixing soybean to maize during crop development (Mudita *et al.*, 2008). Brophy and Hiechel (1989) reported that the soybean released 10.4% of symbiotically fixed N in to the root zone over its growth period. Martin *et al.* (1991) also reported that the elevated yield and protein level observed in maize and soybean intercrop may be a consequence of Nitrogen transfer from soybean to maize.

Now a day the cost of inorganic fertilizer is increasing and the resource poor farmers are forced to use below recommended rate or null. Therefore, technologies that will reduce N fertilizer input by resource-poor farmers in the area are urgently needed. Nitrogen input through biological N_2 fixation (BNF) by grain legumes can help to maintain soil N reserves as well as substitute for N fertilizer requirement for large crop yields. Different growth habit and maturity period soybean varieties have different nitrogen fixation ability. Since late maturing soybean varieties were able to fix more N2 than early and medium varieties, greater N contribution to any cropping system is expected through their roots, litter and harvest residues. Ogoke *et al.* (2003) reported that a positive N balance by soybean crop was reported due to the effect of increased crop duration and N application. Late maturing soybean varieties are, therefore, able to give higher N benefit compared to early and medium varieties for the improvement of the cropping systems. The objective of this experiment was to identify the appropriate combination of Soybean varieties and nitrogen fertilizer rates on yield, yield components and productivity of Soybean and Maize under intercropping at Mechara.

2. Material and Methods

An on station experiment was conducted for in the 2012/13 cropping season at Mechara Agricultural Research Center (MeARC) west Hararghe Zone, eastern Ethiopia. Three varieties of soybean namely; Awasa-95, Cocker-240 and Crowford and maize variety Melkesa-2 were used for the study. Awassa-95 is relatively early to intermediate maturing variety requiring around 120 days reaching physiological maturity depending on the temperature, altitude and moisture availability of the growing locations. It is suitable for production in intermediate rainfall areas. The areas receiving 500 mm rainfall in growing period is conducive for its production. Crowford is early maturing variety, determinate

growth habit and takes 90-120 days to reach physiological maturity. It grows on soils free from excessive rain fall and at altitude ranges from 1300-1700 m. Cocker-240 is a medium maturity class and indeterminate growth habit with a physiological maturity of 121-150 days. It best grows with altitude ranges 1300-1700 m and temperature 23-25°C (Mandafro *et al.*, 2009).The maize variety Melkasa-2 used in this experiment is an open pollinated variety recommended for moisture stress areas that receive annual rainfall of 600-1000 mm. It is early maturing variety that reaches physiological maturity in 130 days after emergence. It was released in 2004 by Melkasa Agricultural Research Center (MoA, 2011).

The three N levels (32 kg N/ha, 64 kg N/ha and 96 kg N/ha) used in maize/soybean intercropping were from DAP and urea. The sole maize received 46kg N/ha from urea and 18 kg N/ha from DAP. Sole soybean varieties received 18kg N/ha and 46 kg P_2O_5/ha from 100 kg DAP ha^{-1}.The rate used for the sole crops was as recommended for production of each crop.

The intercrop of maize and soybean were in 100% of the sole maize population and 53.3% of soybean population was intercropped as additive series between the two maize rows at the same time. Two seeds per hill of both maize and soybean were planted to ensure germination and good stand of the crops and were thinned to one plant per hill after emergence. The plot size for intercropping was 11.25m^2 (3.75 m width and 3m length). The plot size for sole maize was the same as the intercropped with the row spacing of 0.75 m and 0.25 m. Four rows of soybean were planted in maize rows in between plant spacing of 0.05 m. The plot size for sole soybean was 11.25 m^2 (3.75 m width and 3 m length) containing nine rows, 0.4m and 0.05m row spacing and spacing between plants, respectively. The yield data for experiment were collected from the net plot area of 4.5m^2 (2.25 m x 2 m) both for sole and intercropped. The design of the experiment was randomized complete block design in factorial arrangement in three replications.

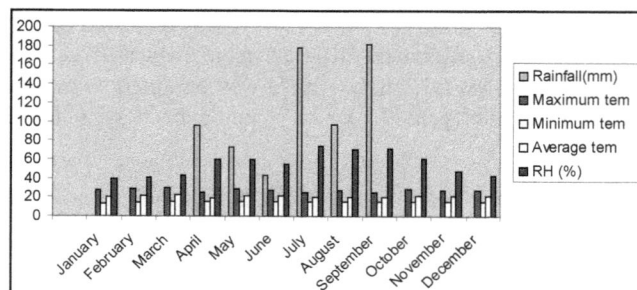

Figure 1. *Weather condition of experimental area during 2012 Source: MeARC weather station, 2012.*

2.1. Soil Condition of Experimental Area

Analysis of soil before planting was done for some physical and chemical properties of soil at Ziway Soil Laboratory of Oromia Agricultural Research Institute (Table 1). The analysis indicated that the soil had low levels of total nitrogen (0.172%) and medium organic matter (2.62%), medium level of available phosphorus (21.3ppm) and high CEC (30.32) as per the criteria developed by Murphy (1968) for Ethiopian soils and Landon (1984) for tropical soils. The pH of the soil was 5.82 showing moderately acidic nature of the soil (Tekalign, 1991). The textural class of the experimental site was silty clay soil.

Table 1. Selected physico-chemical properties of experimental soil.

Soil characteristic	Values
pH (1:2.5 H$_2$O)	5.82
Organic matter (%)	2.62
Total nitrogen (%)	0.172
Available phosphorus (ppm)	21.3
Cation exchange capacity (meq/100g)	30.32

Table 2. Total nitrogen of the experimental plot after harvest in response to the treatments.

Treatments	Total Nitrogen (%)
Maize +Awasa-95+ 32 kg N/ha	0.073
Maize +Awasa-95+ 64 kg N/ha	0.099
Maize +Awasa-95 + 96 kg N/ha	0.114
Maize +Crowford + 32 kg N/ha	0.071
Maize +Crowford + 64 kg N/ha	0.086
Maize +Crowford + 96 kg N/ha	0.099
Maize +Cocker-240 + 32kg N/ha	0.099
Maize +Cocker-240 + 64kg N/ha	0.097
Maize +Cocker-240 + 96kg N/ha	0.085
Sole maize	0.064
Sole cocker	0.099
Sole Crowford	0.085
Sole Awasa-95	0.099
Before planting	0.172

2.2. Post Harvest Soil Analysis

The soil analysis for the samples collected before planting and after harvesting revealed that there was variation in total nitrogen due to variation in cropping practice (Table 2). The soil analysis after harvesting showed that intercropping of maize and soybean resulted in increased total soil nitrogen than sole maize planting. However, all cropping systems reduced total nitrogen compared to total nitrogen of the site before planting (Table 2). The reason for the reduction of total nitrogen could be due to maize and soybean depleted soil nutrients extensively and most of the soil nitrogen was removed through grains and other plant parts of both crops. Other possible losses could be through denitrification, leaching, volatilization and/or their combination. Low soil pH and drought might have affected nodule development and efficiency that ultimately affected the amount of atmospheric nitrogen fixed by soybean.

2.3. Data Collected and Analysis

Data on maize yield components such as number of ears per plant, ear length, thousand kernel weight, grain yield and soybean number of pods per plant,100 seed weight (g), grain yield (kg ha^{-1}) and harvest index (%) were collected. The collected data were analyzed using GenStat Release 13.3 software (Genstat, 2010). Mean separation was carried out using Least Significant Difference (LSD) test at 5% probability level.

3. Results and Discussion

3.1. Maize Yield Components and Yield

Analysis of variance showed that maize stand count at harvest was not significantly affected by main effect of soybean varieties, nitrogen rates and interaction of main effects. Stand count of maize was significantly (p<0.05) affected by cropping system (Table 3). The mean number of stand count of sole cropped maize was higher (21.67/plot) than intercropped maize 20.11/plot (Table 3). The lower stand count in intercropped maize may be due to competition for the same resource with soybean or due to shortage of moisture during early vegetative growth. Similar to this result, Biruk (2007) reported reduction in stand count of intercropped sorghum with common bean varieties.

Table 3. Main effects of the intercropped soybean varieties and nitrogen rates on yield components of maize in maize and soybean intercropping.

Treatments	No. of stand count/plot at harvest	No. of ears per plant	Ear length(cm)	No. of kernels per ear	1000 kernels weight (g)	Harvest index (%)
Soybean varieties						
Awasa-95	20.22	1.044	11.80	396.6	219.1	33.9
Cocker-240	19.78	1.067	12.71	378.2	229.9	37.6
Crowford	20.33	1.133	12.64	398.1	220.8	34.3
LSD (5%)	NS	NS	NS	NS	NS	NS
Nitrogen rates (kgha^{-1})						
32	19.44	1.07	12.24	386.4	226.3	33.8
64	20.67	1.06	12.60	396.0	220.4	37.2
96	20.33	1.12	12.31	390.4	223.8	34.7
LSD (5%)	NS	NS	NS	NS	NS	NS
CV (%)	6.7	11.1	13.8	11.8	8.6	15.4
Cropping system						
Intercropping	20.11b	1.08b	12.39b	391.0	223.3b	35.3
Sole cropping	21.67a	1.26a	14.87a	428.0	294.0a	31.7
LSD (5%)	1.48	0.15	2.19	NS	51.97	NS
CV (%)	5.8	10.8	13.9	10.9	19.1	16.7

Means followed by the same letter(s) in the column are not significantly different at 5% level of significance
NS=not significant

Number of ears per plant, ear length and number of kernels per ear were not significantly affected by main effect of soybean varieties; nitrogen rates and interaction of main effects. However, cropping system had significant effect on number of ears per plant and ear length (Table 3). Sole cropped maize produced significantly more number of ears per plant (1.26) than intercropped 1.08 (Table 3). Similarly, significantly longer ear (14.89cm) was recorded due to sole cropping while shorter ear length (12.39cm) scored due to intercropped maize (Table 3). The reduction in number of ears per plant and ear length in intercropped maize might be due to the reduction in the ear leaf photosynthesis due to competition with soybean that lowers the number of ears per plant. Similar to this result, Wogayehu (2005) and Walelign (2008) reported lower number of ears per plant, ear length and number of kernels per ear of maize from intercropped maize with haricot bean varieties.

Analysis of variance showed that 1000 kernel weight, grain yield and harvest index were not significantly affected by main effect of soybean varieties and nitrogen rates, while grain yield was significantly influenced by interaction of main effects (Table 3 and figure 2). The highest grain yield (2196 kg ha^{-1}) was obtained from combination of maize intercropped with soybean variety Crowford and 32kg N ha^{-1} and the lowest maize grain yield (1352 kg ha^{-1}) was obtained from maize intercropped with soybean variety Awasa-95 at 96 kg N ha^{-1}. Higher grain yield of maize with soybean variety Crowford might be due to good nitrogen fixing abilities related to the higher number of nodules per plant as compared to other varieties, and early maturity of soybean variety Crowford. This might be because intercropping with early maturing legume could lead to increased productivity of the cereal (Rao, 1980).

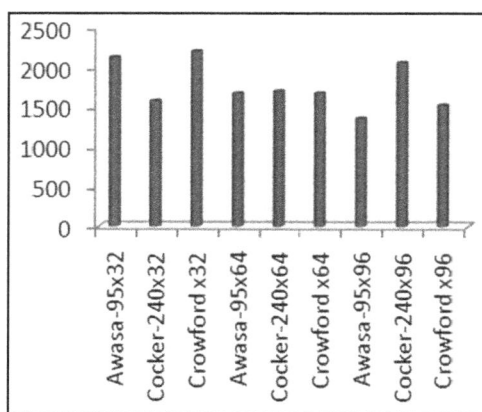

Figure 2. Interaction effect of the intercropped soybean varieties and nitrogen rate on grain yield (kg ha^{-1}) of maize in maize and soybean intercropping.

Grain yield of maize was significantly reduced by 31.7% due to intercropping. Similarly, Wandahwa et al. (2006) found that intercropping maize and soybean reduced the yield of maize probably because of competition for resources.

Even though the difference was not significant, higher

harvest index (35.3%) was recorded from intercropped than sole cropped maize (31.7%) (Table 3). The higher harvest index from intercropped maize could be due to increasing in nitrogen rates in intercropping increased biomass of maize. Similar result was reported by Selamawit (2007) that higher harvest index was from intercropped maize with potato than sole cropped maize.

In this study a severe water stress during early growing period in June (43.3mm) and August (97.7mm) might have contributed for lower yield and yield components of maize.

3.2. Soybean Yield Components and Yield

The analysis of variance showed that number of pod per plant, 100 seed weight and harvest index were significantly (P<0.05) affected by soybean varieties. The highest number of pods per plant (32.22) was obtained from soybean variety Awasa-95 intercropped with maize while the lowest number of pod per plant (26.0) was obtained from variety Crowford (Table 4). Similar to this result, Thole (2007) reported that number of pods per plant of soybean intercropped with maize was significantly reduced by soybean varieties. Number of seeds per pod was not significantly affected by main effect of soybean varieties, nitrogen rates and their interaction and cropping system (Table 4). The number of 100 seed weight was significantly affected by main effect of varieties. The highest 100 seed weight (17.31g) was obtained from soybean variety Crowford intercropped with maize while significantly the lowest 100 seed weight (14.16g) was obtained from soybean variety Awasa-95 (Table 4). The highest 100 seed weight in variety Crowford could be due its larger seed size.

Cropping system significantly (P<0.05) affected 100 seed weight and grain yield. Higher 100 seed weight (15.95 g) was recorded from intercropping (Table 4). This could be the fact that the lower intra species competition between soybean as plant density was lower in intercropping than sole cropped soybean and higher seed weight was recorded from lower plant density (Turk et al., 2003). Similar to this result, Wogayehu (2005) in maize/haricot bean, Biruk (2007), in sorghum/haricot bean and Egbe et al. (2010) from sorghum/soybean intercropping study reported higher 100 seed weight of legume components in intercropping than in sole crop.

Grain yield per hectare of soybean was significantly affected by main effect of soybean varieties and nitrogen rates (Table 4). Yield obtained from plot treated with 96 kg N ha^{-1} (842 kg ha^{-1}) was significantly higher than that of 32 kg N ha^{-1} (586 kg ha^{-1}). In this result, the yield of soybean was enhanced by increased level of nitrogen rates. The response of soybean to increase in grain yield might be the soil was deficit for nitrogen required by crop. In agreement with this result, Wandahwa et al. (2006) reported that the yield of soybean was increased due to increased nitrogen fertilizer in intercropped maize and soybean.

Cropping system highly significantly (P<0.01) affected the yield of soybean. Sole cropping gave significantly higher grain yield (1754 kg ha^{-1}) than intercropping 703 kg ha^{-1} (Table 4).

Lower grain yield of the intercropped soybean might be due to the competition effect exerted by maize component for limited growth factors in intercropping and lower stand count under intercropping. Pal *et al.* (2001) and Muoneke *et al.* (2007) reported similar yield reductions in soybean intercropped with maize and sorghum and associated the yield depression to interspecific competition and the depressive effect of the cereals.

Here in additive intercropping of maize and soybean, the intercropped soybean grain yield per hectare was reduced by 40% as compared sole cropped (Table 4). Comparably, Huxley and Maingu (1978), in cereals and legumes intercropping system, reported that the grain yield of the legume component declined, on average, by about 52% of the sole crop yield, whereas the cereal yield was reduced by only 11%.

Thus, the general observation in this study showed that yields of soybean component were significantly depressed by maize component in intercropping. This is most likely due to competition for soil nutrient and the reduction in transmitted photosynthetically active radiation to the soybean as a result of shading.

The harvest index of soybean was highly significantly (P<0.01) affected by soybean varieties. The highest harvest index (47%) was from soybean variety Crowford and the lowest harvest index was from Awasa-95 (32.7%) which was significantly not different from variety Cocker-240 (36.7%) (Table 4). The highest harvest index recorded for variety Crowford intercropped with maize might be due to the high grain yield to biomass obtained by the variety as a result of high partitioning of dry matter to the grain. Udealor (2002) and Ano (2005) reported that the differences in harvest index might be due to the inherent varietal characteristics, environmental factors and other cultural practices.

Table 4. Effect of intercropped soybean varieties and nitrogen rates on yield components and yield of soybean in maize and soybean intercropping.

Treatments	No. of pods per plant	No. of seeds per pod	100 seed weight (g)	Grain yield (kgha^{-1})	Harvest index (%)
Soybean varieties					
Awasa -95	32.22a	2.49	14.16c	679	32.7b
Cocker-240	28.56abc	2.53	16.07ab	755	36.7b
Crowford	26.0c	2.35	17.31a	676	47.0a
LSD (0.05)	4.67	NS	1.25	NS	8.86
Nitrogen rates (kgha^{-1})					
32	26.33	2.39	15.73	586b	36.1
64	29.67	2.40	15.87	681ab	39.0
96	30.78	2.59	16.23	842a	41.4
LSD (0.05)	NS	NS	NS	212.5	NS
CV (%)	16.2	10.2	7.9	30.2	22.8
Cropping system					
Intercropping	28.9	2.46	15.95a	703b	38.8
Sole cropping	31.8	2.54	14.52b	1754a	33.7
LSD(0.05)	NS	NS	1.21	169.4	NS
CV (%)	19.9	10	9.9	22.4	26.9

Means followed by the same letter(s) within column are not significantly different at 5% level of significance
NS=Not significant

Table 5. Effect of intercropped soybean varieties and nitrogen rates land equivalent ratio (LER) and gross monetary values (GMV) of maize and soybean intercropping.

Treatments	LER			MV		
Soybean varieties	Maize	Soybean	Total	Maize(Birr/Ha)	Soybean(Birr/ha)	GMV(Birr/ha)
Awasa-95	0.70	0.39	1.09	8906	4685	13591
Cocker-240	0.69	0.40	1.08	9213	5208	14421
Crowford	0.69	0.41	1.1	9318	4662	13980
LSD (0.05)	NS	NS	NS	NS	NS	NS
Nitrogen rates (kgha^{-1})						
32	0.76	0.34	1.1	10210	4046	14256
64	0.65	0.39	1.04	8694	4697	13391
96	0.68	0.48	1.16	8533	5811	14344
LSD (0.05)	NS	NS	NS	NS	NS	NS
CV (%)	17.8	20.7	16.3	19.9	38.4	14.7
Cropping system						
Intercropping	0.69b	0.40b	1.09a	9146b	4850.7b	13996.7
Sole cropping	1.0a	1.0a	1.0a	13387a	12102.6a	-
LSD (5%)	0.17	0.089	NS	2878	1169	-
CV (%)	18.5	20.7	16.0	24	22.4	-

Means followed by the same letter(s) within column are not significantly different at 5% level of significance

3.3. Total Land Productivity and Gross Monetary Evaluation

Analysis of variance showed that partial LER of maize and soybean and total LER were not significantly (P<0.05) affected by the main effects of soybean varieties, N rates and their interaction. The highest total LER due to main effect of variety was obtained from soybean variety Crowford intercropped with maize (1.10) while the lowest value was obtained from Cocker (1.08). With respect to the main effect of N, the highest LER (1.16) was obtained due to application of 96 kg N ha^{-1} and the lowest LER (1.04) was obtained due to 64 kg N ha^{-1} (Table 5).

Productivity was improved in almost all intercrops as depicted by LER values greater than one (Table 5). The total land productivity ranged from 108% in Cocker-240 and maize to 110% Crowford and maize intercrop as compared to sole crops. This indicated that intercropping of maize and soybean was advantageous than sole planting of either maize or soybean. The result also indicated that the intercrops are more advantageous in efficiently utilizing land than the sole cropping of either maize or soybean and it would require 10% more land to get the same yield obtained from the intercropping system. This intercropping system resulted in the highest cumulative total yields than either of maize or soybean.

The LERs, greater than one in this experiment might have resulted from morphological differences of these two species and creating various niches for resources such as sun light, nutrients and moisture. The higher LERs in intercropping than mono-cropping were reported by Adeniyan and Ayola (2006), Bingcheng et al. (2008) and Javanmard et al. (2009).

Table 6. *Gross monetary value (Birr ha^{-1}) of maize and soybean under intercropping as influenced by interaction of soybean varieties and nitrogen rates.*

Nitrogen rates (kg ha^{-1})			
Soybean varieties	**32**	**64**	**96**
Awasa -95	15304ab	12922b-g	12548b-h
Cocker-240	12362b-i	13584b-e	17315a
Crowford	15103abc	13667bcd	13168b-f
Intercropped mean			13997a
Sole soybean			9146b
Sole maize			13387a
	Soybean varieties x N rates		Cropping system
LSD (0.05)	3565.3		1877
CV (%)	14.7		19.8

Means followed by the same letter(s) within column and row are not significantly different at 5% level of significance

Unlike main effects of soybean varieties and N rates, the analysis of variance showed that the interaction significantly (P<0.05) affected the GMV in the intercropping system. The highest GMV (17,315 ETB ha^{-1}) and the lowest (12,362 ETB ha^{-1}) were obtained from soybean variety Cocker-240 intercropped with maize at 96 kg N ha^{-1} and 32 kg N ha^{-1} respectively (Table 6). This finding was in agreement with the previous studies on maize-soybean intercropping by Raji

(2007), Thole (2007) and Gani (2012) who obtained higher monetary returns from intercropping maize and soybean as compared to sole maize.

4. Conclusion

Due to increasing population, decreasing land units and soil fertility, integrating legumes in to the cereal production system is a viable option in western Hararghe for food security. The statistical analysis revealed that maize yield components and yield were not significantly affected by main effects of varieties and nitrogen except grain yield which was affected by interaction of main effects. The highest maize grain yield (2196kg ha^{-1}) was for soybean variety Crowford at 32 kg N ha^{-1} and the lowest yield (1352 kg ha^{-1}) was recorded from maize intercropped with soybean variety Awasa-95 at 96 kg N ha^{-1}.The main effect of soybean variety significantly affected yield components of soybean such as number of pod per plant, 100 seed weight and harvest index while N and their interaction had no significant effect on yield and yield components of soybean except grain yield which was significantly influenced by main effect of nitrogen. The highest LER due to main effect of soybean varieties (1.10) was recorded from soybean variety Crowford intercropped with maize while the highest LER (1.16) due to main effect of nitrogen rates was recorded from the highest rate of nitrogen (96 kg N ha^{-1}) and the highest GMV (17315 Birr ha^{-1}) was obtained from Cocker-240 and 96 kg N ha^{-1} while the lowest GMV (12362 Birr ha^{-1}) was from Cocker-240 and 32 kg N ha^{-1}. Awasa-95at lowest rate of nitrogen (32 kg N ha-^{1}) which was not significantly different from Cocker-240 at highest rate of nitrogen (96 kg N ha^{-1}) could be better in intercropping system to maximize yield of both crops as well as total productivity.

Ackowledgements

The author thanks Ahmedziyad Abubaker and Wolansa Mokonin for their support in data collection and Oromia Agicultural Research Institute for financial support.

References

[1] Ano A. O, 2005. Effect of soybean relayed in to yam minisett/maize intercrop on the yield of component crops and soil fertility of yam based system. *Nigeria Journal of. Soil Science*, 15: 20-25.

[2] BirukTesfaye, 2007. Effects of Planting Density and Varieties of Common bean (*Phaseolusvulgaris* L.) Intercropped with Sorghum (*Sorghum bicolor* L.) on Performance of the Component Crops and Productivity of the System in South Gondar, Ethiopia. M.Sc. Thesis. Haramaya University.

[3] Brophy, L. S and G. H. Heichel, 1989. Nitrogen Release from Roots of Alfalfa and Soybean in Intercrops. Direct^{15}N Labeling Methods.

[4] Carruthers K, Prithiviraj BFQ, Cloutier D, Martin RC, Smith DL, 2000. Intercropping maize with soybean, lupin and forages: yield component responses. *European Journal of Agronomy*, 12: 103-115.

[5] Cotteinie, A, 1980. Soil and Plant Testing as a Base of Fertilizer Recommendations. Soils Bulletins, No. 38, FAO, Rome.

[6] Egbe, O. M, Alibo, S. E and Nwueze, I., 2010. Evaluation of Some Extra-Early- And Early-Maturing Cowpea Varieties for Intercropping With Maize in Southern Guinea Savanna of Nigeria. *Agriculture and Biology Journal of North America.* 1(5) 845-858.

[7] FAO, 2000. Fertilizers and Their Use 4th ed. International Fertilizer Industry Association, FAO, Rome, Italy.

[8] Francis, C. A., 1986. Multiple cropping systems. Vol. 1. Macmillan Publishing Co., New York.

[9] Fukai, S. and B. R. Trenbath, 1993. Presses of Determining Intercrop Productivity and Yield of Component Crops. *Field Crops Research*, 34: 247-271.

[10] Gani, O. K., 2012. Effect of phosphorus fertilizer application on the performance of maize/soybean intercrop in the southern Guinea savanna of Nigeria. *Archives of Agronomy and Soil Science,* 58:2, 189-198.

[11] GenStat, 2015. GenStat Release 15, VSN International Ltd.

[12] Ghosh, P. K., 2004. Growth yield competition and economics of groundnut/cereal fodder intercropping systems in the semi-arid tropics of India. *Field Crops Research*, 88: 227-237.

[13] Huxley, P. A. and Z. Maingu, 1978. Use of a systematic spacing design as an aid to the study of intercropping: Some general considerations. *Experimental Agriculture*, 14: 519-27.

[14] Kidane Georgis, 2010. Inventory of Adaptation Practices and Technologies of Ethiopia. Environment and natural resource working paper 38, FAO, Rome.

[15] Mandefro, N., Anteneh, G., Chimdo, A., and Abebe K., (Eds.), 2009. Improved technologies and resource management for Ethiopian Agriculture. A Training Manual. RCBP, MoARD, Addis Ababa, Ethiopia.

[16] Martin, R. C., Voldeng, H. C. and Smith, D. L. 1990. Intercropping corn and soybean in a cool temperate region: yield, protein and economic benefits. *Field Crops Research.* 23: 295–310.

[17] MoA (Ministry of Agriculture), 2011. Crop Variety Registry, Issue Number 14, Addis Ababa.

[18] Mudita, I.I., Chiduza, S.J. Richardson-Kageler and F.S. Murangu, 2008. Performance of Maize and Soybean Cultivars of Varying Growth Habit in Intercrop in Sub humid Environments' of Zimbabwe. *Journal of Agronomy*, 7(3): 227-237.

[19] Muoneke C.O, Ogwuche M.O, Kalu B.A., 2007. Effect of maize planting density on the performance of maize/soybean intercropping system in a guinea savanna agroecosystem. *Afr. J. Agric. Res.*, 2: 667-677.

[20] Murphy, H.F., 1968. A report on fertility status and other data on some soils of Ethiopia. Expt. Bull. No. 44. College of Agriculture, Haile Selasie I University, Alemaya, Ethiopia. 551p.

[21] Ogoke I.J., Carsky R.J., Togun A.O. & Dashiell K., 2003. Effect of P fertilizer application on N balance of soybean crop in the Guinea savanna of Nigeria. *Agriculture Ecosystem and Environment*, 100: 153-159.

[22] Olsen, S.R., C.V. Cole., F.S. Watanabe and L.A. Dean, 1954. Estimation of Available Phosphorus in Soils by Extraction with Sodium Bicarbonate. USDA Circular, 939: 1-19.

[23] Raji, J. A., 2007. Intercropping soybean and maize in a derived savanna ecology. *African Journal of Biotechnology.* 6 (16): 1885-1887.

[24] Selamawit Getachew., 2007.Effect of Plant Population and Nitrogen Fertilizer on Growth and Yield of Intercropped Potato (*Solanumtuberosum* L.) and Maize (*Zea mays L.)* at Haramaya, Eastern Ethiopia. M.Sc. Thesis, Haramaya University.

[25] Tekelign Tadesse, 1991. Soil, Plant, Water, Fertilizer, Animal Manure and Compost Analysis. Working Document No. 13. International livestock Research center for Africa (ILCA), Addis Ababa.

[26] Thole, A., 2007. Adaptability of soybean varieties to intercropping under leaf stripped and detasseled maize. MSc Thesis. University of Zimbabwe.

[27] Turk M.A., A.M. Tawaha and M.K.J. El-Shatnawi, 2003. Response of Lentil (*Lens culinaris*medic) to plant density, sowing date, phosphorus fertilization and ethephonapplication in the absence of moisture stress. *Journal of Agronomy and Crop Sciences,* 189: 1-6.

[28] Udealor A., 2002. Studies on the growth, yield, organic matter turnover and soil nutrient changes in cassava (*Manihotesculenta* Crantz)/vegetable cowpea (*Vignaunguiculata* L. Walp.) mixtures. Ph. D. Dissertation, University of Nigeria, Nsukka, Nigeria.

[29] Walelign Worku., 2008. Evaluation of haricot bean (*Phaseolus vulgaris* L) Genotypes of diverse growth under sole and intercropping with maize in southern Ethiopia. *Journal of Agronomy*, 7(4): 306-313.

[30] Walkley, A and I.A. Black, 1934. An Examination of Digestion of Degrjareff Method for Determining Soil Organic Matter and Proposed Modification of the Chromic Acid Titration Method. *Soil Science*, 37: 29-38.

[31] Wandahwa P, Tabu IM, Kendagor MK, Rota IA. 2006. Effect of intercropping and fertilizer type on growth and yield of soybeans. *Journal of Agronomy.* 5(1): 69–73.

[32] Willey, R.W., 1979. Intercropping-its importance and research needs. Competition and yield advantages. *Field Crops Research*, 32: 1-10.

[33] Wogayehu Worku., 2005. Evaluation of Common Bean (*Phaseolus vulgaris* L.) Varieties Intercropped with Maize (Zea mays L.) for Double Cropping at Alemaya and Hirna areas, Eastern Ethiopia. M.Sc. Thesis. Haramaya University.

Properties of Mine Soils in a Forested Hilly Terrain of South eastern Nigeria

E. U. Onweremadu[1, *], E. I. Uzor[1], Egbuche C. T.[2], L. C. Agim[1], D. J. Njoku[3], A. C. Udebuani[4]

[1]Department of Soil Science and Technology, Federal University of Technology, Owerri Nigeria
[2]Department of, Forestry and Wildlife Technology, Federal University of Technology, Owerri, Nigeria
[3]Department of Environmental Technology, Federal University of Technology, Owerri, Nigeria
[4]Department of Biotechnology, Federal University of Technology, Owerri, Nigeria

Email address:

uzomaonweremadu@yahoo.com (E. U. Onweremadu)

Abstract: Soils of a mine site at Leru, Abia State, Nigeria were characterized and classified for proper usage .A transect survey technique was employed in which a traverse was cut to link soils affected by mining to the unaffected soils in the area. Soil profile pits were dug described and sampled using standard techniques. Routine laboratory analyses were conducted on soil samples for selected soil properties. Soil data were subjected to mean statistic. Results indicated increasing sandiness and hulk densi1 in the epipedons closest to the mine site. Mean value of soil moisture in un-mined site (149.8 g kg^{-1}) was higher than those of mine soils (144.2 g kg^{-1} for middle land unit and 137.0g kg^{-1} for pedon closest to the mine site). Mine soils were younger with silt-clay ratio ranging from 3.7 to 7.0 while unaffected soils were older (silt-clay ratio = 0.6). Low values of calcium-magnesium ratios (Ca/Mg < 3.0) were reported. Soil pH progressively increased towards un-mined site. The soil profile proximal to the mine site was shallower (depth <1 00cm) when compared with other soil profiles, and had a lithic contact at 90 cm depth. Based on field and laboratory analysis soils were classified as Lithic Dystrudepts (5 metres away from mine site), Typic Haplanthrepts (25 metres away from mine site) and Typic Hapludults (2 kilometres away from mine site).

Keywords: Classification, Degradation, Mining, Soils, Topography

1. Introduction

Mining activities disrupt partially or totally the original characteristics and qualities of soils and re-set the pace of soil formation. Mining influences soil physical and structural properties [1], disturbs farmland forests and waterways [2], increases water stable aggregates [3] and initiates differences in water stable aggregation among soil horizris [4]. Overburden in soils of mine sites influence soil texture, soil colour and soil subsurface pH [5]. [6] reported that unconsolidated materials resulting largely from landfills, mine soils. Rubble garbage dumps and dredging generate fresh anthropgeomorphic parent materials on which new soils develop. The nature of soils formed depends on the type of parent material [7]. Mined soils can have poor water retention resulting from high coarse fragment content, lack of fine earth, and poor soil structure which allow water to drain quickly from the soil profile [8]. High levels of soluble salts in minerals inhibit water and carbon dioxide uptake: and inactivate enzymes affecting protein synthesis. C metabolism and photophosphorylation [9]. Soil nitrogen, organic carbon and phosphorus have been reported as growth limiting factors on mined soils, especially within 10 years after disturbance. Organic matter and total nitrogen are used as good indicators of N-availability in mine soils [10]. In southeastern Nigeria, there is an increasing demographic pressure on available soil resources [11] leading to soil stresses [12] (Reich et al., 2001) and land degradation [13] (Onweremadu, 2008; [14] Onweremadu et al.. 2010). Unprofitable agricultural enterprise has led to diversion of socio-economic activities towards non-agricultural ventures in Southeastern Nigeria. Because quick returns are obtained in milling rock minerals like coals, rocks and salts, it has become a major socioeconomic activity in Nigeria. These activities have attracted extractive and non-extractive industries such as quarrying industries, most of which do not consider the impact on the ecosystem despite the need for environmental quality control. At Leru in Umunneochi, Abia State Nigeria, sand and gravel mining coupled with quarrying are socio-economically attractive, causing more native populations to engage in it. This study was therefore undertaken to characterize soil properties of the affected soils by these mining activities, and classify them using USDA soil Taxonomy.

2. Materials and Methods

2.1. Study Area

The study was conducted at Leru in Umunneochi Local Government Area in Abia State, Southeastern Nigeria. Leru mine lies between Latitudes 5° 30' and 6° 08' N, and Longitudes 7° 20'and 7° 50'E. The mine sites are sparsely populated and occupies over 5000 ha of farmland. The soils are derived from upper coal measures or Nsukka formation [15] and belong to the Eastern Nigeria highlands [16]. The area lies within the humid tropics, with a mean annual rainfall range of 2000-2250 mm and mean annual temperature ranging from 27-28 °C' [17]. Leru has a rainforest vegetation which has been depleted b> anthropogenic activities such as agriculture, stone mining and quarrying. The rainforest is less dense compared with southern locations, but comprises varying levels and species of plants arranged in storey. Prominent plant species include oil palms (Elaeis guineensis), oil bean tree (Pentaclethra macrophyllum), maize (Zea mays), cashew (Anarcadium occidentale), cassava (Manihot esculenta), Pineapple (Ananas cornosus), yams (Dioscorea Species), and others. Mining of earth minerals and agriculture are the major socio-economic activities of the area.

2.2. Field Sampling

Transect soil survey technique guided field sampling. A transverse was cut to link three identified land units, namely soils 5 meters away from the mine site, soils 25 metres away from the mine site and soils 2 km away from the mine site, based on observed macro morphological changes.. Profile pits were dug in each land unit and described according to [18] procedure. Soil samples were collected based on the degree of soil horizon differentiation, thereafter, they were bagged and transported to the laboratory where they were air-dried and sieved using 2-mm sieve preparatory to laboratory analyses. Core soil samples were collected for bulk density determinations from delineated horizons.

2.3. Laboratory Analyses

Particle size distribution was determined by hydrometer method [19] (Gee and Or. 2002). Bulk density was measured by core method [20] (Grossman and Reinsch.

2002). Total porosity was calculated as follows:

$$\text{Total porosity} = 100\% - \left(\frac{BD}{PD} \times \frac{100}{1}\right)$$

Where BD = bulk density
PD = particle density (assumed to be 2.65 Mg/kg)
Available water capacity was computed as the difference between the moisture retained at 0. 10 and 15 bars tensions.

Soil pH was determined on a 1:1 soil/water sample [22]. Soil organic carbon (SOC) was estimated according to the procedure of [23]. Total nitrogen was determined by Kjedahl digestion [24]. Available phosphorus was obtained using Bray P No .2 method [25]. Exchangeable acidity was measure titrimetrically [26]. Exchangeable basic cations were extracted using IM Nl-I1OAC, and exchangeable Ca and Mg were determined by EDTA titration, while Na and K were measured photometrically [27]. Elemental ratios of CIN and Ca/Mg were computed to assess fertility of the mine soils.

2.4. Data Analysis

Soil data were analyzed using descriptive statistics: mean and statistical tests of difference were conducted using t-test at 5 % level of significance between mine and natural soils.

3. Results and Discussion

3.1. Macromorphological Properties

Morphological properties of the soils are presented in Table 1. The Soils show thicker A-horizon nearest to the mine. Effective soil depth increased with increased distance from mine site while colors of A horizon became darker. Differences in soil thickness could be a result of human activities like spoil layering more than pedogenic processes. Soils 5 meters away from the mine site were redder (Reddish brown 5YR4/6) moist as opposed to darker colors (Dark reddish brown (YR3/2)) obtained from soils 2 kilometers away from the mine site at surface horizon. Color changes in horizons were likely a result of spoil layering. Removed strata during mining were transported sideways and downwards, giving rise to heterogeneity of soils in terms of soil morphological properties. Root development was improved in soils far away from the mine site. Granular structures predominate in soils distal to the mine site, indicating more pedogenic activities away from the mine site. However, the thicker A-horizon (21 cm) in soils closest (5-meters away) the mine site compared to non-mine site (10 cm) could provide greater volume of rooting zone for crops. In a similar study. [28] found that mine soils have deeper root zones, higher hulk densities and weaker soil structure compared with native soils. Greater thickness in A-horizon is attributable to high population of grasses near the mine.

Table 1. *Morphological Properties of the Mine Soils*

Horizon	Depth (cm)	Soil colour (moist)	Structure		Consistence	Boundary	Roots
5 meters away from mine site (Lithic Dystrudepts)							
A	0-21	Reddish Brown (5YR4/6)	Weak Subangular blocky	Fine	fr	Cs	Many very fine roots
BC	21-68	Reddish brown (5YR4/8)	Weak Subangular block	Coarse	f	d	Few medium and large roots
C	68-90	Orange (5YR 6/6)	Weak	Coarse	fr	-	Very few large roots
25 meters away from mine site (Typic Haplanthrepts)							
A	0-11	Reddish brown (5YR4/8)	Moderate granular	Medium	vfr		Many vey fine roots
						C	
Bw₁	11-49	Bright reddish brown (5YR 5/6)	Moderate	coarse	fi	D	Many fine root
Bw₂	49-80	Bright reddish brown (5 YR 5/8)	Moderate medium subangular blocky		vfi	D	Many fine roots
2E/B	80-88	Orange (6/6)	Moderate course subangular blocky	coarse	vfi	D	Very few large roots
2Btxb	88-120	Orange (5YR 9/8)	Moderate prismatic	Coarse	efi	-	Very few large roots
2 Kilometers away from mine site (Typic Hapludults)							
A	0-10	Dark reddish brown (YR3/2)	Granular medium		fr	Cs	Many very fine root
E	10-26	Light brown reddish (YR6/4)	Medium sub angular blocky		fi	D	Common fine and few medium large root
Bt₁	26-65	Yellowish red (7YR4/4)	Medium sub angular blocky		fi	D	Few medium large root
Bt₂	65-120	Yellowish red (7.5YR5/4)	Medium subangular blocky	Course	fi	D	Few medium large root
BC	120-150	Yellowish red (7.5YR6/6)	Medium course sub angular blocky		f	-	Very few large root

fr = friable, fi = firm, vfr = very friable, efi = extremely firm c = clear, s = smooth, D = diffuse

Table 2. *Selected physical properties of the soils*

Horizon	Depth cm	Total sand (kg⁻¹)	Silt (g kg⁻¹)	Clay g (kg⁻¹)	SCR	Texture	BD (Mg M⁻³)	TP (g kg⁻¹)	MC g (kg⁻¹)
5 meters away from mine site (Lithic Dystrudepts)									
A	0-21	910	80	10	8.0	S	1.40	47.2	133
Bw	21-68	890	100	10	10.0	S	1.58	41.4	141
C	68-90	880	90	30	3.0	S	1.67	36.9	41.37
	Mean	893.3	90.0	16.7	7.0	-	1.53	42.6	137
25 meters away from mine site (Typic Haplanthrepts)									
A	0-11	890	100	10	10.0	S	1.38	47.9	142
Bw1	11-49	900	80	20	4.0	S	1.50	43.3	145
Bw2	49-80	880	80	40	2.0	S	1.52	42.6	146
2E/B	80-88	860	100	40	2.5	SL	1.61	39.2	140
2Btxb		820	30	150	0.2	SL	1.63	38.4	148
	Mean	870	78	52	3.7	-	1.52	42.2	144.2
2 kilometers meters away from mine site (Typic Haplanthrepts))									
A	0-20	855	50	100	0.5	SL	1.29	51.3	150
E	20-26	870	20	110	0.2	SL	1.35	49.1	125
Bt1	26-65	850	10	140	0.1	SL	1.40	47.2	158
Bt2	65-120	860	20	120	0.2	SL	1.42	46.4	155
BC	120-150	880	80	40	2.0	LS	1.56	41.1	141
	Mean	862	36	102	0.6	-	1.40	47.0	149.8

S= sand, LS= loamy sand, SL=sandy Loam BD = bulk density, TP= total porosity, MC= moisture content, CV = coefficient of variation, SCR = silt clay ratio, G. mean= grand mean, G. CV = grand coefficient of variation.

3.2. Soil Physical and Chemical Properties

The soil physical properties (Table 2) indicate higher bulk densities ranging from 1 .29 4g M3 in so that is 2km away from mine area to 1 .67 Mg M3 in soil that is 5 meters away from the site and lower total porosities in soils proximal to the mine with values ranging from 36.9 to 51.30 g/kg. This could he due to overburden effect and coarseness of the soils. It was reported that such soils do not retain enough water, possibly emanating from high sandiness [29]. Particle size distribution showed dominance of sand (820 910 g/kg) and silt (30 -100 g/kg) over clay. The dominance of these two particle sizes over clay implies that crushing of parent materials and rock fragments genetically and sequentially produced sand, then silt, and these sizes are transformed into clay with intense weathering and pedogenesis. Very high silt-clay ratios ranging from 8 - 10 were obtained in soils closest to the mine site, indicating that they are young. [30] reported that mine soils show either some development (Inceptisols) or exhibits little or no development (Entisols).

The soil chemical properties are shown on Table 3, with soil acidity increasing towards the mine site. Higher acidity (5.6) of soils proximal to the mine site is a property probably inherited from the overburden parent materials, that are acid-producing. Even where these materials contain basic cations, they are easily leached due to high rainfall amount, intensity and duration characteristic of the study site leaving a preponderance of acidic cations. [31] reported that carbonates contained in overburden strata in mine sites are readily leached thereby creating an acid environment. It is possible that the oxidation potential of the parent materials counts in influencing pH of mine soils. [32], found that mine soils forming in partially oxidized sandstone overburden have an initial surface p1-1 of 5.5, whereas mine soils forming in un-oxidized sandstone, and siltstone overburden had an initial pH of 7.5. There was no trend in the distribution of organic matter in a spatial orientation, although it decreased with depth in all the profiles. Highest values 23.12 g kg was recorded in soils, 25 meters away from the mine site. However, total nitrogen value were higher in native soils, implying that greater leaching predisposed by high acidity of mine soils may have affected their nitrogen content. Carbon-nitrogen ratios (14.9) of native soils are close to values typical of West African soils with C/N 12-14. However, these values were very low in mine soils. It implies that C/N ratios of mine soils have not stabilized to typical values in the West African biome. Calcium-magnesium ratio is an index of soil fertility [33], with values tying below 3.0 indicating poor fertility of soils in all land units. [34] suggested the use of calcium-aluminum ratios as a better indicator of calcium nutrition. The Ca:Mg ratios below 3.0 lead to unavailability of calcium and available phosphorus [35].

The soils show low available phosphorus in soils closest to the mine site which is attributable to high acidity of the soils, and consequent fixation of phosphorus (Table 3). In some soils, calcium can he derived from rock fragments [36], but the rock fragments in this mine site are from sandstones, which are naturally acidic and calcium-poor. There were significant (p = 0.05) changes in the distribution of OM, TN, C/N, Ca/Mg ,clay SCR and MC while non-significant differences were recorded in BD, pH and Avail. P in the study site (Table 4).

Table 3. Selected chemical properties of the soils

Horizon	Depth cm	pH water	OM g kg^{-1}	TN (gkg^{-1})	SCR	Avail P mg kg	C/N	Ca/Mg
	5 meters away from mine site (Lithic Dystrudepts)							
A	0-21	5.9	21.0	1.3	33.1	S	5.3	0.90
Bw	21-68	5.4	20.9	1.0	26.1	S	8.6	0.91
C	68-90	5.7	19.3	0.8	21.7	S	14.0	0.92
	Mean	5.6	20.3	0.6	27.1	-	9.3	0.91
	25 meters away from mine site (Typic Haplanthrepts)							
A	0-11	5.5	25.9	1.8	37.6	S	3.9	0.91
Bw1	11-49	5.8	23.6	1.7	28.9	S	5.2	0.94
Bw2	49-80	5.7	23.8	1.6	7.8	S	7.2	0.94.
2E/B	80-88	5.7	24.2	1.1	15.9	SL	8.5	0.90
2Btxb		5.7	18.1	0.9	13.2	SL	5.2	0.90
	Mean	5.7	23.12	1.4	22.7	-	6.2	0.92
	2 kilometers meters away from mine site (Typic Haplanthrepts)							
A	0-20	6.6	24.3	2.3	46.4	SL	10.6	2.9
E	20-26	6.0	11.8	11.8	41.3	SL	11.9	2.0
Btl	26-65	6.2	9.2	9.2	40.6	SL	13.6	3.0
Bt2	65-120	6.5	7.6	7.6	40.3	SL	14.9	1.8
BC	120-150	6.4	5.2	5.2	40.1	LS	12.0	1.6
	Mean	6.3	11.6	11.6	41.7	-	12.6	2.3

S= sand, LS= loamy sand, SL=sandy Loam, BD = bulk density, TP= total porosity, MC= moisture content, CV = coefficient of variation, SCR = silt clay ratio, G. mean= grand mean, G. CV = grand coefficient of variation.

Table 4. Statistical tests of difference for selected properties in mine and natural soils (p = 0.05).

Soil property	Calculated t– values
pH water	1.862[NS]
OM (g kg^{-1})	3.940*
TN (g kg^{-1})	2.939*
Avail. P (mg kg^{-1})	2.617[NS]
C/N -	2.739*
Ca/Mg -	2.983*
SCR -	2.886*
Clay ((g kg^{-1})	3.623*
MC (g kg^{-1})	2.932*
B.D. (Mg m^{-3})	0.768[NS]

OM=organic matter, TN=total nitrogen, Avail. P =available phosphorus, C/N=carbon-nitrogen ratio, Ca/Mg =calcium-magnesium 1atio, SCR =silt-clay ratio, MC=moisture content, BD =bulk density * =significant at 5% level of probability, NS=not significant

4. Conclusions

The study revealed differences in some morphological and physiochemical soil properties among geographically-associated mine soil units. Soils closest to the mine site were least differentiated based on distinctness of soil horizons while most distal soils exhibited pronounced soil formation as indicated by Bt (argillic) horizon. Bulk density and moisture content decreased away from the mine site. Soil pH and calcium-magnesium ratio decreased in soils proximal to the mine site.

References

[1] Shukla, M.K, Lal, R., Underwood , J. and Ebinger, M. 2004. Physical and hydrological characteristics of reclaimed mine soils in southeastern Ohio. Soil Sd. Soc Arn.J., 68:1352.1359.

[2] Rodrigue,J.A. and J.A.Burger.2004.Forest soil productivity of mined land in the Midwestern and Eastern Coalfield regions.Soil Sci. Soc .Am.J.,68:833-844.

[3] Thomas, K.A. and Sencindiver, J.C., Skollusem , J.G. and Gorman, J.M. 2000a. Soil horizon development on a mountain surface mine in southern vest Virginia Greenlands 30:41-52 Rodrigue, 3.A . and Burger , J.A. 2004 . Forest soil productivity of mined land in the Midwestern an d eastern coalfield regions. Soil Sci. Soc. Am. J., 68:833-844

[4] Gorman, J.M. and Sencindiver, J.C. 1999. Changes in minesoil physical properties over a mine year period. Proceeding of the Annual National meeting of the American society of surface mining and Reclamation at Scottsdale, Az, pp. 245-253.

[5] Haerring, K.C; Daniels, W.L. and Roberts, J.A.. 1993. Changes in mine soil properties resulting from overburden weathering J. Environ. Qual.

[6] Koose , A 2000 . Pedogenesis in the human environment. Proceedings of first International Conference on soil of Urban Industrial Traftic and mining Areas (SUTTOMA) at University of Essen, pp. 24 1-246.

[7] Khotchanin, K., Thanasuthipitak, P. and Thanasuthpipitak, T. 2010. characteristics of' trapiche blue sapphires from southern Vietnam. Chiang Mai J.Sci.. 39(1): 64- 73.

[8] Thurman, N.C. and Sencindiver, J.C.. 1986. Properties, classification and interpretation of minesoils at two sites in West Virginia. Soil Sci. Soc. Am. .1. 50:15 1-185:

[9] Taiz, L. and Zeiger, E 1991. Stress physiology. In: plant physiology, Benjamin cummings PubI. Co. California. Pp. 362-3 64.

[10] [10] Bendfeldt, E.S; Burger, J.A. and Daniels, W.L.. 2001. Quality of amended mine soils after sixteen years. Soil Sci Soc. Am. J., 65:1736-1744.

[11] Onweremadu, E.U., Izuogu, O.P. and Akamigbo, E.O.R.. 2010. Aggregation and pedogenesis of seasonally inundated soils of a tropical watershed. Chiang Mai J.Sci, 37 (I): 74—84.

[12] Reich, P.F., Number, S.T. Alrnaraz, R.A. and Eswaran, H. 2001. Land resource stresses and desertification in Africa. Agro-Science, 2(2): 1-10.

[13] Onweremadu, E.U. 2008. Evaluating soil structure and hydraulic conductivity by land use in Nigeria. Soil Surv. Horiz; 49:6-11.

[14] Onweremadu, E.U., Izuogu, O.P. and Akamigbo, E.O.R.. 2010. Aggregation and pedogenesis of seasonally inundated soils of a tropical watershed. Chiang Mai J.Sci, 37 (I): 74—84.

[15] Orajaka, S.O. 1975. Geology. In: Ofornata, G.E.K. (ed.) Nigeria in maps: Eastern States. Ethiope Publishing House, Benin City. Pp. 5-7.

[16] Ofomata, G.E.K. 1975. Relief. In: Ofomata, G.E.K. (ed.) Nigeria in maps: Easterm States. Ethrope publishing House, Benin City. pp. 5-7.

[17] Monanu, P.C. 1975, Temperature and Sunshine. In: Ofornata, G.E.K. (ed.). Nigeria in maps: Eastern states. Ethiopic Publishing House, Benin city. pp. 16-1 8

[18] FAO (Food and Agriculture Organization). 2006. Guidelines for soil profile description. 3 editions, FAO Rome. 7Opp.

[19] Gee, G.W. and Or, D. 2002. Particle size analysis. In: Dane, J.H. and Topp, GC. (ed.) Methods of soil analysis, part4. physical methods. SSSA Book Senes Nos, ASA and SSSA, Madison, W.1. pp. 255-293.

[20] Grossman, R.B. and Reinsch, T.G. 2002. Bulk density and linear extensibility. In Dane, J.H. and top, G.C. (eds). Methods of soil analysis. Physical methods. Soil Sci. Soc. Am. Book Series, AsA and SSSA, W1.pp. 20 1-22

[21] Foth, H.D. 1984. Fundamentals of soil science. 7thi edition: John Wiley and Sons, New York. 435 pp.

[22] Watson, W.E. and Brown., J.R. 1998. pH, And lime requirement. In: Recommended chemical soil test procedures for the north central region North Central Regional Research Publ. 221, Missouri Agri. Exp. Stu, Colombia. Pp 13-16.

[23] Nelson, D.W. and Sommers, L.E. 1982. Total carbon, organic carbon and organic matter. In: page. A.L.; Miller RI-I. and Keeaey, D.R. (eds). Methods of soil analysis. 2uid ed. Agron. Monogr. ASA and SSSA, Madesion, Wl.pp. 539-579.

[24] Brernner, J.M. 1996. Nitrogen-total. In: Sparks, D.L. (ed.) Methods of soil analysis, 2'ed:'Agrom—Monogr; ASA-andSSSA; Madison;WI:Pp. 1085-1 12 I: -

[25] Olsen, S.R. and Sommers, LB. 1982. In: Page, A.L; Miller, R.H. and Keeney, DR. (eds). Methods of soil analysis, 2' ed. Agron. Monogr. ASA and SSSA. Madison, WI. pp. 403-430.

[26] Mclean, E.O. 1982. Soil pH and lime requirement. In: page, A.L., Miller, R.H. and Keeney, D.R. (eds) Methods of soil analysis. 2nd ed. Agron. Monogr. ASA and SSSA, Madison, WI.pp.199-234.

[27] Jackson, M.L. 1958. Soil chemical analysis. Prentice Hall Englewood cliffs Nev Jersey. 498 pp.

[28] Schafer, W.M., Nielsen, G.A. and Nettleton, W.D.. 1980. Minesoil genesis and morphology in a spoil chronosequence in Montana. Soil Sci. Soc. Am. J.. 44:802-807.

[29] Daniels, W. 1999. Creation and management of productive mine soils. Povell project Reclamation: Guidelines for surface —mined land in southwest Virginia l2Ipp.

[30] Thomas, K.A., Sencindiver, iC.. Skousen, J.G and Gorman, J.M.. 2000b. Soil horizon development on a mountaintop surface mine in southern West Virginia. http://www.wvu .edu/-agexten/landrec/soilhori.htm.

[31] Sobek, A.A.,. Skousen , J.G and Fisher, S.E.. 2000. Chemical and physical properties of overburden and minesoils. pp. 77-104. In: R.I. Barrihisel et al. (eds) Reclamation of drastically disturbed lands. Agron. Monogr. 41. ASA, CSSA and SSSA, Madison, WI

[32] Ahn, P.M. 1979. West African soils. Oxford Univ. Press, Oxford.Great Britain.332pp. Andrews, J.A., Johnson, J.E. . Torbert, J.L Burger, J.A. and Kelting. D.L.. 1998, Mine soil and soil properties associated with early height growth of eastern white pine, J. Environ. Qual., 27: 192-199 (Ahn. 1979).

[33] Landon. T.R. 1991. Booker tropical soil manual: A handbook of soil survey and agricultural land evaluation in the tropics and subtropics. Longrnan Scientific & Technical, U.K. 474 pp.

[34] Jandi, R; Alewell, C. and C. Prietzel, C. 2004. Calcium loss in central European forest soils. Soil Sci Soc. Am. J., 68: 588-595.

[35] Landon. T.R. 1991. Booker tropical soil manual: A handbook of soil survey and agricultural land evaluation in the tropics and subtropics. Longrnan Scientific & Technical, U.K. 474 pp.

[36] Kohler, M., von Wilpert, K. and Hildebrand. E.E. 2000. The soil skeleton as a source for the short term supply of base cations in forest soils of the Black Forest (Germany). Water, Air, Soil pollut; 122: 37-48.

3

Determination of Appropriate Planting Time for Dekoko (*Pisum sativum var. abyssinicum*) Productivity Improvement in Raya Valley, Northern Ethiopia

Berhane Sibhatu[1, *], Hayelom Berhe[2], Gebremeskel Gebrekorkos[1], Kasaye Abera[1]

[1]Department of Agronomy, Ethiopian Institute of Agricultural Research, Mehoni Agricultural Research Center, Maichew, Ethiopia

[2]Land and Water Research Process, Ethiopian Institute of Agricultural Research, Mehoni Agricultural Research Center, Maichew, Ethiopia

Email address:

berhane76@gmail.com (B. Sibhatu), hayelomberhe631@yahoo.co.uk (H. Berhe), gebremeskel12@gmail.com (G. Gebrekorkos), kasayeab123@gmail.com (K. Abera)

*Corresponding author

Abstract: Dekoko is highly appreciated by the local people for its taste and high market value. However, productivity of Dekoko is limited by improper planting time. An experiment on Dekoko planting time was, therefore, conducted in 2013 and 2014 cropping seasons to determine the appropriate planting time of Dekoko that maximizes its productivity under rain fed conditions. Treatments comprised combinations of four planting time (dry planting about 5-7 days before the beginning of main rain season, when the rain fall amount received greater or equal to 10 mm at once or cumulative, when the rain fall amount received greater or equal to 20 mm at once or cumulative and when the rain fall amount received greater or equal to 30 mm at once or cumulative) were carried out in Randomized Complete Block Design (RCBD) with three replications. The analyzed result showed that days to maturity, number of pods plant^{-1}, grain and biomass yields were significantly influenced ($P<0.05$) by planting time. Dekoko matured late during dry planting. Dekoko planted when the rain fall amount received is greater or equal to 20 mm at once or cumulative gave high (21) number of pods plant^{-1}. Similarly, the maximum grain (533.53 - 638.00 kg ha^{-1}) and biomass (1635.23 - 1820.06 kg ha^{-1}) yields were produced during planting time when the rain fall amount received is greater or equal to 20 mm at once or cumulative, while the minimum values were due to dry planting. It is, therefore, concluded that planting of Dekoko when the rain fall amount received is greater or equal to 20 mm at once or cumulative can be recommended for the growers in the study area to improve Dekoko productivity. Moreover, further research works on different varieties along with different soil moisture levels, planting dates and soil types can be a step forward to identify best sustainable technology on the growth and yield improvements of Dekoko.

Keywords: Dekoko, Planting Time, Yield, Yield Components

1. Introduction

Cool-season food legumes (CSFLs), which are largely produced in Ethiopia, are mainly produced by subsistence farmers and serveas supplementary protein sources and soil fertilityrestorers. Among the CSFLs, a pea variety locally called Dekoko (*Pisum sativum* var. *abyssinicum*) is a unique crop developed and cultivated in Ethiopia [1]. It is restricted to highland regions of Ethiopia (South Tigray and North Wollo) [2].

Dekoko has highly appreciation for its taste and obtains a premium price in local markets compared to field pea or 'Ater' (*Pisum sativum var. sativum*) [2]. Farmers and consumers call it as the "Dero-Wot of the poor". This may be to express for its good taste and high nutritional value. Most often, the dry seeds of Dekoko are decorticated and split ('split peas') before boiling. According to [3], in Ethiopia the annual consumption per person of field pea including Dekoko seeds is estimated at 6 -7 kg. Because of its favorable amino acid profile, it can be a suitable complementary protein source for a cereal based diet. Moreover, its early maturation can make it an important crop in areas where the growing season is too short for other cool season food legumes (CSFLs) and yield losses caused by terminal droughts are common [2].

Appropriate planting time has a promising impact in improving the productivity of legumes. The time from sowing to seedling establishment is of considerable importance in crop production as it has major impacts on crop germination, growth, final yield, and post harvest grain quality [4]. Planting time is adjusted with the availability of soil moisture. Soil moisture supply is an important environmental factor controlling germination and seedling establishment [5]. High seed emergence and seedling establishment contribute directly to the crop yield [6].[7] reported that rapid seed emergence along with fast plant growth and early maturity substantially contributes to high chickpea yield under drought conditions. One of the first physiological disorders taking place during seed germination under dry conditions is a decrease in water uptake by the seed due to low water potential of the germination medium. Slow or sporadic germination and emergence generally result in fewer and small plants, which are more vulnerable to different biotic and abiotic stresses [4]. According to [8], lack of adequate soil moisture in the seedbed is a major hindrance to the establishment of chickpea crop. This is because inadequate soil moisture can reduce seed germination, slow down seedling growth and diminish yield in rainfed crops. Thus, soil moisture is important for early seedling emergence. Similarly, [9] reported that high soil moisture make early seedling emergence of chick pea genotypes; the fast emerged seedlings were taller, produced more branches, developed larger leaf areas and accumulated more above ground biomass as well as specific leaf area than the slow emerging genotypes. According to [10], planting dates should be based on the onset of the rainfall of the growing season in each agro-ecology/location. It is advisable to plant early whenever there is enough moisture in the soil to benefit from higher soil fertility present at the beginning of the rainy season and to achieve physiological maturity before the end of the rainy season.

Dry lands experience unreliable and erratic rainfall that is always inadequate for crop germination, growth and development to reach maturity. With adequate soil moisture chickpea germination percentages are reported to be high leading to high crop yields. With low soil moisture regimes as it is in arid and semi arid lands (ASALs), poor crop germination is experienced leading to poor crop stand and hence low crop yields [11].

There is, therefore, need to explore various technologies that can ensure early and uniform crop germination that will enhance optimum crop stand and establishment that will lead in to optimal crop yields in the ASAL environments. Dekoko, which can grow in the ASAL environments, is the most neglected pulse crop in the research area. A research has not yet been done on yield improvement and development of

management practices for Dekoko in the study area. As a result of this, its productivity is low because of mainly poor agronomic practices. Hence, this study was aiming to determine the optimum planting time of Dekoko under varying initial amount of rain fall for its productivity improvement.

2. Materials and Methods

2.1. Description of Experimental Area

The experiment was carried out under rain fed conditions in 2013 and 2014 cropping seasons at Mehoni Agricultural Research Center testing site. It is 678 km north of Addis Ababa (the capital city of Ethiopia). The area is situated at an altitude of 1578 meter above sea level (m.a.s.l) having a mean annual rainfall of 750 mm and its average minimum and maximum annual temperature is 18°C and 25°C, respectively. It lies at latitudes of 12°41'50" N and longitudes of 39°42'08" E. The textural class of the soil is Clay loam with a pH value of 7.9 [12].

2.2. Field Experimental Design and Procedures

The experiment was consisting of four treatments of planting dates. The treatments included dry planting about 5-7 days before the beginning of main rain season, when the rain fall amount received greater or equal to 10 mm at once or cumulative, when the rain fall amount received greater or equal to 20 mm at once or cumulative, when the rain fall amount received greater or equal to 30 mm at once or cumulative. The amount of rain fall was measured in the metrological station of the research site. The treatments were arranged in a randomized completed block design (RCBD) with tree replications having a plot size of 6m x 5m. The spacing between blocks and plots was 1.5m and 0.5m, respectively. Urea and Triple super phosphate (TSP) were used as source of N and P, respectively. Full dose of P (20 kg P ha^{-1}) in the form of P_2O_5 was applied at planting as broad casted application method. Similarly, 23 kg ha^{-1} of N was applied as a starter during planting time of each treatment. Local variety of Dekoko was used as a test crop. The other crop management practices like weeding, thinning and chemical spraying were applied uniformly for all plots as per recommendations in field pea.

2.3. Data Collection and Statistical Analysis

Data on days to 90% maturity, plant height (cm), pod number plant^{-1}, seed number pod^{-1}, grain yield (kg ha^{-1}), biomass yield (kg ha^{-1}) were collected and analyzed. The data were collected from the net harvestable area of 5.5 m by 4.5 m. Five plants from the net plot area were pre tagged to collect data of plant height, pod number plant^{-1} and three pods per each of these plants with a total of fifteen pods were considered to determine seed number pod^{-1}. Dry matter was measured using electronic sensitive balance after the net plot area plants had been harvested and oven dried at 70°C till constant dry weight was attained. Similarly, shelled seed yield was taken at 10.5% adjusted moisture level using electronic sensitive balance from the harvested plants of net plot area.

The collected variables were subjected to the analysis of variance using the SAS software version 9.1 [13] and significance difference among the treatment means was computed with least significant difference (LSD) at 5% probability level as cited in [14].

3. Results and Discussion

3.1. Days to 90% Physiological Maturity

Days to 90% physiological maturity was significantly influenced due to planting time in both cropping seasons (Table 1). Dry planting of Dekoko took more time (86.67 days) than the others. It was significantly superior (P<0.05) to the others. The other treatments were significantly similar to each other. Time taken to maturity was remarkably prolonged with a decline of soil moisture content. In other words, it has been noted that on leaving the sown seeds for a considerable period at low soil moisture content, the time to emergence becomes maximally prolonged which in turn results delay of maturity. This finding was in agreement with the findings of [9] who noted that emergence of chick pea genotypes was significantly influenced due to soil moisture at which low soil moisture prolonged seedling emergence.

Table 1. Effect of planting time on mean values of days to 90% maturity and plant height of Dekoko.

Treatments	Days to 90% Maturity			Plant height (cm)		
	2013	2014	Mean	2013	2014	Mean
Dry planting	86.67a	85.67a	86.17	41.33	56.67	49.00
When rain fall ≥10 mm	81.33b	82.00b	81.67	36.67	61.67	49.17
When rain fall ≥20 mm	81.33b	82.00b	81.67	36.33	63.00	49.67
When rain fall ≥30 mm	80.67b	81.67b	81.17	39.00	62.33	50.67
CV (%)	2.58	1.75		12.93	6.03	
LSD (0.05)	4.25	2.90		NS	NS	

Means with the same letter (s) in the same column are not significantly different at P<0.05; NS= Non-significant; LSD= least significant difference; CV= Coefficient of variance

3.2. Plant Height

According to Table 1, plant height was not significantly affected by planting time in both cropping seasons. Generally, the plant height was ranged from 49 cm to 51 cm. However, plant height significantly differed (P< 0.05) due to soil moisture difference at planting time at which the shortest plant height of pea was obtained under limited soil moisture content [9]. Moreover, a substantial reduction in soybean plant height as a result of a short period of water stress has been reported by [15].

3.3. Number of Pods Plant[-1]

Concerning to pods plant[-1], it was significantly influenced

(P<0.05) due planting time in both cropping seasons. In 2013 cropping season, the highest number of pods plant[-1] (20.67) was obtained from the time of Dekoko planting when the rain fall amount received is greater or equal to 20 mm at once or cumulative, and it was statistically at par with the other treatments excluding dry planting. However, the lowest number of pods plant[-1] (17.00) was gained when Dekoko was planted at dry condition. similarly, in 2014 cropping season, the highest value (21.33) was obtained from the time of Dekoko planting when the rain fall amount received is greater or equal to 30 mm at once or cumulative, and it was statistically at par with the other treatments excluding dry planting. Like to 2013, Dekoko planted during dry soil condition, the number of pods plant[-1] got reduced by 15.61%. This ascribed to the essential of water for the maintenance of the turgidity necessary for cell enlargement and plant growth as well as its major constituent of physiologically active tissue which brings well performance of the plant to produce its yield components [16].

Table 2. Effect of planting time on pod number plant[-1] and seed number pod[-1] of Dekoko.

Treatments	Pod number plant[-1]			Seed number pod[-1]		
	2013	2014	Mean	2013	2014	Mean
Dry planting	17.00b	18.00b	17.50	3.67c	4.33	4.00
When rain fall ≥10 mm	19.00ab	19.67ab	19.33	4.33bc	4.33	4.33
When rain fall ≥20 mm	20.67a	21.00a	20.83	5.33a	5.33	5.33
When rain fall ≥30 mm	19.00ab	21.33a	20.17	4.67ab	4.67	4.67
CV (%)	5.50	5.34		9.80	18.56	
LSD (0.05)	2.08	2.13		0.88	NS	

Means with the same letter (s) in the same column are not significantly different at P<0.05; NS= Non-significant; LSD= least significant difference; CV= Coefficient of variance

3.4. Number of Seeds Pod[-1]

Planting time caused significant effect on the number of seeds pod[-1] of Dekoko in 2013 cropping season. Based on this result, the highest number of seeds pod[-1] (5.33) was recorded from the time of planting when the rain fall amount received is greater or equal to 20 mm at once or cumulative, and it was statistically similar with the time of planting when the rain fall amount received is greater or equal to 30 mm at once or cumulative. On the other hand, the lowest number of seeds pod[-1] (3.67) was recorded due to dry planting. This might be due to the fact that water is the solvent in which salts, sugars and other solutes move from cell to cell and organ to organ which in turn results in formation of seeds, and water is a reagent in photosynthesis for production of more dry matter and yield. This was confirmed to the report of [16]. In 2014 cropping season, number of seeds pod[-1] was not significantly varied due to planting time where seeds pod[-1] of Dekoko ranged from 4.33 to 5.33.

3.5. Grain Yield

Concerned to grain yield of Dekoko, planting time caused

significant effect (P<0.05) on it in both cropping seasons. Accordingly, in 2013cropping season, the highest grain yield (533.53 kg ha⁻¹) was obtained during planting time when the rain fall amount received is greater or equal to 20 mm at once or cumulative, and unlike to dry planting, it was statistically at par with the other treatments (Table 3). Nevertheless, the lowest grain yield (448.47 kg ha⁻¹) was gained due to dry planting. A similar trend was occurred in 2014 cropping season where the highest value (638.00 kg ha⁻¹) was recorded when the rain fall amount received is greater or equal to 20 mm at once or cumulative, while the lowest (545.33 kg ha⁻¹)

was obtained due to dry planting of Dekoko (Table 3). This was attributed to the role of moisture in early vegetative growth and enhancing leaf area which in turn results in accumulation of more yields. This confirmed to the previous findings of [7] that showed that rapid seedling growth has been found to be associated with well established seedling and subsequent early growth in chickpea which in turn contributes favorably to high yield under drought conditions. Likewise, [17] reported that reduced soil moisture during the early seedling stage diminishes leaf growth which in turn would result in a reduction of both dry matter and yield.

Table 3. Mean values of grain and biomass yields of Dekoko as influenced by planting time.

Treatments	Grain yield (kg ha⁻¹)			Biomass yield (kg ha⁻¹)		
	2013	2014	Mean	2013	2014	Mean
Dry planting	448.47b	545.33b	496.90	1386.78b	1428.95b	1407.86
When rain fall ≥10 mm	493.02ab	563.33b	528.18	1496.60ab	1580.22ab	1538.41
When rain fall ≥20 mm	533.53a	638.00a	585.77	1635.23a	1820.06a	1727.65
When rain fall ≥30 mm	518.31a	600.33ab	559.33	1536.97ab	1730.78a	1633.87
CV (%)	5.41	5.08		5.26	7.85	
LSD (0.05)	53.83	59.56		159.04	257.17	

Means with the same letter (s) in the same column are not significantly different at P<0.05; LSD= least significant difference; CV= coefficient of variance

3.6. Biomass Yield

Referring to Table 3, the treatments exhibited substantial variations for biomass yield. In 2013, the highest biomass yield (1635.23 kg ha⁻¹) was produced during planting time when the rain fall amount received is greater or equal to 20 mm at once or cumulative, and unlike to dry planting, it was statistically at par with the other treatments. It gave 17.92% more biomass yield than dry planting at which the lowest result was obtained. Correspondingly, in 2014, the maximum biomass (1820.06 kg ha⁻¹) was produced during planting time when the rain fall amount received is greater or equal to 20 mm at once or cumulative (Table 3). Nevertheless, the lowest value was obtained due to dry planting which was 21.49% lower than the maximum biomass yield. This could be due to soil moisture stress during seedling emergence which retarded vegetative growth of the plant. This result is consistent with the work of [9] who reported who reported that biological yield of chick peas was significantly influenced due to soil moisture at which low soil moisture declined the yield. Similar findings have been also reported by [11] who demonstrated that water deficit decreased dry matter accumulation (biological yield) and grain yield per unit area of chick pea.

4. Conclusions

Appropriate agronomic management of Dekoko improved its productivity. Identifying optimum planting time is one of the agronomic practices that enhanced Dekoko yield and yield related traits. Accordingly, the findings of this experiment showed that number of pods plant⁻¹, grain and biomass yields were significantly affected by planting time of Dekoko. The highest pods plant⁻¹ was obtained during planting time when the rain fall amount received is greater or equal to 20 mm at once or cumulative. Moreover, the maximum grain and biomass yields were produced during

planting time when the rain fall amount received is greater or equal to 20 mm at once or cumulative in both cropping seasons. However, the lowest values of these variables were recorded due to dry planting. Generally, early planting is advisable whenever there is enough moisture in the soil to achieve physiological maturity before the end of the rainy season. It is, therefore, concluded that planting time of when the rain fall amount received is greater or equal to 20 mm at once or cumulative can be recommended for the growers in the study area to improve Dekoko productivity. Moreover, it can recommend from the findings that further investigation on different varieties along with different soil moisture levels, planting dates, soil types and Integrated Pest Management (IPM) techniques can be a step forward to identify best sustainable technology on the growth and yield improvements of Dekoko.

Acknowledgements

The authors are thankful to Ethiopian Institute of Agricultural Research (EIAR) for funding this research project. The authors are also sincerely acknowledging the Mehoni Agricultural Research Center for providing all farm facilities.

References

[1] Haddis, Y., Hussein, M., Berhanu, A., Birhanu, A. "Association of traits with yield in Dekoko (*Pisum sativum* var. *abyssinicum*) accessions in the highlands of southern Tigray, Ethiopia." *African Journal of Agricultural Research*, 10(12): 1480-1487, 19 March 2015.

[2] Yemane, A., and Skjelvag, A. O. "The physic-chemical features of Dekoko (*Pisum sativum* var. *abyssinicum*) seeds." *J. Agron and Crop Sci.*, 189: 14-22, 2002.

[3]   Sentayehu, A. "Assessment of Nutrient Contents of Different Field Pea Genotypes (*Pisium sativum*L.) in South west Ethiopia." Department of Plant Sciences, Jima University, Jimma, Ethiopia, 2009.

[4]   Harris, D., Breese, W. A., and Kumar Rao, J. V. D. K. "The improvement of crop yield in marginal environment using "On farm" seed priming: Nodulation, nitrogen fixation and disease resistance." *Australian Journal of Agronomy*, 56 (11): 1211-1218, 2005.

[5]   Tylor, A. G., Motes, J. E. and Kirkham, M. B. "Germination and Seedling Growth Characteristics of Three Tomato Species Affected by Water Deficits."*Hort. J.*, 107: 282-285, 1982.

[6]   Maiti, R. K. and Moreno-Limon, S. "Seed and Seedling Traits in Bean (*Phaseolus vulgaris* L.) and Its Relation to Abiotic Stress Resistance." *Legume Res.*, 24 (4): 211-221, 2001.

[7]   Gupta, S. N. "Studies on Genetic Variability for Drought Resistance in Chickpea."*Seed Res.*, 25 (1): 19-24, 1985.

[8]   Sharma, R. A. "Influence of Drought Stress on the Emergence and Growth of Chickpea Seedlings." *International Chickpea Newsletter*, 12: 15-16, 1985.

[9]   Hosseini, N. M., Siddique, K. H. M., Palta, J. A.. and Berger J. "Effect of Soil Moisture Content on Seedling Emergence and Early Growth of Some Chickpea (*Cicer arietinum* L.) Genotypes." *J. Agric. Sci. Technol*, 11: 401-411, 2009.

[10]   Kidane, G. *Dryland Agriculture Production System in Ethiopia*. Addis Ababa, Ethiopia: Ethiopian Institute of Agricultural Research, 2015, pp. 116.

[11]   Kamithi, K. D., Kibe, A. M. and Wachira, F. "Effect of different initial soil moisture on desi chickpea ICCV 95107 (*Cicer arietinum* L.) dry matter production and crop growth rate." *International Journal of Environmental and Agriculture Research* (IJOEAR), 1 (6): 1-10, October 2015.

[12]   Haileslassie, G., Haile, A., Wakuma, B., and Kedir, J. "Performance evaluation of hot pepper (*Capsicum annum* L.) varieties for productivity under irrigation at Raya Valley, Northern, Ethiopia." *Basic Research Journal of Agricultural Science and Review*, 4 (7): 211-216, July 2015.

[13]   SAS (Statistical Analysis System) Institute. SAS User Guides, Version 9.1. North Carolina, USA: SAS Inc. Cary, 2004.

[14]   K. A. Gomez and A. A. Gomez. *Statistical Procedures for Agricultural Research, 2nd edition*. New York. John Viley and Sons Inc., 1984, pp. 121-35.

[15]   Momen, N. N., Carlson, R. E., Shaw, R. H. and Arjmend, O. "Moisture Stress Effects on Yield Components of Two Soybean Cultivars." *Agro. J.*, 71: 86-90, 1979.

[16]   Kramer. P. J. "Water stress and plant growth."*Agron. J.* 55: 31-5, 1963.

[17]   Constable, G. R. and Hern, A. B. "Agronomic and Physiological Responses of Soybean and Sorghum Crops to Water Deficits. I. Growth, Development and Yield."*Aust. J. Plant Physio.*, 5: 159-167, 1978.

4

Rapid Risk Reduction Strategies Using Some Horticultural Plants in a Changing Atmosphere among Urban and Peri-Urban Centres of the Atlantic Coast in Nigeria

E. U. Onweremadu[1, *], A. C. Udebuani[2], Egbuche C. T.[3], Ndukwu B. N.[4]

[1]Department of Soil Science and Technology Federal University of Technology, Owerri, Nigeria
[2]Department of Biotechnology, Federal University of Technology, Owerri, Nigeria
[3]Department of Forestry and Wildlife Technology, Federal University of Technology, Owerri, Nigeria
[4]Department of Agricultural Extension, Federal University of Technology, Owerri, Nigeria

Email address:
uzomaonweremadu@yahoo.com (E. U. Onweremadu)

Abstract: There are irregular global changes in climatic attributes. Nigeria is not left out in the unpredictable atmospheric variability especially in its coastlands. The situation has led to varying forms of environmental challenges, calling for rapid risk reduction responses. This paper suggested four technologies namely, vegetable intercrop production, improved fallow systems, biomass technology and night-soil technologies as efficacious in sequestering atmospheric carbon directly or indirectly. These technologies are easily adaptable in the agro ecological zone following its characteristic multifloristic structure and climatic peculiarities as well as demographic attributes. Coastland climate change adaptation and irrigation experimental stations should be established in the area to evaluate efficacy of these technologies.

Keywords: Climate Change, Risk Response, Settlement, Resilience Vulnerability

1. Introduction

Nigeria belongs among the least developed countries (LDCs) with high degree of vulnerability to climate change. Perhaps, Nigeria ranks among the least endowed in will power, resources and technology to minimize the adverse effects of global warming and climate change despite low contribution to the problem. [1] contended that location of poor and LDCs is contributory to their vulnerability to the changing atmosphere, reporting that Nigeria has a potential loss of 11% of her gross domestic product by 2060 with a 2 ^{0}C rise in temperature. It is known that LDCs contribute to climate change through land use and forestry activities. Changes in land use emitted an estimated 1.6 billion t C/Yr during 1990s [2]. Tropical deforestation for agriculture, mining, fuel wood, industrialization and urban expansion are essentially responsible for a good volume of emitted carbon dioxide in Nigeria. A substantial share of Nigerian economy is dependent on climate-sensitive natural resources [3], making Nigeria high vulnerable to risks associated with climate change [4].

Niger Delta region in Nigeria has a coastland area of about 6000 Km^2 in a country of about 8000KM^2 of total mangrove area [5]. A rise in sea level implies an inundation of the coastlands and displacement of human and other terrestrial inhabitants with consequent loss of lives, property and infrastructure. Unprecedented disasters such diseases, total loss and submergence of farms, homes, coastal landslides, etc. are recorded. In this regard, modern engineering practices have brought significant achievements via seismic building codes and flood management systems. Robust disaster risk reduction and preparedness systems have been put in place especially in advanced countries. Yet, there is general lack of awareness on how inhabitants contribute to disaster occurrence and what they can do within their resources to adapt, avoid or minimize them especially in this era of global climate change. There is a change in perception that climate change disasters are largely anthropogenic. A systematic assessment, reduction and management of a good number of risks associated with climate change can be done at communities semi peri-urban, urban and rural settings. The coastlands in Nigeria in addition to aquatic resources are endowed with timber, fuelwood, pulpwood (paper and match making), fruits (*Rativenear Dacryodes edulis; Cola; Cola lepidota*), spices, food condiments and sweeteners (*Irvingia, Monodora, Piper guineense*), drug plants (Bitter Kola: *Garcinia kola; Neem Azadiraditha Indica*), fibre (Raphia Horticultural Crops (Bananas, Plantain), tuber crops (Yam, Cocoyam), tree crops (Oil Palms, Rubber), forest trees (Teak, Mahogany, Iroko) and cereals (Maize, Swamp Rice). Some of these agricultural resources are seasonally available, implying climate sensitivity. In addition to this, there is a decline in the agricultural resource base of the coastlands owing to increasing urbanization and conflictive land use types and practices. Seasonal availability of major fruits and vegetables are shown in Table 1 [6]

Table 1. *Seasonal availability of major fruits and vegetables in Nigeria*

Crop	Availability Period
Fruits	
Mango	March – July
Pineapple	November – March
Plantain/Banana	October – January
Citrus	September – November
Guava	November – January
Pawpaw	October – March
Vegetables	
Okra	July – October, January – April
Pepper	April – November, February – June
Amaranthus	July – November
Melon	July – November
Corchons	July to November

Adapted from Babatola (2001)

Crude oil exploration activities in the area with attendant oil spills has led to declining biodiversity, impeded farming activities and an attitudinal change where inhabitants lean towards crude oil fallouts for livelihood. The growth of urban and peri-urban towns in Nigeria is at an exponential rate, with a high tendency in crude oil-rich coastlands of the country. Since increasing urbanization implies reducing available space for agriculture and forestry spaces with attendant increase in greenhouse gases and heightened household expenditure on food items, it becomes necessary to adopt strategies that with capture emitted GHGs especially in the urban and peri-urban centres. The strategies suggested in this paper include;
- Vegetable/inter crop production
- Use of improved fallow systems
- Biomass technology
- Night soil technology

Vegetables intercropped with two or more arable crops on the same piece of farmland such that the periods overlap is long enough to include vegetative stages which will absorb emitted CO_2 from the atmosphere. Vegetables intercropped with legumes will ensure optimum ground cover, optimum use of sunlight, more efficient root growth, spreads risks of crop failure over more crops due to multiple cropping and reduces effects of pests and diseases, a majority of which are crop

specific. With increasing population in urban and peri-urban centres, tuber crop production leads to nutrient mining, moreso with shortened fallow periods. Future of tuber crop production (yam, cocoyam, cassava) depends on development of cropping systems that enhance soil fertility within short fallow periods. [7] noted that such cropping systems must ensure adequate nutrients content in the soil as well as promote large quantity of biomass. Total forest areas under forest cover, chemical and physical properties of some coastland have been documented by [8] and [9]. Although, staple crops such as maize (*Zea mays*), plantains (*Musa paradisiaca*), banana (*Musa sapientum*), even fruits such as pawpaw (*Carica papaya*), and tubers (yams: *Dioscorea spp*)are grown, urban and peri-urban populations tend to be more adapted to vegetable crop production. Vegetables grown outside urban and peri-urban centres cannot reach the market on time [10] and advantages of vegetable crop production are so many and include short shelf life [11], low transportation costs, independence from middlemen and fast access to market [12], Vegetable crop production ensures continuous production even during dry periods with simple irrigation techniques. Vegetable crop production is preferred in situations when land tenure is insecure. Urban and peri-urban small-scale farmers in Nigeria earn little or nothing from staple crops (cassava, banana, maize, yams and plantain) due to competition from rural producers [13], thereby promoting the role of vegetable production as income earner for urban dwellers.

Table 2. *Some Chemical Properties of Coastland Soils at Koko, Delta State*

Sample	pH Water	OC g/kg	TN g/kg	Av.P mg/kg	Ca cmol kg	Mg	Na	ECEC
1	4.70	12.0	1.2	8.76	0.5	0.4	0.1	10.3
2	6.66	13.0	1.1	3.59	0.6	0.6	0.1	5.5
3	4.79	9.8	1.0	3.30	0.9	0.3	0.1	5.2
4	4.72	9.7	1.0	2.60	0.7	0.3	0.1	4.6
5	5.20	10.0	1.0	1.29	0.5	0.6	0.1	3.6
6	5.54	9.9	1.0	5.89	0.6	0.6	0.2	7.7
7	5.26	10.9	1.1	3.30	0.6	0.5	0.1	5.2
8	5.43	11.6	1.2	2.41	0.7	0.4	0.11	4.2
9	5.22	11.0	1.1	1.06	0.6	0.4	0.1	2.7
10	5.63	9.9	1.0	4.30	0.6	0.4	0.1	6.0
Mean	5.32	9.8	1.1	3.65	0.6	0.5	0.1	5.5

Adapted from: Imadojemu (2011)

Table 3. *Some Physical Properties of Coastland Soils at Koko, Delta State*

Sample	Sand (g/kg)	Silt (g/kg)	SCR
1	912	58	1.93
2	807	47	1.02
3	792	108	1.08
4	737	67	0.34
5	812	163	6.54
6	832	138	4.60
7	802	62	0.46
8	772	137	1.43
9	882	68	1.36
10	842	88	1.23
Mean	819	94	1.99

SCR=silt clay ratio

Table 4. Standard Set of 25% of Total Land Area Under Forest Cover

State	Total land area (ha)	Area of forest reserve (ha)	Standard set of the total land area (ha)	Deficiency % (ha)	% deficiency (ha)	% available (ha)
Lagos	3393900	6773	98475	91602	23.26	1.74
Ogun	1608600	275362	402150	126788	7.88	17.12
Ondo/Ekiti	2045100	305541	511275	205734	10.06	14.94
Osun	979100	91268	237275	146007	15.39	9.26
Oyo	2784800	169173	696200	527027	18.93	6.07
Southwest	7781500	848217	1945375	1097158	14.10	10.9

Source: Faleyimu et al. (2009).

Table 5. Effect of Ash on Growth and Yield of Amaranthus

Ash (t/ha)	Fresh mater yield/plant (g)	Seed weight per plant (g)	Root length (cm)	Root weight (g)
0	100.7	33.3	17.8	10.5
2	279.5	85.6	20.8	27.8
4	168.5	92.8	18.8	25.0
6	161.1	92.1	18.8	18.4
LSD$_{0.05}$	26.1	7.5	2.6	4.5

Adapted from Ojeniyi et al. (2001).

Table 6. Effect of Wood Ash on Celosia

Ash (t/ha)	Plant height (cm)		Number of leaves	
	5WAP	6WAP	5 WAP	6 WAP
0	8.5	9.7	13.1	17.2
2	11.2	13.2	13.7	17.7
4	11.5	13.4	15.5	19.5
6	13.1	15.1	17.1	20.0
LSD$_{0.05}$	13.3	15.3	17.3	21.3
	1.6	6.2	4.2	2.5

WAP = Weeks after planting
LSD = Least significant difference
Adapted from Ojeniyi et al. (2001)

2. Assessment

Urban and peri-urban vegetable production can utilize organic urban wastes, a majority of which are burnt in incinerators, releasing CO_2 contained therein. Recycling these organic wastes by incorporating directly or after composting can be an effective and sustainable way of improving soil fertility and reducing disposal costs. Cities are known to have nutrient surpluses [14] but these wastes are dumped into water bodies. Incorporation of organic wastes in vegetable crop production builds soil macro aggregates hence aggregate stability which reduces risk of coastland and riverbank erosion while increasing water storage capacity. Farmers prefer application of municipal wastes on staple crops for fear of contaminating intensive vegetable production [15] but pre-treatment using composting, microbial decomposition and the use of water ferns before utilization [16] can improve its quality. Compost storage reduces pathogen levels (*Ascaris and Trichuris* eggs) markedly especially with increasing pH [17]. Municipal wastes are often criticized for high heavy metal load but this depends on soil and crop factors as well as source of the wastes. Consumers' exposure to heavy metals depends on choice of crop, nature of pollution and preparation method [18], compost fortification as using wood ash to increase soil pH thereby reduces solubilization and availability of heavy metals [19]. (Table 5 – 8)

Table 7. Effect of Wood Ash on Yield Components of Okra (MH-47-4)

Ash (t/ha)	Number of pods		pod weight (g)		Plant height (cm)		Top root length (cm)	
	1999	2000	1999	2000	1999	2000	1999	2000
0	2	3	60.3	51.9	45.9	41.1	12.4	10.2
1	4	4	82.3	67.9	54.5	53.0	13.4	13.4
2	5	4	92.9	94.4	55.5	53.8	13.0	13.8
3	5	5	96.7	101.6	59.7	59.1	14.1	14.2
4	5	5	101.5	103.1	62.7	64.7	16.3	15.0
5	6	6	101.6	109.3	70.0	64.9	14.7	14.4
LSD$_{0.05}$	1.3	1.2	25.6	32.0	8.5	8.4	NS	4.3

NS = Not significant Source: Ojeniyi et al.(2001)

Table 8. Effect of Wood Ash on Leaf Nutrients of Vegetables

Treatment	N	P	K	Ca	Mg
	g/kg				
Amaranthus					
Control	21	1.6	18	2.0	0.5
250kg urea	35	4.0	28	2.4	0.6
2t/ha as	30	3.5	26	3.0	0.9
2t/ha ash + 62.5 kg urea	44	3.7	25	2.8	0.7
2t/ha ash + 125 kg urea	45	3.2	25	2.9	0.8
2t/ha ash + 187.5 kg urea	36	3.0	23	2.6	0.1
LSD$_{0.05}$	0.8	0.9	0.4	0.03	0.01
Celosia					
0t/ha ash	38.8	34.2	20.9	20.8	32.1
2 t/ha ash	40.8	40.3	34.9	35.0	34.3
4 t/ha ash	48.7	43.2	33.1	34.1	35.4
6 t/ha ash	55.7	53.32	48.3	45.8	46.3
8 t/ha ash	49.6	54.2	47.3	56.8	48.3
LSD$_{0.05}$					
Okra					
0t/ha	31	1.5	14.5	4.5	16.7
1 t/ha	34	1.9	15.8	5.1	17.6
2 t/ha	41	2.1	16.1	5.8	17.9
3 t/ha	42	2.3	23.5	6.7	30.0
4 t/ha	46	2.6	24.1	7.2	21.6
5 t/ha	40	2.9	25.7	10.2	22.9
LSD$_{0.05}$					

Source:Ojeniyi et al.(2001)

Municipal solid wastes were used in reclaiming crude oil polluted soils of the Niger Delta region of Nigeria at rates equivalent to 1.2 – 3 t/ha [20]. It increases crop yield, improves soil properties [21] and increases nitrogen fertility in drastically disturbed low N soils [22]. However, application of industrial compost to agricultural soils may increase total concentrations of heavy metals without increasing their phytoavailability [23]. Raising biomass in situ using leguminous trees and/or shrubs improves soil nutrient status. Legumes such as *Leucaena leucocephata*, *Sesbania sesban* and *Gliricidia sepium* can generate abundant biomass in coastland soils of Nigeria. Their rapid and luxuriant growth will ensure soil cover and utilize CO_2 in the process of photosynthesis.

In peri-urban centres, natural fallow lengths may be non-existent, suggesting the use of improved fallows which regenerate soil fertility more rapidly. Two and three year *Sesbania sesban*-based fallows have proved highly effective in soil fertility regeneration in Zambia. [24] reported that maize grain yield following a 3-year *Sesbania sesban* fallow without N-fertilizer in Chipata, Zambia were 2.27, 5.59 and 6.02 t/ha after 1, 2 and 3 years, respectively compared with control plots with 1.6, 1.2 and 1.8 t/ha, respectively.

Green manuring involving the cultivation forage or legumes with high N-content enhances soil fertility and fits into adaptation strategy for climate change. Plants like *Crotalaria juncea* (Sunn hemp) can be used, and incorporated while succulent into the soil. Greening the landscape keeps the soils under cover at all times.

Vegetable crop diversification is an adaptive strategy as it ensures existence of vegetation at all times on coastland farms in urban and peri-urban areas. Multiple cropping arrangements accommodate vegetable crops such as okra (*Abelmoschus esculentus*), fluted pumpkin (*Telfairia occidentalis*), Gnetum (*Gnetum africanum*), Amaranth (*Amaranthus caudatus/Amaranthus cruentus*), jute (*Corchorus olitorius)*, spinach (*Celoisia argentea*) Indian spinach (*Basella alba*), Bitter leaf (*Vernonia amygdalina*) waterleaf (*Talinum fruticosum*) and Moringa (*Moringa oleifera*). Under humid tropical climate, the coastland urban and peri-urban centres of Nigeria, these vegetables grow luxuriantly and vegetatively thereby increasing their capacity to use gaseous carbon (CO_2) and yield profitably thereby reducing poverty. There is mutualistic interaction between these plant types which enhance great, survivability under plant association. Engagement of farmers in multiple vegetable productions on available land space in urban centres guides against crop failure and tendency for climate – unfriendly practices.

In Southeastern Nigeria, human waste has been considered as a valuable source for soil fertility regeneration consequent upon this, heavy feeders such as plantain (*Musa paradisiaca*), banana (*Musa sapientum*), yams (*Dioscorea* spp) and the likes are planted on soils proximal to *abandoned* and old pit latrines. It becomes worthwhile to ignore the social stigma attached to its use.

It is often argued that night-soil use is unhealthful. But, a look at World Health Organization (WHO) guidelines for safe use of wastewater, excreta and grey water convinces one to have a rethink. The WHO [26] states as follow:

- Where faecal matter and other organic materials are composted at ambient temperature, the end-product of such an aerobic composting process does not smell and has good properties as a soil conditioner and slow-release phosphorus fertilizer.
- To minimize health risks from using night-soil as fertilizer, WHO makes various statements and recommendations. Where it is difficult to increase the temperature of the compost heap, WHO recommends prolonged storage to ensure safety. With ambient temperatures of $2 - 20^0C$, they note that storage times of one and a half to two years will eliminate material pathogens; will reduce viruses and parasitic protozoa below risk levels.
- In addition, WHO recommends various pre cautions to control exposure to risk. Precautions for those handling night-soil include wearing personal protection such as boots, gloves, facemasks, and using tools or equipment not used for other purposes.
- At the time of applying the night-soil compost to the field, if the quality cannot be guaranteed, it is recommended to use close to the ground application, working the material into the soil, and covering it. In addition, children should be kept away from all areas where night-soil is prepared, treated or has been applied.
- Finally, WHO notes that domestic and personal hygiene is important. Technology alone cannot stop transmission of diseases, and communities must be aware of good hygiene practices. If treatment recommendations are followed, coupled with good general community hygiene, the risks to be people who collect and use night-soil as well as those consuming fertilized products will be reduced to acceptable levels.

With declining organic matter content of tropical soils following reduction in vegetal sources via deforestation, night-soil composting becomes a reliable source to build up soil fertility. Some scholars [26]; [27] reported benefits from using night-soil. Night-soil has been found to be the best source of nutrients compared to other organic sources as it tends to give a quick response, especially when used as top dressing. In addition to this, an application of night-soil sustains the soil for over three years. It is currently a common practice in Tanzania where farmers buy contents of old toilets, and this is encouraging in this era of organic agriculture and climate change.

3. Conclusion

Climate change is obvious and its attendant problems cannot sustain ecosystem balance, food security and human health. The situation is more intractable in the fragile coastland ecosystems of Nigeria, especially in their urban and peri-urban centres. Following the peculiarities of the study area in question, the paper opined for the use of vegetable intercrop production improved fallows, biomass and night-technologies in reducing risks of atmospheric changes. It is recommended that rapid response experiments be conducted within the ecosystem for indepth knowledge on the suggested adaptive and irrigitative technologies.

References

[1] Mendelsohn, R., A. Dinar and I. Williams. 2006. The distributional impact of climate change on rich and poor countries. Environ. Dev. Econs. 11(2):159-178

[2] Bolin and Sukumar, 2000. Global perspective Chapter 1. In Watson R.T; I. R. Noble; , B. Bolin, N.H Revindrath, D.J. Vennde and D.J. Doklean eds. Landuse, Landuse Change and Forestry, Special Report of the Intergovernmenmental Panel on Clinate Change. Cambridge University Press. Cambridge, England. Pp 23-51

[3] IPCC (Intergovernment Panel on Climate Change) 2002. The report of the working group 1 of the Intergovernmental Panel on Climate Change, Survey for Policymakers.

[4] Boko, M.I. Niang, A. Nyong, C. Vogel, A. Githeko, M. Medany, B. Osaman-Elasha, R. Tabo and P. Yanda. 2007. Climate change 2007: Impact, adaptation and vulnerability in Africa. Contribution of Working Group II to the Fourth Assessment Report of the Intergovernmental Panel on Climate Change. Parry M.L., O.F. Canziani, J.P. Palutikof, P.J. Vander Linden and C.E Hansen (eds.) Cambridge, U.K. Cambridge University Press. Pp. 433-467.

[5] Onofeghara, F.A. 1990. Nigeria Wetland: An overview In: Akpata, TV. and Okali, D.U.U. (eds.) Report of the man and the biosphere (MAB) National Committee, Nigeria. UNESCO National Commission, Federal Ministry of Education.

[6] Babatola, J.O. 2001. Postharvest technology of horticultural crops as a means of improving dietary intake and socioeconomic empowerment of youths in Nigeria. Proc. Of the 35th Ann. Conf. of the ASN at UNAAB, Set. 16-20, 2001, pp. 9-19.

[7] Ano, A.O. and G.C. Orkwor. 2006. Effect of fertilizer and intercropping with pigeon pea (Cajanus cajan) on the productivity of yam minisett (Dioscorea rotundata) based system. Niger. Agric. J. 37:65:73.

[8] Imadojemu, P.E. 2010. Properties of soils as affected by different sources of waste in Koko, Wani, Delta State, Nigeria. An M.Sc. Thesis of the Fed. Univ. of Technology Owerri, Nigeria. 75.

[9] Faleyimu, O.I., B.O. Agbeja and B.O. Oni 2009. National stipulated forest cover: Mirage or reality (Southwest Nigeria experience). Proceedings of the International Conference on Global Food Crisis held at FUTO, Owerri, Nigeria Apid 19-21, 2009. pp. 228-231.

[10] Ratta, A. and J. Nasv. 1996. Urban agriculture and the African food supply system. African Urban Quarterly 11:154-161.

[11] Tixier, P. and H. de Bon. 2006. Urban horticulture In: an

[12] de Neergaard, A., A.W. Drescher and C. Kouame. 2009. Urban and Peri-urban agriculture in African cities. In: Shacklefon, C.M., M.W. Pasquini and A.W. Dresher (eds.). African indigenous vegetables in urban agriculture. Earthscan, London pp. 35-64.

[13] Ezedinma, C. and C. Chukuezi. 1999. A comparative of urban agricultural enterprises in Lagos and Port Harcourt, Nigeria. Environ. & Urbaniz., 11:135-144.

[14] Khai, N.M., P.Q. Ha and I. Oborn. 2007. Nutrient flows in small-scale peri-urban vegetable farming systems in Southeast Asia. A case study in Hanoi. Agric Ecosyst. Environ., 122:192-202.

[15] Eaton, D. and T. Hilhorst. 2003. Opportunities for managing solid waste flows in the peri-urban interface of Bamako and Ougadougou. Environ & Urbaniz., 15:53-64.

[16] Gaye, M. and E. Diallo. 1997. Community participation in the management of the urban environment in Rufisque (Senegal). Environ. & Urbaniz., 9:9-29.

[17] Vinneras, B. 2002. Possibilities for sustainable nutrient recycling by faecal separation combined with urine diversion unpublished Ph.D. Thesis, Acta Universtatis Agricultural sueciac, Agrana, Swedosh Univ. of Agric. Science, Sweden. http://diss-epsilon.slu.se/arlhwe/00000332

[18] Nabulo, G., H. Oryem-Origa and M. Diamond. 2006. Assessment of lead, cadmium and zinc contamination of roadside soils, surface films and vegetables in Kampala City, Uganda. Environ. Res., 1011:42-52.

[19] Ojeniyi, S.O., O.P. and A.A. Aroitoilu. 2001. Response of vegetables to woodash fertilizer. Proc. Of the 35th Ann. Conf. of the ASN at UNAAB, Setp. 16-20, 2001. pp. 39-43.

[20] Ayolagha, G., F.S. Kio-Jack and N.O. Isirimah. 2000. Remediation of crude oil polluted soil using municipal solid compost for soybean production in the Niger Delta. Proc. Of the 26th Ann. Conf. of Soil Science Society of Nigeria, Ibadan, Oyo State. Pp. 161-169.

[21] Ingrid, W., F. Martinez and G. Cueva. 2006. Plant and soil responses to the application of composted MSW in a degraded, semi-arid shrubland in Central Spain. Compost Sci. Util., 14(2):147-154.

[22] Claasen, V.P. and J.L. Carey. 2004. Regeneration of nitrogen fertility in disturbed soils using composts. Compost Sci. Util., 22(2):180-186.

[23] Rodd, A.V., P.R. Warman, P. Hickleton and K. Webb. 2001. Comparison of N. fertilizer source separated MSW compost and semi-solid beef manure on the nutrient concentration in boot stage barley and: wheat tissue. Can. J. Soil Sci., 82:33-43.

[24] Kwesiga, F. and S.M. Chisumpa. 1992. Multipurpose trees for the eastern province of Zambia: An elthnbotanical survey of their use in the farming systems. AFRENA Report No. 49 ICRAF, Nariobi, Kenya.

[25] WHO (World Health Organization).2006. Guidelines for the safe use of wastewater, excreta and greywater. Vol. IV: Excreta and greywater use in agriculture. Geneva, Switzerland. (2006)

Veenhuzen, R. (ed.) Cities farming for the future, RUAF Foundation, IDRC and IIRR, Leusden, NI. Pp. 313-346.

[26] Dranger, J.O. 1998. Urine blindness and the use of nutrients from human excreta in urban agriculture. Geojcurn., 45"201-208.

[27] Johnson, H., A.R. Stinzing, B. Vinneras and E. Salomon. 2004. Guidelines on the use of urine and faeces in crop production EcoSanRes., 1-35.

5

Impacts of Tourism on the Coastal Environment of South China Sea: Terrestrial Perspective

Egbuche C. T.[1,*], Nwaihu E. C.[1], Umeojiakor A. O.[1], Zhang Jia'en[2], Okechukwu Ukaga[3]

[1]Department of Forestry and Wildlife Technology, School of Agriculture and Agricultural Technology, Federal University of Technology Owerri, Nigeria
[2]College of Agriculture, South China University of Agriculture Guangzhou, China
[3]College of Food, Agriculture and Natural Resources, University of Minnesota Extension USA

Email address:
ctoochi@yahoo.co.uk (Egbuche C. T.)

Abstract: At recent times, Asia and the Pacific have been rated very high as major tourism destinations. This transdisciplinary (nature and social science concepts) study reveals that tourism activities generate pressure on forest ecosystems and coastal biodiversity. Dongguan forest (site) park attracted impact on terrestrial ecosystem with impact base respondent (9 000), while Shenzhen (18 000), Guangzhou (18 000) and Zhuhai (12 000) districts showed much higher values on general impact. A coastal activity survey and impact base of visitor response model evaluation was used. A general impact and causative platform were identified as driving fundamental (direct/indirect) factors in the region. Subjective and observed broad impacts were presented however; trampling had a major terrestrial impact on both ecosystems. Further investigation is needed to evaluate the economic impact of tourism in the region using industrial, urban and tourism potential factors after the 2008 Olympic Games in China. Strategies for appropriate action and government regulations are recommended concepts of terrestrial and coastal conservation planning and land use.

Keywords: Coastal Environment, Coastal Conservation, Terrestrial Ecology, South China Sea, Tourism Impact

1. Introduction

This paper is based on broad field investigations conducted on tourism potential impacts on the coastal environments of Guangdong province in the Peoples Republic of China. It addresses the issues of tourists' interaction with terrestrial, coastal ecosystems and the associated impact of such human interactions. We focused on the impact of tourism causative and platform in Dongguan forest park from a natural science perspective and in designated coastal districts from social science perspective. The increased recreational use of forest parks, coastal areas, and other protected areas has endangered areas of great ecological interest due to the negative effect of tourism platforms on local ecosystems and plant communities. [1] documented that tourism and outdoor recreation could lead to negative impacts on the environment due to pressures that stretch the natural capability of the ecosystems. Uncontrolled conventional approach to tourism poses a potential threat to the natural environment and biodiversity thereby tourism conducted along coastlines/beaches is bound to influence the ecosystems natural capacity [2]. Under tourism destination consequences, impact scenarios on platforms were considered under coastal (intertidal and forest park of individual (single) tourists and tour groups (group package) effect on plant species). We adopted the conceptual idea of natural science in taxonomic identification of forest species (amounting to 99 species) in the Dongguan forest park and applied social science methods to enumerate coastal ecotourism activities (9 subjective outdoor activities and impact base) in Guangdong province along the coast of the South China Sea. The potential positive effects of coastal ecotourism and forest park visits was considered, but overall tourism left evidence of negative environmental impact such as generating waste, pollution, and trampling of terrestrial ecosystems. Coastal recreation and forest parks have long been considered popular destinations for tourists and attract many tour groups [3]. Tourism and outdoor recreation is a human activity that may catalyze pollution and generate waste. It allows people to utilize, compete for, and modify natural resources. This study targets the impact of tourism activity when human population (consumers and potential beneficiaries) is concentrated in a certain geographical region (producer). Field assessments include the identification of forest park species, coastal tourism activity, and an assessment of the general impact on coastal environment/vegetation. Direct observations were employed which is study site scaling methods to document associated impact of tourism. Single site and group spot sites were identified as tourists destinations where [4] identified tourist destinations include public services and facilities, as well as physical and natural attractions, so tourist destinations are composed of (i) attractions, (ii) facilities and services, (iii) infrastructure, (iv) Hospitality and (v) cost variables. Destination tourism products comprise a set of tangible and non-tangible components based on an activity at the destination. [5] reported impact of tourism with varied activities that range from relatively innocuous pursuits to extremely popular sports all result in measurable deterioration of the world's coral ecosystems despite good management practices. Furthermore, [6], conducted visitor surveys for varieties of reasons but on the background that we considered coastal and forest parks as fragile ecosystems. Tourism activities that are conducted within forest parks and coastal districts may not substantially be of long-distance and duration [7] but may have significant ecological impact, especially from mass and packaged tour groups.

Increasing economic and industrial development both in developed and developing nations has created a higher volume of worldwide demand for recreation, leisure activities, and created the potential for even more tourism. Visits to seaside attractions are more prevalent in fast growing economic and industrial regions like Guangdong province exhibits critical issues of spatio-temporal development of international seaside tourism [8]. This concept results in development of tourism infrastructures, utilities and expanded urban growth, that could stretch (conurbation) towards coastal regions. Other important environmental impact of tourism are the conflicting and serious issues involving ownership and of who controls potential tourism properties/facilities, the local people, regional and national administration, and multiple policies/regulations such as land use competition, natural environment and infrastructures, [9]. Individual and group tourism in coastal areas lead to illegal structures and juxtapose seaside growth with garbage and "shanty towns". Generally, tourism is linked to contamination and degradation of ecologically valuable plant species (Such as mangrove), reduction of insect species, and seaside modification [10] and [11]. Coastal regions and their management, has been identified as posing

multidisciplinary challenges [12]. Specifically, this study aimed to assess the following issues: (1) To understand the impacts of tourism on vegetation and (2) To assess tourism activity impact on coastal region like the South China Sea. This study objectively considers the direct impact of tourism on two levels: (1) that the outdoor recreation activities of both individual tourists and packaged tour groups have a multiple effect on terrestrial ecosystem and (2) that seaside (coastal) recreation and activities often generate pollution and waste.

2. Study Area

Guangdong province is located in the southern part of China and has a long coastline along the South China Sea. The province has a total area of 179,766 square kilometer. Key tributaries are rivers such as the Xijiang River, the Beijiang River, the Dongjiang River, the Hanjiang River, the Rongjiang River and the Moyangjiang River, which together form the Zhujiang River (Pearl River) as in fig. 1. The region has both tropical and sub-tropical climatic zones. The Tropic of Cancer runs through Guangdong Southwestern province. The region is generally warm and rainy all the year. The average annual rainfall is more than 1 500 mm, (rainy season is from April to September). The average temperature ranges from 19 to 26°C. The province is endowed with a long coastline of about 3 368 km along the South China Sea and additional coastline around an archipelago of 759 small islands (excluding Hong Kong, Macao and Dongsha Islands). The total land area of the archipelago accounts for only 1 649 km^2 and constitutes only 1. 85% of the total land area of the province.

Fig. 1. *South China Sea and geographical location of Guangdong province*

2.1. Data Source and Collection

"Tourism Hot Spots" that attracts high volume of tourists for the activity impact survey include a) Dongguan Forest Park (Impact Survey on vegetation) and b) Popular tourist areas in Zhuhai, Xinhui, Jianmeng city, areas in Guangzhou and Shenzhen in coastal areas as shown in fig.1 and 2. Then, primary data were collected through a field survey using Dillman's total design method [13]. Detailed questionnaires were administered to coastal tourists on their activities while on their stay at the destinations.

2.2. Direct Impact of Individual Tourists and Packaged Group Tours on the Study Sites

Direct impact data was collected from 1 200 visitors to the Dongguan Forest Park and observed tourists activities in vulnerable coastal areas in the province. Two separate surveys were created with similarity in design, but with different objectives, tailored to identify tourist's activities that are associated with the area.

2.3. Coastal Outdoor Activity Identification

Coastal areas and beaches are popular for swimming and other water sports. This study is in agreement that beaches constant changing environment because of weather (storm impact), beach nourishment, pollution and development. Considering this background, each visitor was asked to identify/enumerate specific activities being conducted while on the visit. The enumeration forms were used to create activity profile that was assessed to identify the environmental impact associated with each activity. Figure 2 shows viable site surveys along the coastal areas - Zhuhai (300), Xinhui (100), Jianmeng city (100), Guangzhou (200) and Shenzhen (300) and Dongguan Forest Park (200) and the environment impact on each site.

2.4. Environmental Impact Assessment

Utilizing field activity response datasheet, direct impact was derived from each location by cumulative tourist's responses. Regional impact is the total spot across the study region (constant), spot impact is the number of hotspots adopted in each district (multiplier), and the summation produces the estimated total impact. This application has been utilized mostly in evaluating economic impact with IMLAN software. We adopted a simplified approach with this application because it allows for a direct field variable multiplication for every district and avoids overestimations unless the spot impact (multiplier) changes. Induced effects are not considered because of the nature of outdoor visits which last for a specific (short time). This model has been widely used by [14] of the Department of Agricultural Economics at Oklahoma State University. Their application was extensively used with multiple multipliers for the economic evaluation of participants' impact on the Summer Games held each May on the campus of Oklahoma State University [15]. We thereby applied the following simplified formula:

$$LAC = \sum DI * RI * MI \Rightarrow TI$$

Where:

LAC is location based on activity categories

$\sum$ is the summation of observed variables

DI is direct impact based on an active tourist's response

RI is the total number of assessed spots along the region (constant)

MI is the regional impact based on the hotpots of a district

TI is the grand total of tourism impact in the region

2.5. Trampling Impact of Tourism on Vegetation

On-the-spot- effect of tourists trampling and cutting in some common forest communities in Dongguan Forest Park was prominent, although such action was identified in other coastal districts of the region. The enumeration of species was conducted and established a spread sheet that identified 99 plant species. Vulnerability to plant disturbance among plant communities were in terms of (1) direct trampling and cutting of plant shoots and flowers, (2) changes in ground vegetation cover of the site, (4) soil compaction, and (4) litter.

2.6. Analysis of Data

Microsoft and other computer based packages were used in describing field and on spot observed data. The classification and division of plant species into groups facilitate the description of community assemblages and overlap within the grid system across the entire site. The variations were identified using the grid ordination, thus providing a complete vegetation data assemblage. [16].

3. Results

3.1. Direct Observed Impact of Tourism

Direct field observation and assessment within the regional (districts) designated tourism hotspot inventory survey using cumulative factors showed Shenzhen district (18 000), Guangzhou (18 000) and Zhuhai (12 000), respectively showed significantly high environmental impact from tourism based on recreational impact platform, geographical comparative advantage, and urban/industrial development base factors. This is shown in table 1, below, which ordinarily portrays specific districts to regional impact base value. Table 1 informs us that various categories of environmental impact were recorded and well observed in each study district exhibit associated and peculiar activity impact. However, Dongguan still maintains a high impact level in respect to plant species impact and industrial influences.

Table 1. *Direct location impact of tourism along South China Sea of Guangdong province*

Location/ district along the region	Impact inventory/ Category (LAC)	Regional impact-hotspot (RI)	Spot impact (Multiplier) (MI)	Direct impact by response (DI)	Total impact value (TI)
Dongguan district	Solid wastes, noise, trampling, species degradation, transport	20	3	150	9000
Xinhui district	Wetland-impact, trampling,-solid wastes,- fishing /boating modification	20	3	100	6000
Shenzhen district	Emissions, noise, beach-litter, solid-industrial-wastes, transport,-coastal modification, overcrowding	20	4	300	18 000
Jiameng city district	Hot-spring degradation, plant species, population, solid wastes, modifications	20	3	100	6000
Guangzhou district	Industrial noise, pollution, coastal degradation, modification of landscape, microbiological, ecological and environmental hygiene and insecurity	20	3	300	18 000
Zhuhai district	coastal modification/aesthetic-quality,- overcrowding, resource degradation	20	4	150	12 000

LAC – Location/ Impact inventory category based on environmental platform. DI – Direct impact based on visitors' activity response. RI – Total number of tourism hotspots along the districts (constant). MI - Multiplier impact base on a district potential attraction of visitors. TI – Grand total impact

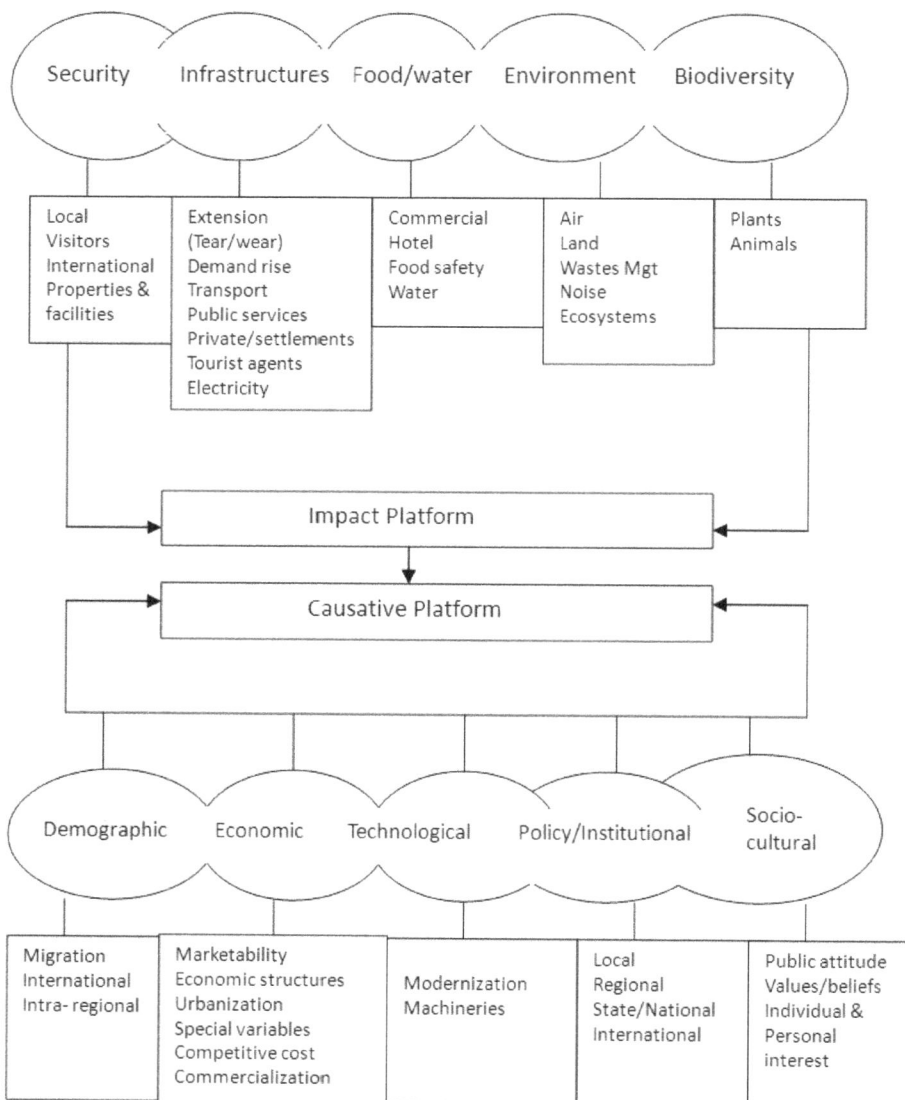

Fig. 2. *Causative/impact platform of broad cluster of underlying driving and fundamental platform underpin the proximate impact of tourism identified directly and indirectly along the region.*

3.2. Causative/Impact Platform of Broad Cluster

Field assessment of tourism among ecosystems indicated broad causative and subjective impact (fig.2) that is associated with direct and indirect impact of tourism activities. Security of tourists is an important factor. Movement of people and properties originating from local and international platforms requires measures to protect the environment. Infrastructure is critical for effective tourism practices ranging from food consumption, utilization of public utilities, energy and waste removal to conservation of scarce resources. The constant wear and tear on the environment caused by tourist activities hampers rehabilitation and modifies the environment. The utilization of commercial and environmental properties becomes a source of degradation to the ecosystem and pollution to the environment. The natural biodiversity especially among fragile ecosystems has been directly impacted by tourism practices. In generally, causative platforms like demographic characteristics that involve local, regional and international structures are considered. Economic interplay within urban and rural interface leading to commercialization of economic goods brings into play competition and the high cost of special variables. Technological development becomes a catalyst for equipment, machinery, and modernization to support tourism activities.

3.3. General Impact on the Coastal Environment of the Region

Subjective platforms identified in forest parks and coastal regions (Table 2) underpinned the depletion of coastal natural and water resources, and local, land/vegetation. The platforms attract pollution in coastal areas. Furthermore, marine development affected depletion of the coral reefs (subjective platform) identified in the region. Field observation revealed greater demand and consumption of coastal and sea resources, such as fishes and activities (boat, swimming etc). There are observed scramble for local food, artisan materials, high demand for technical facilities. These observed factors results to contamination, scenic landscape modification. We observed higher demand for housing, both for commercial and accommodation which correspondingly increases generation of waste, emission from public services and transportation. Generally, all these factors increase impact on both lands, coastal and marine ecosystems.

Table 2. General impact on the coastal environment of the study region

Subjective platform of the ecosystem	Observed broad impact
Depletion of coastal natural and water resources	•1 Increased consumption of especially scarce resource •2 Overuse of water resources •3 Degradation of water supplies and increase waste of in fresh water.
Coastal local resources	•1 Generates pressure on local resources like energy, food, and other raw materials. •2 Greater extraction and transport physical intensity. •3 High demand for tourists' facilities and technology.
Land/vegetation depletion	•1 Degrade coastal wetland resources, forests, wetland and coastal wildlife. •2 Increase construction of coastal tourism facilities and modifications such as creating scenic landscapes. •3 Direct impact –like developing coastal accommodation, infrastructures and building materials. •4 Coastal forestry suffers deforestation, trampling and clearing of wetlands.
Coastal pollution impact	•1 Coastal industries •2 Air emissions and noise •3 Solid waste and littering •4 Release of sewage – sewage pollution results to public health problems (human and coastal animals). •5 Cause discharge of oil and chemicals •6 Visual pollution/modification, aesthetic pollution resulting to lack of planning and encourage sprawl of facilities.
Marine Development	• Breakwater channels and change in currents and coastlines
Coral reefs	• May alter and destabilize fragile marine ecosystems, shoreline, anchoring, ships grounding, diving, yachting and cruising

Based on www.uneptie.org/pc/tourism/sus-tourism/env

3.4. Physical Impacts – Impact of Trampling in Forest Park and Coastal Regions

The effect of trampling was critically observed in Dongguan forest park and also observed to be a common impact factor on the soil in coastal areas, as shown below in table 3. The observations were notable in most areas trampled upon by visitors and collection of flowers and shoots. Species composition and coastal soil porosity impact were seriously observed in the region as supported by table 3.

Table 3. showing trampling impact on coastal vegetation and soil/environment

Trampling on coastal vegetation	Trampling on soil and coastal environment
Breaks and bruises vegetation, especially stems and shoots	Loss of soil organic matter
Cause reduction in plant vigor	Reduce coastal soil macro porosity
Results in poor and reduction of regeneration	Results in poor air and water permeability
Reduces and loss of ground cover	Increase water run off
Reduces and changes species composition	Accelerate coastal erosion

4. Discussion

4.1. Relationship Between the Environment and Tourism Activities

There exists a complex relationship between natural and manmade environments in regard to outdoor recreation and coastal tourism. This investigation revealed that tourism, by both induced and direct interaction, can have adverse environmental impacts. Considering the broad definition of environment, both impact and causative platform (fig.4) include natural and man-made physical features and historical sites. It is perceived interrelationships between tourism activities, infrastructure, industrial development, and environment sustainability calls for specialized discipline, such as coastal ecology and recreational development, [17]. The impacts are linked to construction of associated infrastructures ranging from low to intensive projects needed for coastal and outdoor recreation. Properly managed tourism can sustain local economies without damaging the environment, protecting and thereby enhancing opportunity for the future [18] as well as respecting the limits imposed by ecological communities [19]. The environment–tourism relationship in any platform must be sustainable tourism. Parks and coastal areas must primarily be defined in terms of sustainable ecosystems. [20]. Tourism aims at providing people with a variety of leisure and recreational activities related satisfaction, which the industry promotes and provides in return for financial reward such activities include travel, shopping, cruising, sun bathing, yachting, and diving, as well as dining and residing in restaurants and hotels. In many coastal areas, tourism is the largest sector of the economy while construction of hotels, apartments, and other tourist infrastructure is becoming the dominant form of development. Tourism activities have far-reaching impacts upon the human (cultural, social) resources of tourist destinations as well as on the physical (vegetation, wildlife, water, etc.) resources, but must be conducted for present tourists and host regions while protecting and enhancing opportunity for the future [21]. In all we strongly identified the use of vehicular and other group movement patterns that have been investigated by [22]. This type of mobility generates attendant environmental waste. We agree that it is environmentally friendly to walk and trek along beaches/coastlines and forest parks.

4.2. Environmental Impact of Tourism on Coastal Regions of Guangdong Province

Southeast Asia is an area of extraordinarily high biodiversity. This high diversity in land and sea is a result of three major factors: the overlap of independently - evolved varieties of species, high rates of local speciation, and differentially-high survival rates temporally and spatially heterogeneous habitats [23]. [24] documented that coastal regions can be defined on a transdisciplinary challenge which we considered in utilization and management (1) as a geographic area where land and sea interact through natural processes, but we can also consider it (2) as a zone of economic activity based on the exploitation of coastal and marine resources, (3) as a socio-cultural entity, with specific traditions and values or (4) as an institutional entity with administrative boundaries. This has made the region at large and China specifically, a major tourism destination. Notably, both coastal and interior parts of China show immense prospects for growth in tourism. We identified that, in table 1, based on our single multiplier impact ratings, Shenzhen, Guangzhou, Zhuhai and Dongguan were rated high. This was attributed to multiple activity base and geographical position, not only factors generated by tourism factors in the region. It is agreed and related efforts to help coastal regional authorities inculcate ecosystem-based management approaches in sustaining coastal resources with attention to pressures and population influx. Notably, modifications to the coastal environment due to tourism tend to degrade environmental resources. On the other hand, this can also contribute to protection of the coastal environmental due to increased awareness of the need for better maintenance and conservation of the coastal environment. A biodiversity conservation approach generates awareness and becomes a tool to finance the ecological conservation of coastal regions at large, especially the natural tourism potential, which is applicable in the study region. The negative impact of tourism occurs when the pressure from visitors (level of users) becomes greater than the environment's ability to cope (carrying capacity) within natural acceptable limits of change.

This pressure concept is often referred to as the "Wear and Tear Effect" of tourism facilities and coastal resources. Negative environmental impacts on man - made and natural coastal tourism potential may be worsened through uncontrolled conventional tourism practices. In the case of coastal regions like the coast of the South China Sea in Guangdong province, enormous pressure leads to coastal and sea erosion, increased pollution, effluent discharges into the sea, loss of natural habitats, increased pressure on endangered species, and increased vulnerability of coastal wetland plant species. Further negative impact includes stress and strain on water resources and compels the local people or communities to scramble for coastal resources, especially critically scarce resources. [25] supported that our field impact inventory as in fig. 4 and table 1 are critically experiencing: 1) increased pressure from automobile emissions, resulting in significant air pollution and other wastes; 2) increased pressure from use of natural/manmade facilities resulting to anthropogenic influence such as sand beach quality; 3) population increase and concentration leading to overcrowding and degradation of coastal resources; 4) significant deterioration and alteration of biodiversity and natural ecosystem; 5) increase in environmental management, insecurity, uncontrolled building, and change to the natural landscape (like sand dunes); 6) An increase in noise, air, water, solid wastes pollution; 7) degradation of aesthetic quality due to litter from cannel flooding and careless disposal like polyethylene, paper, cigarette packs, beer bottles/cans of drinks; 8) decrease microbiological ecological quality and general environmental hygiene – caused by increased refuse and contamination; 9) serious impact on plant species through flower collection, cutting and trampling.

4.3. Impact of Trampling and Waste Generation by Tourists in the Coastal Region

Trampling is a human induced factor, [26]; identified in this study and its impact that is predominant on coastal vegetation and soil environment as shown in table 4. The damages done by trampling and associated outdoor activities by tourist do results in alteration of ecosystems. Field observation confirmed tourists collecting branches and flowers along the tourist destinations in the area. These observations underscore the direct negative environmental impact of tourism there. Trampling affects coastal vegetation and soil, and generates further impact on tidal flat in fauna [27]. These activities by tourists are described as "recreational disturbance". The impact of trampling was considered low at most sites but fragile ecosystems like the littoral active zone have been documented as heavily impacted. [28]. However, [29] confirmed that the potential damage due to trampling has mostly been studied in the short term (two years). We followed used routes marked on tourist maps to observe the practical effect of trampling as associated with tourism such as shoot/flower collection and off-path hiking. These activities impact the least vulnerable plant communities. Waste is associated with human activities generally as most field (tourist hot spots) environments were littered with plastic materials, food wrappers, and all kinds of drinks bottles. However, the placement of waste bins is a way of dealing with this problem, but some tourists still continue to litter rather than put waste in such bins. [30] documented that beaches and coastlines are impacted more by litter pollution, and however, the presence of waste collectors and waste management trucks in most of the spots we visited was encouraging. Observed impacts are linked to construction of associated infrastructure ranging from simple to intensive projects needed for coastal and outdoor recreation. Based on this observation, it is evident that coastal regions have attracted outdoor enthusiasts and tourists for some time now as a result of available coastal tourism facilities. The coastal region is currently experiencing a rapid increase in development of infrastructures and in urban development. Tourist destinations include an amalgam of industries such as transportation, accommodation, food and beverage services, recreation, entertainment, and travel agencies [31]; [32]. Much of the discussion on sustainable development and tourism has been in the context of the environment in which tourism occurs [33] [34]. There is an argument that tourism preference could be accounted for, to some extent, by global climate change, especially as with regard to tourism destination preferences.

5. Conclusion

This multidisciplinary study provides a good understanding of the long term, immediate benefits and natural potential of coastal and environmental properties along the South China Sea in Guangdong province. The environmental impact of tourism results in modifications that becomes relevant for harmonized local, regional and national authorities for the development of appropriate strategies and updated tourism management techniques to achieve sustainable coastal/forest ecosystems. Further investigation is required to evaluate the annual and seasonal economic impact at the district and regional level in order to build major tourism - related industry profiles and assess their impact on the provincial economy. This is paramount and strategic in terms of the post Olympics China, industrial output, employment, and income and taxation. This paper strongly suggests that there should be outlined appropriate actions and strategies and responsive adoption from various government levels as implemented in Canada. Furthermore, there should be regulation and control, and attention paid to the concepts of fragile ecosystem and sensitive areas, in all areas of land use and coastal developmental planning where tourism plays a major role in the economy.

Acknowledgments

This study was supported by two key research programs of Guangdong Province (2005A30402003, 2006A36702003) and China Scholarship Council Beijing Award

References

[1] United Nations Environmental Programs (UNEP) (www.uneptie.org/pc/tourism) (http://www.uneptie.org/pc/tourism/library/home.htm) – *webp Planning and managing ed.age: recreation and Tourism: Impacts of Tourism*. Visited on the 2008/December.

[2] Pearce, D. G.; Kirk, R. M. (1986): Carrying capacities for coastal tourism. *Industry and Environment (UNEP Paris)* 9(1), 3–7

[3] Dimanche, F., M.E. Havitz et al. (1991). "Testing and involvement profile scale in the context of selected recreational and tourists activities" Journal of Leisure Research 23(1): 51-66.

[4] Kozak M, Rimmington M. (1998). Benchmarking: destination attractiveness and small hospitality business performance. *International Journal of Contemporary*

[5] Davenport John and Julia L. Davenport (2006) The impact of tourism and personal leisure transport on coastal environments: A review: *Estuarine, Coastal and Shelf Science* 67. Elsevier

[6] Leones, J., B. Colby, and K. Crandell. (1998). Tracking expenditure of the elusive nature tourists of Southeastern Arizona. Journal of Travel research 36: 56-61.

[7] Burger, J., (2002) Tourism and ecosystems. In: Douglas, I. (Ed.), Causes and Consequences of Global Climate Change. *Encyclopedia of Global Environmental Change, vol. 3. John Wiley & Sons, Ltd, Chichester*, pp. 597e609.

[8]   Gormsen, E. (1981): The spatio-temporal development of international tourism, attempt at a centre-periphery model. *Etudes & Mémoires* 55, Centre des Hautes Etudes Touristiques, Aix-en- Provence, 150–169 (Burger 2002)

[9]   Gormsen, E. (1988): Tourism in Latin America, spatial distribution and impact on regional change. *Applied Geography and Development* 32, 65–80

[10] Kreth, R. (1985): Some problems arising from the tourist boom in Acapulco and the difficulties in solving them. Mainzer Geograpische Studien 26, 47–59 *Hospitality Management* 10(5): 184–188.Kreth (1985)

[11] Uthoff, D. (1996): From traditional use to total destruction. Forms and extent of economic utilization in the Southeast Asian Mangroves. *Natural Resources and Development* 43/44, 58–94.

[12] Dronkers J., de Vries I. (1999) Integrated coastal management: the challenge of transdisciplinarity. *Journal of Coastal Conservation* 5, 97-102. 99. Doi: 10.1007/BF02802745

[13] Dillman, Don A. (1978) Mail and telephone surveys – The Total Design Method. New York, NY: Wiley and Sons.

[14] Mike Woods and Suzette Barta (2002): Estimating Impacts of Tourism Events: *Methodology and a Case Study – National Extension Tourism Conference: "Changing Faces – Changing Places*. University of Michigan Traverse City, Michigan

[15] Barta, Suzette, Woods, Mike D , Trzebiatowski, Susan, and Cain, Derek, (2002) "The Economic Impact of Special Olympics Oklahoma on the Economy of Stillwater, OK." *Oklahoma Cooperative Extension Service*, AE-02135,

[16] McCune, B., and J. B. Grace, (2002). Analysis of ecological communities. *MjM Software Design, Gleneden beach, Oregon.*

[17] Frankenberg, D., Pomeroy, L.R., Bahr, L. & Richardson, J. (1971). Coastal ecology and recreational development. In *The Georgia coast: issues and options for recreation* (ed. C.D. Clement), II, 1-49. The Conservation Foundation, Washington, D.C.

[18] World Tourism Organization. 1993. *Sustainable Tourism Development: Guide for Local Planners*. Madrid: WTO. *Southern Africa* 19(1): 123–141.

[19] Payne, R. (1993). Sustainable tourism: Suggested indicators and monitoring techniques. In *Tourism and Sustainable Development: Monitoring, Planning, Managing*, ed. 1993: 154-5

[20] Woodley, S. (1993). Tourism and sustainable development in parks and protected areas. In Tourism and Sustainable Development: Monitoring, Planning and Managing ed. 1993: 94).

[21] World Tourism Organization. 1993. *Sustainable Tourism Development: Guide for Local Planners*. Madrid: WTO. *Southern Africa* 19(1): 123–141.

[22] Brodhead, J.M. & Godfrey, D.J. (1977) Off-road vehicle impact in Cape Cod National Seashore; disruption and recovery of dune vegetation. *Int. J. Biometeor.* 21, 299-306.

[23] McManus, J.W. (1985).Marine speciation, tectonics, and sea-level changes in Southeast Asia. Proceedings of the 5th International Coral Reef Congress, Tahiti. 4:133-138.

[24] [12] Dronkers J., de Vries I. (1999) Integrated coastal management: the challenge of transdisciplinarity. *Journal of Coastal Conservation* 5, 97-102. 99. Doi: 10.1007/BF02802745

[25] Bhaskar Nath (1998) *Environmental Management in Practice: Managing the Ecosystem*, Routledge Publishers Bhaskar Nath (1998)

[26] Burden R. F. and P. F. Randerson (1972). Quantitative Studies of the Effects of Human Trampling on Vegetation as an Aid to the Management of Semi-Natural Areas, *British Ecological Society*

[27] Chandrasekara W.U., Frid C.L.J. (1997) Effects of human trampling on tidalflat infauna. Aquatic conservation-Marine and Freshwater Ecosystems 7, 299-311 *Conserv,* 71, 223-230 (Chandrasekara, and Frid (1997).

[28] Heath, R. (1987). *Impact of trampling and recreational activities on the littoral active zone - a literature review.* Univ. of Port

[29] Roovers P, K. Verheyen, M. Hermy, and H. Gulinck (2004): Experimental trampling and vegetation recovery in some forest and heathland communities. *Applied Vegetation Science, BIOONE Online Journals.*

[30] Bowman D, Manor-Samsonov N., Golik A., (1998) Dynamics of litter pollution on Israeli Mediterranean beaches: a budgetary, litter flux approach. *Journal of Coastal Res.,* 14, 418-432

[31] Poonyth D, Barnes JI, Suich H, Monamati M. (2002). Satellite and resource accounting as tools for

[32] Smith and Massieu, 2005, pp. 865– 866). Much of the discussion on sustainable development and tourism has been in the context of the environment in which tourism occurs (Eagles 1994;

[33] Eagles, P.F.J. (1994) Understanding the market for sustainable tourism. In *Linking Tourism, the Environment and Sustainability,* ed. S.F. McCool and A.E. Watson, pp. 23-33. Ogden, UT: USDA (General Technical Report INT-GTR-323).

[34] McCool, S.F. (1994). Linking tourism, the environment, and concepts of sustainability: Setting the stage. In *Linking Tourism, the Environment and Sustainability,* ed. S.F. McCool and A.E. Watson, pp. 3-7. Ogden, UT: USDA (General Technical Report INT-GTR-323).

6

Experimental Program and Technical Assistance in Alpine Cheesemaker Huts of Friuli Venezia Giulia (Italy)

Simona Rainis[1], Ennio Pittino[2], Giordano Chiopris[2]

[1]Crita S.c.a.r.l. - Research Center for Technological Innovation in Agriculture, Via Pozzuolo, Udine, Italy

[2]Ersa - Regional Agency for Rural Development of Friuli Venezia Giulia, Via Del Montesanto, Gorizia, Italy

Email address:

s.rainis@libero.it (S. Rainis), ennio.pittino@ersa.fvg.it (E. Pittino), Giordano.chiopris@ersa.fvg.it (G. Chiopris)

Abstract: In Friuli Venezia Giulia (Italy), traditionally transhumance and mountain grazing characterize the activity of zootechnic breeders during the summer period. In these alpine pastures ("malga") there are typical cheesemaking huts (called "casere") with an important activity of dairy productions that represents an important sector of the agriculture in this Region. The main products are alpine cheeses and smoked ricotta cheese, with flavors and aromas unique and inimitable, due to the old autochthonous recipes, the "art" of the cheesemakers and the environmental conditions. In the present paper, the experimental area taken into account is represented by the Friulian mountains in Italy. The trial consisted in the technical assistance related to the productive pathways of the alpine cheese and alpine ricotta cheese, performed by ERSA - the Regional Agency for Rural Development of Friuli Venezia Giulia (Italy). 800 dairy operations, 418 of mountain cheese and 382 of smoked ricotta cheese respectively, were followed in 20 productive units. Amount of milk transformed, acidities of the milk and whey, temperatures, durations of the processing and microclimatic parameters were registered. Suggestions and advices were given to the cheesemakers in order to improve the quality of the products. Analyzing the period taken into account, it can be observed that important goals along all the productive pathway were obtained. This type of experimental and technical assistance can be a good practice for all the farms in marginal areas.

Keywords: Alpage, Alpine Cheese, Alpine Ricotta Cheese, Marginal Areas Development, Mountain Agriculture

1. Introduction

Pastures of Friuli Venezia Giulia are situated at the extreme Eastern edge of the Italian slope of the Alpine arc. The farming landscape system in this region is characterized especially by livestock husbandry (almost the 77% of the agricultural holdings) of bovines and ovi-caprines. In this marginal area, traditionally, transhumance and mountain grazing mark the activity of zootechnic breeders during the summer period, thanks to the fact that this mountainous portion of Italy presents a significant climatic, geographic and agricultural diversity. For definition, an alpine pasture ("*malga*") is a holding where livestock are moved over summertime from the lowland permanent farms to exploit the meadows, as shown in Figure 1.

Figure 1. An example of "malga", productive unit in the Alps during the summer.

In these high altitudes there are traditional cheesemaking huts (called "*casere*") so there is an important activity of cheesemaking without the need to carry the milk back down to the valley [1]. This seasonal dairy production represents an important sector of the agriculture in this Region [2], in fact this peculiar activity constitutes a fundamental train for the local farmers, because it allows the social and financial progress of the autochthonous population and in general it ensures the tutelage of their identity and culture patrimony. In this delicate and fragile historical moment, the new paradigm for a sustainable development emphasizes the role of extensive breeding and low impact and eco-friendly activities as the alpage. In these specific areas, the mountain grazing and all the activities related to it contribute to preserve and also to exploit the peculiarity of the landscape and its aesthetic value, enhancing the recreational-ecological quality of the surroundings and its tourist appeal [3]. The alpiculture favorites the development of marginal areas, the production of safety food and environmental protection [4]. It is evident that the supporting of the activities that take place in this mountain productive units represents the tutelage of the typical products, the increasing of the farming profitability and the inclusion of new young strengths actively committed in the developing of their future [5].

Centuries of old traditions and methods insure that alpine cheeses and smoked ricotta cheese are unique from other types of similar products. Their flavors and aromas are typically described as nutty, fruity, spicy, floral, herbal, grassy and/or buttery. These inimitable tastes and smells are created by the skilled hands of the cheesemakers, the traditional recipes perfected over generations and the high-quality, butterfat milk from cows grazing mainly on lush, seasonal plants and grasses found up and down the mountainside [6]. Since the animals are fed almost on exclusive mountain pastures with little microclimates, each product tastes different than the next, due to the strong alpine herbs that grow differently in every area. Furthermore the regions developed their own unique style due to herd sizes, remoteness and local preferences [7].

In this context, a relevant experimental design of technical assistance for the dairy sector in alpine lands is performed form the 2002 by Ersa, the Regional agency for rural development in Friuli Venezia Giulia (Italy). This action consists in several technical visits in the cheesemaking huts in order to follow the processing pathway of dairy production, ensuring a high level consultancy. The suggestions and advices are related chiefly to milk quality and productive technology. Together with the farms, it is an important moment also of evaluation of the estate and functionality of all the productive unit, of the animal welfare and so on. In every visit, they also register the productive and technological parameters (as the quantity of processed milk, the acidity, the temperatures, etc.).

The aim of the present paper is to present an overview of the data collected during these 14 years of assistance, in order to describe the traditional modality of dairy production in the alpine pastures of Friuli Venezia Giulia (Italy) and to describe the actual trend of this compartment.

2. Methodology

The project area taken into account is situated in the Friulian mountains, an Italian region.

In the present work the result of 14 years of experimental activity related to the productive pathways of the alpine cheese and alpine ricotta cheese is presented. This technical assistance was performed by the technicians of ERSA - the Regional Agency for Rural Development of Friuli Venezia Giulia (Italy). The joining to this support program is voluntary and it is open to all the cheesemaker huts of the area. 800 dairy operations, 418 of mountain cheese and 382 of smoked ricotta cheese respectively, were followed in 20 productive units.

For each one, apposite technical sheets were filled up, in order to note down the key parameters and possible criticisms during the processing as dairy products (Annex 1 and 2).

The acidities were registered by the employment of an acidimeter, with the Soxhlet-Henkel method and expressed in °SH/50ml.

The microclimatic parameters were monitored by an hygrometer and a thermometer, expressed respectively in % of relative humidity and in °C. In the case of the measurement of a trend, it was employed the logger, to determine these values at determinate intervals.

The concentration of brines were measured by a densimeter and expressed in the scale of Baumé (°Bé).

In this contribution, to stress out how the compartment is modifying, average data of the main parameters from 2002 to 2015 were taking into account. In particular, the comparison between the 2002 (the year of the begging of the experimental period), the 2009 (considered the mid-term) and the 2013, 2014 and 2015 (the last ones, that allows to observe how the present situation is, if there are some stabilizations and where are the possible ameliorate margins for improvement) was chosen to be explained.

All the collected data were then elaborated, to have a clear and exhaustive framework of the typical productions methodology. It was also possible to evaluate the evolution of the sector during the years.

It was performed using SPSS for Windows (version 7.5.21, SPSS Inc., Chicago, IL). Normality of data distribution and homogeneity of variance were tested using Shapiro-Wilks test. Data were also processed by principal component analysis carried out using The Unscrambler X version 10.2 (Camo Software AS, Oslo, Norway). Data were mean centered, variables were weighted with 1/SD, and the full cross-validation method was used.

3. Results and Discussion

3.1. Dairy Production

The following data are referred to the dairy production, from the raw material collection till the final storage of cheese and smoked alpine ricotta (Figure 2). All these steps can affect considerably final result of the processing.

As regards the daily average amount of milk transformed, it can be observed that the medium size cheesemaker huts are the predominant (for example in 2015 they represent the 53% of the total productive units). Small size ones reach the 30-38%, quite constant in all the period analyzed. The larger typologies showed a decreasing trend (19%, 22%, 20%, 34% and 14% respectively in 2002, 2009, 2013, 2014 and in 2015) (Figure 3).

Figure 2. Milk in the boiler before the starting of the processing steps.

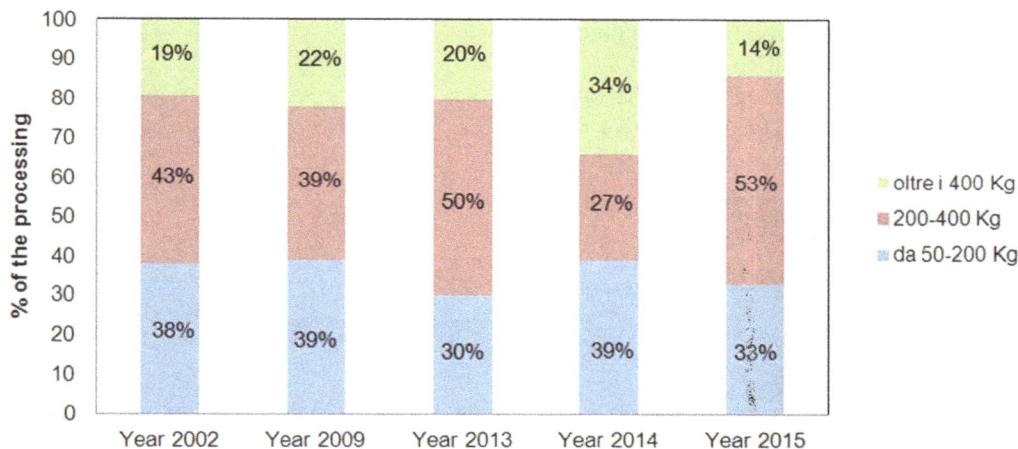

I modify the format of the graphs

Figure 3. % of processing divided by the productive dimensions (comparison between years 2002-2009-2013-2014-2015) (P<0.05).

This values reflect perfectly the situation of the downhill farms. In fact, in Friuli Venezia Giulia, the dimensions of the agricultural holdings are year by year more oriented towards the medium size [8].

Particular attention was paid towards the rearing system and animal welfare. The herds were always suggested to control the sanitary situation of their cows before stocking the alpine meadows, in order to avoid the presence of ill animals that could affect the quality of milk. Following an ancient *motto* that says "*a good cheese starts from the stable*", a part the importance to avoid dairy cows with mastitis, the milking processing must be absolutely done with the maximum care, from the hygienic and physiological point of view. A preliminary work consists in the filtration of the milk that can allow the reduction of the largest part of contamination. An important step then is the storage of the milk of the previous days, that must be done considering especially the modality of the raw material cooling and the containers employed for this delicate phase. The cheese makers were always encouraged to maintain milk at a temperature between 8 and 14°C, collecting it in bins perfectly sanitized and sealed. In the investigated farms, actually not always the storage temperatures of milk were respected, with obvious negative effects on the cheese production. The same attention should be put also for the storage rooms, that have to be always clean and tided up.

The acidity of fresh milk, of stored one, of their mix in the boiler, of whey after the curd milling and of final whey were check out [9]. This parameter allows the evaluation of the sanitary estate of the cows, the freshness of the raw material, its possible alteration, its maturation level and its attitude to the dairy transformation. Variations in organoleptic qualities are mainly due to technological factors and among them to the kinetics of acidification during cheesemaking, that can characterize a so-called disgenesic milks (low acidity, slow and difficult coagulation, low acidification rate and inadequate creaming activity).

In Figure 4 it is shown the comparison of the acidities measured in storage milk. In 2002, the average value observed was 4,01°SH/50 ml, the following years were 3,34°SH/50 ml, 3,38°SH/50 ml, 3,47°SH/50 ml and 3,48°SH/50 ml respectively. As regards the acidity in the fresh milk (data not shown), the values were almost constant for all the years taken into account and it was about 3,40°SH/50 ml.

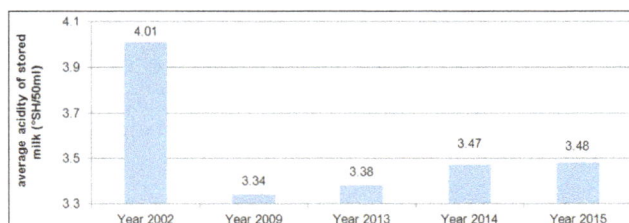

Figure 4. Acidity of the stored milk (comparison between years 2002-2009-2013-2014-2015) (P<0.05).

Throughout the amelioration of the sanitary estate of the herd and the more attention paid towards the quality of the milk (first of all during the storage phase of the raw material), taking into account the suggestions given by the ERSA technicians, it was possible to get the ideal values for this parameters (3,5÷3,9°SH/50 ml) for the dairy transformation [10, 11].

The cheese makers were helped also to learn the use of the instruments for a preliminary milk evaluation, as the acidimeter, the pH-meter (this instrument was introduce just in the two last years) and the "Leucocytest Roger Bellon" (the California Mastitis Test), which is useful for an early evaluation of the possible presence of mastitis in the herd, to attend immediately on the infect and to exclude the not suitable milk. They were encouraged to register all the values by themselves, in order to create a personal archive very useful to improve further processing [5].

(local name: formadi di mont, Figure 5)

Figure 5. *Typical autochthonous alpine cheese.*

A very interesting element to be considered, especially for its relevant ecological impact, is the type of combustible material employed during the dairy processing. The utilization of wood in this alpine structure was almost the 76% in the 2002, while this percentage unfortunately decreased to 69%, 65%, 56% and 69% respectively in 2009, 2013, 2014 and 2015, with in parallel, obviously, the increment of the use of other synthetic fuels. This choice has a great relevance from the quality of the products, but also it is important for the economic and environmental point of view. In fact, the heating by wood contribute to impart to cheese and ricotta a unique bouquet of tastes, smells and perfumes, emphasizing the particular style of the Alpine cheesemaker who has produced. There are also implications on the biodiversity of the surroundings, because the collecting of firewood implies a care of the forest. This activity is considered precious and necessary for the tutelage and wise control of the mountain land. Last but not least, the natural resource permits a saving from the monetary point of view [12, 5, 13].

The modality of typical dairy processing in alpine chalets are described below.

3.2. Alpine Cheese

The autochthonous productive pathway that characterizes the dairy tradition in this Region is shown in Figure 6.

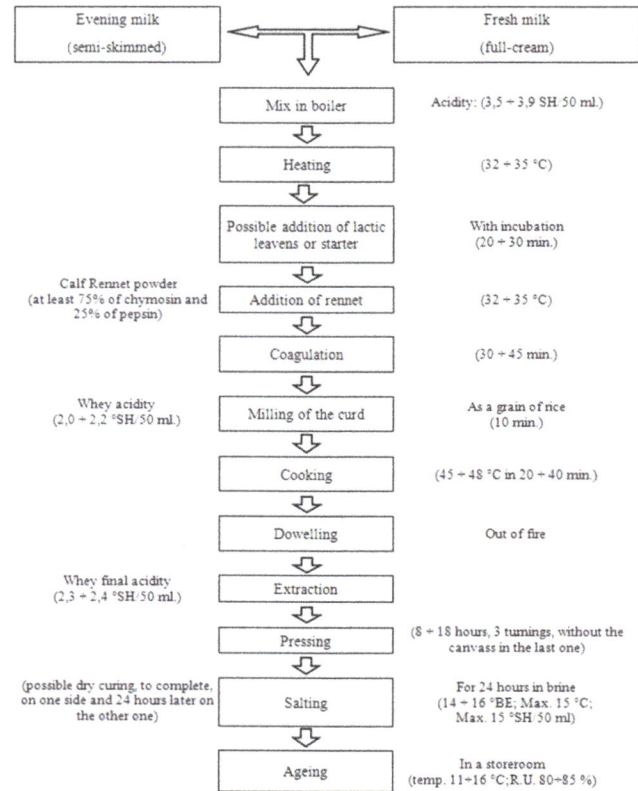

Figure 6. *Technological diagram for the alpine cheese.*

Own characteristics of this alpine cheese include:
- usually made from raw cow's milk heat treated (cooked), but not fully pasteurized;
- a semi-firm to hard texture with a dense paste;
- natural brushed/rubbed rind;
- large sized wheels;
- often presence of holes or "eyes". [6].

Lucey *et al.* [14] observed that the "art" or "science" of cheesemaking, that affects the physical properties of cheese (i.e., body/texture, melt/stretch, and color), is all about managing five key factors: initial milk composition, rate and extent of acid development, moisture content, curd manipulation and maturation conditions.

The cheese ripening is always affected by the interaction between milk (specific) and the traditional technology applied to the transformation process (non-specific). Also, the environment can be quite definite and not reproducible. The factors of typicality are:
- species and/or breed (genetic factors);
- the general environment and the pedo–climatic conditions;
- animal management system and feeding (that influence the protein and fat content, renneting properties, fatty acid composition and other chemical flavors);
- milk treatments and processing;
- the original *microflora*;
- the ripening procedures. [15].

This particular environmental obviously has a great influence on the sensory quality of dairy products, as confirmed by Martin *et al.* [16]. In particular, the botanical composition of the pasture have been recognized as a factor in flavor enrichment of cheese not only *per se*, as terpenes, sesquiterpenes etc., but also for compounds of microbial origin. Several studies have shown that the floristic diversity of grazed herbage can influence the sensory characteristics of raw milk cheese, especially for the direct or indirect effect of a complex blend of volatile compounds present in the different grass species [17]. Going into deep, Dovier *et al.* [18] underlined the differences between lowland and highland cheese, while Coloumb *et al.* [19] investigated also the positive correlation between the composition of fodder consumed by the cow and the altitude at which the cow grazes.

The renneting properties, as well as its clotting time, the coagulum strength and the syneresis rate are of major importance for the rheological characteristics of curd and then for the chemical-physical and structural properties of cheese. They influence whey drainage, which affects the moisture content of the curd, fundamental for the appropriate start of the cheese ripening processes, which are in turn linked to fermentation activity [20, 21].

The absence of the pasteurization ensures the maintenance of the incomparable fragrance of smells and tastes that the surrounding environmental of alpine pastures confers to the crude material. In these types of dairy products, it is essential to keep unaltered the constitutive *microflora*, with various microorganism of different species and with distinctive features. It can be affirmed that a major factor of success is the correct management of bacteriological traits, because, as reported by Agabriel and colleagues [22], the characteristics of ripened cheeses depend both on the cheesemaking technology and on the chemical and bacterial composition of milk. The microbes are able to multiple themselves actively, acidify and produce important substances that confer the desired flavor. Furthermore, these peculiarities could be exalted with the employment of the natural lactic leavens during the curdling, that don't modify the typical characteristics of milk, enhancing the differences and the authenticity among the productive units. This can be useful also in the case that the raw material presents microbiological, chemical or physical imbalances. In particular, the natural lactic leavens are used to:

a) increase the initial acidity of milk and to allow the curd to acidify. The measure of the °SH/50 ml value of the raw material determines the quantity of natural lactic leavens to be employed;

b) prevent coliform bacteria, since, being rich in lactic acid bacteria, they hinder the development of these unwanted microbes and limit the formation of tiny holes in the texture of the product;

c) modulate the fermentation and reduce defects in cheese [23].

Actually, in the last years, there is an increasing trend to employ industrial starters. In fact, in 2002, the 6% of the cheese makers used a synthetic starter, while the 42%, the 49%, the 47% and 31% were the data observed in the

following years taking into account (Figure 7). The tendency for the natural lactic leavens was: 9%, 26%, 14%, 30% and 38% (always considering the years 2002, 2009, 2013, 2014 and 2015). The technicians always encouraged the cheese makers towards an own production, even if it requires time, precision and a particular accuracy (and this is the reason why a lot of producers tends to prefer the industrial ones, that are ready to be employed), because it is the only way to have a guaranty about the peculiarity and authenticity of the product. Furthermore, the group of Franciosi et al. [24] stated that the use of commercial starters in raw milk cheeses may modify the characteristics of the cheese microbiota, in particular lowering their biodiversity.

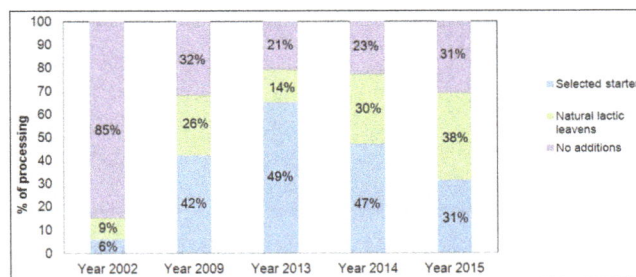

Figure 7. % of productive units that employs different starters (comparison between years 2002-2009-2013-2014-2015) (P<0.05).

The *microbiome*'s role is also related to food safety and to health issues. Many of the *bacteria* that can be found on the rind prevent the spread of potentially dangerous pathogens by excreting inhibitors against them, such as *listeria monocytogenes* [25]. Other undesired germs can be recognized pathogens like *Brucella spp.*, *Staphylococcus aureus*, *Myobacterium tubercolosis*, some *Enterobacteriaceae*, etc.; those that are unsuitable for cheese production like *clostridia* spores or simply competitors for the desired *lactobacillum* activity. For this reason, in order to optimize the bacterial quality, the cheese maker were suggested to respect the procedures of prophylactic and hygienic means of eradicating human transmissible diseases, to ensure the minimization of the animal fecal contamination and to guarantee the cleanliness in the milking plant and of the cheesemaker hut. The first reduction in active *bacteria* and spores can be successfully obtained through the physical effect of fat globules rising during natural creaming [21].

A general amelioration of the health and sanitary aspect was allowed also by the modernization of the milking process, with the introduction of small milking parlors directly on the pastures.

Going into deep of the technological stages, it is worth of consideration the size of the curd after the milling. It can be affirmed that gradually, from 2002 till 2015, the dimensions were reduce as a rice grain (data not shown). This dimension was strongly recommended during the technical visits because it allows to decrease the internal humidity of the cheese, thanks to a higher level of extraction of the whey. In this way, it is facilitated the further phase of the storage, in fact, there are less molds and it is preserved best physical parameters of the cheese wheels in the storage room.

Precise instructions were given about the timing and the temperatures to be respected during all the processing. In the comparison of the 5 years examined, the duration of the total processing in the boiler has reduced (from 107 minutes in 2002 to 93 minutes in 2015, as reported in Table 1). This aspect has an important advantage, because with a shorter manufacture the acidity of the whey can be contained, enhancing the following preparation of the ricotta cheese (that employs the residual whey of the previous processing of the cheese). A brief processing time allows also more free time for the cheese maker that can focus on the care of the subsequent steps as for example the pressing of the cheese [13].

In general, as it can be observed in Table 1, the average value registered respected the reference range.

Table 1. Average duration of the total processing pathway (min.) (comparison between 2002-2009-20013-2014-2015) (P<0.05).

Technological parameters	2002	2009	2013	2014	2015	Reference range
Temperature of calf rennet powder addition (°C)	34.6	34.8	34.5	34.6	34.9	32.0 ÷ 35.0
Duration of the coagulation (min.)	41	37	34	35	34	30 ÷ 45
Duration of the curd milling (min.)	11	13	15	14	13	10 ÷ 15
Duration of the curd heating (min.)	30	27	25	23	24	20 ÷ 40
Temperature of heating (T°C)	47.8	46.9	46.3	46.5	46.5	45.0 ÷ 48.0
Duration of the doweling (min.)	24	23	22	26	22	15 ÷ 30
Total duration of the processing (min.)	107	101	98	98	93	90 ÷ 110

In the dairy technology, the modality of salting has a prominent role, in fact, it:

- completes the whey extraction,
- enhances the conservation of the final product,
- delays the acidification and the microbial colonization,
- favorites the rind formation,
- stimulates the protein solubilisation,
- confers the perfect flavor.

This step consists in dipping the cheeses in brine, made by a mix of water and salt. For this reason, this solution must be constantly monitored in order to avoid its degradation and fermentation, caused by the contamination with impurities and whey. These undesired processes are triggered by high temperatures. As Ponce De Leon-Gonzalez *et al.* [26] observed, this process requires a careful monitoring of the saturated solutions to obtain cheeses of good quality. During the experimental trial, the parameters of acidity (°SH/50 ml), salt concentration (°Bé) and temperature (°C) were always registered in the appropriate tubs of the brine. In particular, the acidity should be under the value of 15°SH/50 ml and salt concentration has to be in the range of 14÷16°Bé, otherwise it is urgent to renew the brine. Furthermore, the temperatures of the solution must be kept under the 15°C and the ratio among the volumes of cheeses and salt concentration must be 1:5. In the course of the surveys, sometimes, the data observed were not satisfying the required value for an optimal salting phase, affecting seriously the final quality of the cheese. So the technicians tried to stress out the importance of caring about this important step of the dairy process, in order to avoid a brine that can confer to the product unpleasant smells of ammonia, ruin the rind and bring it an unusual color [27, 28].

The last phase of the processing consists in the storage that is performed in appropriate maturing warehouses (known as "*celârs*"). In the past, some building reconstructions or some *ex novo* stores weren't projected respecting the ideal features for an optimal cheese stocking phase. Traditionally, this room was oriented towards the North, at the basement and with thick walls made up by rocks and lime to guarantee the following conditions:

- constant temperature and humidity, due by a thermo-hygrometric exchange, also during very high thermal day/night temperature gap;
- a correct ventilation, throughout the windows wisely provided.

During the summer 2014, two loggers were simultaneously settled down in order to compare the microclimate parameters in a traditional store and in a modern one. Figures 8a and 8b show that, in the typical building (red line in the graph), the internal conditions are better for cheese storage because there are no sudden variations of temperature and humidity, as observed, instead, in the new one (blue line). The tendencies noted were analogous to previous measurements done in 2004, with the same technique and in similar conditions (data not shown). Abrupt modifications of the values enhance the development of defects in the cheese [29]. In the structures with this type of inconveniences, it was suggest to install a dehumidifier or a conditioning system and to manage the opening of the windows to assure a good level of ventilation.

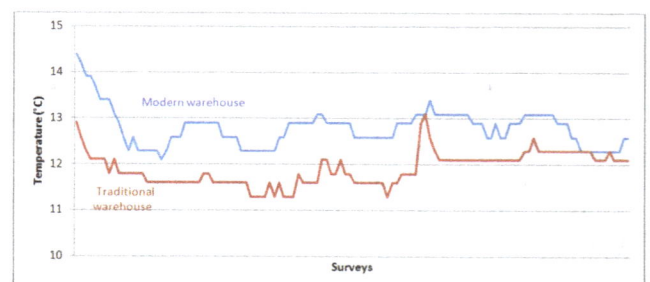

Figure 8a. Comparison of the thermal pattern observed in two warehouses (118 surveys in 5 days).

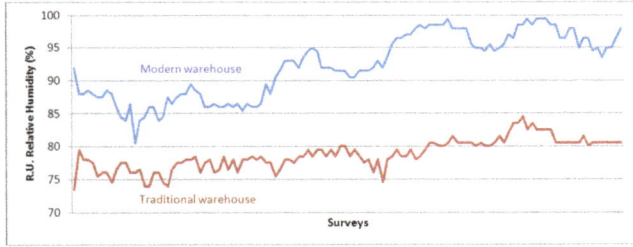

Figure 8b. *Comparison of the hygrometric pattern (R.H. %) in two warehouses (118 surveys in 5 days).*

3.3. Alpine Ricotta Cheese

The second dairy product is the alpine ricotta cheese that is considered traditional because it has a documented history of more than 25 years. It symbolizes a style of cooking using autochthonous flavors of this Region [5]. In Figure 10, it is exemplified the technological diagram for this production. The whey, resulted from a previous cheese processing and still rich in some proteins, is flocculated with the addition of an acidifier. A further heating is necessary to curdle the mixture into a suspension of finely divided curd particles and to allow the fixture to stand while the curd particles rise to the surface. The curd is skimmed off by hand ladling into a suitable receptacle (traditionally canvass bags), as residual whey is drained off below them, to drip. The Ricotta curd is very delicate and will readily disintegrate and re-mix with the residual whey body if subjected to shock from rough handling during ladling. The following steps are the pressing, the salting and the smoking one.

(traditional name: scuete fumade di mont, Figure 9)

Figure 9. *Typical autochthonous alpine ricotta cheese.*

Technological diagram for the alpine ricotta cheese

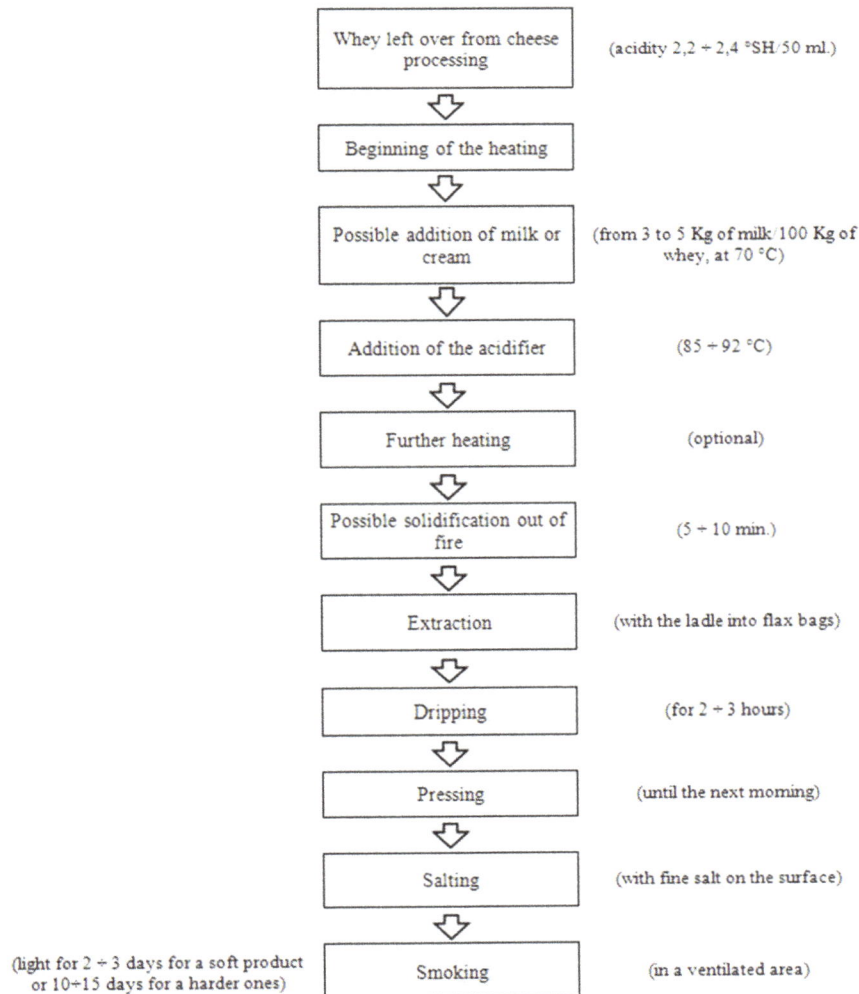

Whey left over from cheese processing	(acidity 2,2 ÷ 2,4 °SH/50 ml.)
Beginning of the heating	
Possible addition of milk or cream	(from 3 to 5 Kg of milk/100 Kg of whey, at 70 °C)
Addition of the acidifier	(85 ÷ 92 °C)
Further heating	(optional)
Possible solidification out of fire	(5 ÷ 10 min.)
Extraction	(with the ladle into flax bags)
Dripping	(for 2 ÷ 3 hours)
Pressing	(until the next morning)
Salting	(with fine salt on the surface)
Smoking	(in a ventilated area)

(light for 2 ÷ 3 days for a soft product or 10÷15 days for a harder ones)

Figure 10. *Alpine ricotta cheese.*

During the technical visits, all the values were registered in the schedules (Annex 2).

A very important aspect is the use of the acidifier. During the years the habits had some modifications. In 2002 the cheese makers mainly employed preferably magnesium sulfate (65%). The 16% opted for the citric acid, while the 19% of them used some different products (referred as *other* in the Figure 11) that are related to a cultural heritage or a family tradition (for example: the vinegar or the *siç*, a particular mix of whey, beech bark and sorrel leaves). Analyzing all the experimental period, it can be observed the rise of the citric acid (from 16% in 2002 to 33% in 2015) and the reduction of magnesium sulfate (from 65% in 2002 to 38% in 2015). Furthermore, in 2009, there was the introduction, in some productive units, of a new industrial compound called "*Sali mix*". This substance is becoming more and more present, in fact, the % of the processing utilizing it increased from 5% to 17% in 2002 and 2015, respectively.

During the trial program, some recommendations were provided about the correct way of employing the acidifiers, their dose, the ideal temperature at which adding them. These aspects are essential to obtain a high quality ricotta cheese.

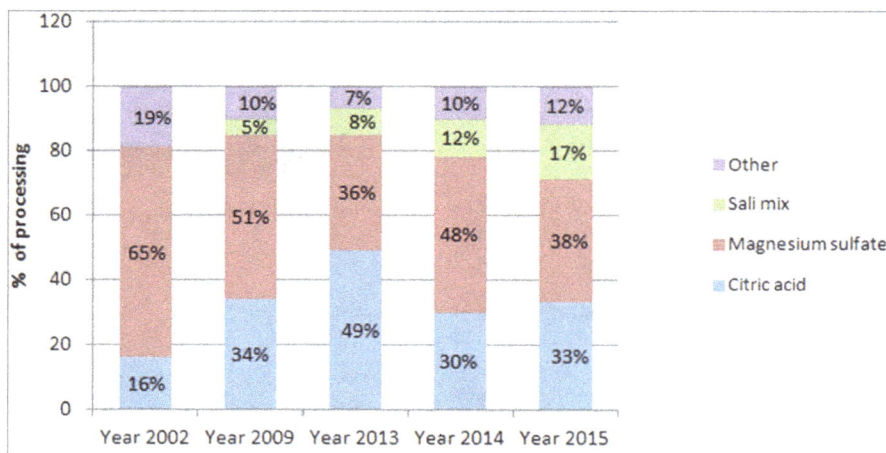

Figure 11. *Type of acidifier employed for alpine ricotta cheese processing (comparison between years 2002-2009-2013-2014-2015) (P<0.05).*

Considering the technique of adding milk, cream or buttermilk to the whey, it can be pointed out that also in this case there were evident modifications in the custom and practice of the cheese makers. In 2002, the 93% of them employed the whey pure to produce this dairy. Year by year, they tended to enrich the solution with milk, cream or buttermilk (as reported in Figure 12). In 2015, buttermilk was also mixed during the productive pathway of the alpine ricotta cheese. This new habit is related to the need to satisfy the tastes of the modern consumers that prefer a product more soft, creamy, doughy and lightly smoked. These addition modify the structure of this dairy, because it becomes more tender and moist. These features, as underlined by the technicians, determine a less shelf-life, in fact, it is important to consume these ricottas quickly to avoid the formation of bad smells and of a chalky texture [30]. In general, it can be pointed out that in this way, the identity of the product losses its peculiarity, with the risk to standardize the inimitable alpine ricotta cheese, as Chiorpis *et al.* [13] always emphasized.

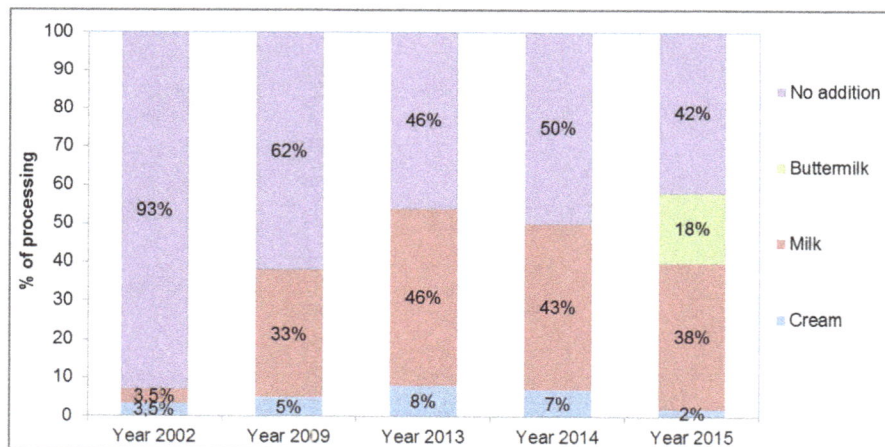

Figure 12. *Technological addition of milk, cream or buttermilk to prepare the alpine ricotta cheese (comparison between years 2002-2009-2013-2014-2015) (P<0.05).*

From the technological point of view, it was suggested to the cheesemakers to reduce the duration of the heating preferably at 30-40 minutes, in order to improve the rising of the albumin (principle protein in the whey, after the cheese production).

A correct salting step is very essential, because this practice is relevant since it enhances the expulsion of the whey and it prolong the preservability of the final product. Different are the methodology:

- dry salting: rubbing the surface of the ricotta with fine salt. Than stand it for a day, before smoking it.
- salting the whey: this modality consists in salting the whey before starting the processing. The procedure presents some defects, because it is expensive for the great amount of salt necessary and it is also not suitable for the further employment of this liquid for swine feeding.
- before the dripping of the canvas bags: the ricotta is put into tubs with sodium chloride and mixed. In this way a homogenous distribution of salt is guaranteed but there is the risk to ruin the texture of the final product.

Ponce De Leon-Gonzalez et al. [26] recovered that the final result isn't affected by the modality of salting.

The ending phase of the all processing is represented by the smoking. Usually it lasts 3/4 days for soft ricottas, while it is prolonged till 10/15 days for harder ones. This duration is essential in order to enhance a gradual and uniform result, with abnormal colorations of the rind or too much intense taste.

4. Conclusions

Some reflections can be done about the perspective of qualitative improvement of this typical dairy products, that can, in a wider perspective, be an important train for the developing of all this marginal mountain area. In fact Alpine cheese and ricotta are emblematic and a symbol of the mountain of Friuli Venezia Giulia (Italy) [31]. The summer grazing of mountain pastures with lactating dairy cows is still a widespread practice in all the alpine regions and has positive repercussions on the quality image of many local dairy products [32]. Bovolenta [33] underlined the need to development models which examine the sustainability and multi-functionality of livestock production systems. The alpage can be considered a tool for improving the appeal of the landscape, the benefit of local cheese products and the diversification of this type of marginal agriculture in tourism [34]. This seasonal activity represents a reality with strong symbolisms and heritagization where nature and human presence join together in an intimate and harmonic way, creating attractiveness for the tourists, in a sort of "open air" museum [35].

In the future, typical dairy products will only be able to maintain their markets or develop them unless they are capable of holding their commercial ground and adapting to the market's needs and demands without losing their specificity, originality and authenticity (Bertoni et al., 2001). The new extraordinary success of niche food and the market demand for organic and traditional foods (for instance raw milk cheese of certified origin and zero-mileage food or farmers' market foodstuffs) are a symptom of a widespread consumer desire for better quality, which would be found in the peculiarities of local ones. Nowadays, the search for authenticity and preference for food with low ecological-impact, seem to find in the alpine environment a natural candidate for a repository of culture and bio-diversity [2]. In the specific context analyzed in the present paper, the efforts of Ersa recently were oriented also towards a quality liability system (quality mark: Aqua) which could guarantee artisan traits linked to the locality and cultural roots, as well as good quality features based on the standards required to satisfy consumer expectations [36]. Nevertheless, the unique features of these production systems have to be taken into account to avoid any excessive changes that could contribute to the loss of these 'inimitable treasures'. As observed by Bertoni et al. [21] it is otherwise important to evaluate carefully the controversy about whether and to what extent technology can be innovated. They affirmed that tradition does not exclude innovation, because the cultural heritage is not to be interpreted with the meaning of static and unchangeable. The techniques involved in traditional technology, which must be respected, can be modern. The most important objectives has to absolutely high safety and quality, that not always are related to typicality. The transformation of traditional Alpine cheese-making systems obviously requires a certain amount of standardization of traditional food processing, in order to offer local products on a wider market basis and in the logic of local/global dimension. It is interesting to notice, as suggested by Grasseni [2], that, in this way, autochthonous strategies for re-valuing intangible patrimony (local histories, material culture and landscapes) are mobilized.

During the 14 years of technical assistance, the cheesemakers that participated to the experimental project reached important goals along all the productive pathway. This support was dedicated with particular care to encourage and to guide the new generations, on the way to make the best technical and management choices. Accompanying the young in this productive chain that involves the artisanal culture is of particular relevance. The most important issue was that diversity and rootedness could become a commodity and they were always suggested to pursue an economic efficiency and a high quality level of the products [36]. The further programs of Ersa are oriented to continue in this activity, that can be considered a "good practices" to be spread, in order to keep on following the development of this compartment.

Acknowledgements

The authors wants to thank all the cheesemakers that take part to the experimental project of technical assistance, during these fourteen years.

Annex 1: Schedule for Alpine Cheese Date of Collection: __/__/__

General information

Cheesemaker hut	
Cheese maker	
Type of boiler	
Type of heating (wood or other sources)	

Milk

Total amount of milk (Kg)	
of bovine (Kg)	
of goat (Kg)	
Temperature of milk storage (°C)	
Number of milkings	

Technology	Yes or No	Dose (g/Kg)	Acidity	
			(°SH/50 ml)	(pH)
Natural lactic leavens				
Indirect inoculum (*)				
Starter				
Calf rennet powder				

Acidity	Preserved milk		Fresh milk		Luecocytest
	(°SH/50 ml)	(pH)	(°SH/50 ml)	(pH)	(+, -, ±)
Bovine milk					
Goat milk					
	(°SH/50 ml)	(pH)			
Mixed milk in the boiler					
Whey after the curd milling					
Final whey					

Operations in the boiler	Start time	Duration (min)	Temperature (°C)
Addition of the starter or of the natural lactic leavens			
Duration of the incubation			
Addition of the calf rennet powder			
Duration of the coagulation (min)			
Beginning of milling			
Type of milling (Rice - Fat rice or Wallnut)			
Duration of milling			
Break (Yes or No)			
End of the milling and beginning of the heating			
Duration of the heating			
End of the heating			
Duration of the cooking out of fire			
End of the heating out of fire and extraction			
Total duration of the processing			

Final treatments	Number of turnings of the cheese loaves		Duration (hours)		
Pressing phase					
Break at room temperature (Yes or No)					
Parameters of the brine	Acidity		Concentration	Temperature	Duration
	(°SH/50ml)	(pH)	(°Bé)	(°C)	(hours)
Dry cured (Yes or No)					
Storage parameters	Relative umidity (%)			Temperature (°C)	

* The Indirect inoculum is made heating of fresh milk until 90°C, than cooling it down to 45°C. Then there is the addition of the starter, that is incubated for 6÷9 hours to reach a final acidity of 10÷12 °SH/50 ml.

Annex 2: Schedule for Alpine Ricotta Cheese Date of Collection: __/__/__

General information

Cheesemaker hut	
Cheese maker	
Type of boiler	
Type of heating (wood or other sources)	

Whey

Total amount of whey (Kg)		
Acidity	(°SH/50 ml)	(pH)

Technology

Disacidification of the whey (Yes or No)	

Addition	Type	Dose	
		(g/Kg)	l/Kg
Acidifier			
rock salt			
milk			
cream			
buttermilk			
other			

Operations in the boiler		Start time	Duration (min)	Temperature (°C)
Beginning of the heating				
Addition of	rock salt			
	milk/cream/buttermilk, other			
	acidifier			
End of the precipitation and extraction				
Duration of the heating				
Duration of the precipitation (outcrop)				
Total duration of the processing				

Final treatments

Purging by gravity (hours)	
Pressing (hours)	
Dry curing (Yes or No)	
Smoking phase	

References

[1] J. Forrest, "Alpine-cheese", available online at: http://www.seriouseats.com/2009/02/-101.html.

[2] C. Grasseni, "Re-inventing food: Alpine cheese in the age of global heritage", in Anthropology of food, August 2011 (available online at: http://aof.revues.org/6819).

[3] S. Rainis, S. R. S. Cividino, and F. Sulli, "Analysis of the criticizes for zootechnics development in a mountain area of Italy", in Journal of Life Science, Vol. 7, No. 8, 846-855, August 2013.

[4] FAO, "Restarting from Nature for the Agricultural Development, FAO Introduces a New model of Agricultural Development to Produce More with Less", 2011, available on line at: http://www.fao.org/news/story/it/item/80136/icode/.

[5] Chiorpis G., and Rainis S., "L'assistenza tecnica dell'ERSA negli alpeggi del Friuli Venezia Giulia: tredici stagioni tra sperimentazione e tradizione" in Notiziario Ersa, pp. 15-23, February 2014.

[6] J. Meier, "What is alpine cheese" available online at: http://cheese.about.com/od/cheese_varieties/fl/What-is-Alpine-Cheese.htm, 2016.

[7] J. Hazard, "Alpine cheeses", available online at: http://culturecheesemag.com/cheese-iq/alpine-cheeses Style Highlight: Alpine Cheeses, 2013.

[8] Istat. 6° censimento generale dell'agricoltura in Friuli Venezia Giulia. Dati definitivi. Regione Autonoma del Friuli Venezia Giulia, July 2013. Available online at: http://www.regione.fvg.it/rafvg/cms/RAFVG/GEN/statistica/.

[9] O. Guneser, Y. K. Yuceer, "Characterization of aroma-active compounds, chemical and sensory properties of acid-coagulated cheese: Circassian cheese International", in Journal of Dairy Technology, Vol. 64, Issue 4, pp. 517–525, August 2011.

[10] V. Bottazzi, "Microbiologia e biotecnologia lattiero-casearia". Il Sole 24 Ore Edagricole, Bologna, 1993.

[11] O. Salvadori del Prato, "Trattato di Tecnologia Casearia". Edagricole, Bologna, 1998.

[12] S. Gunasekaran, and M. Mehmet Ak," Cheese Rheology and Texture", CrC Press, December 2002.

[13] Chiorpis G., Pittino E., and S. Rainis, "Assistenza lattiero-casearia in malga, i nuovi orizzonti del 2014", in Notiziario Ersa, pp. 43-49, December 2015.

[14] J. A. Lucey, M. E. Johnson, and D.S. Horne, "Invited Review: Perspectives on the Basis of the Rheology and Texture Properties of Cheese", in Journal of Dairy Science, Vol. 86, Issue 9, pp. 2725–2743, September 2003.

[15] J. B. Coulon, A. Delacroix-Buchet, B. Martin, and A. Pirisi, "Relationships between ruminant management and sensory characteristics of cheeses: a review", in Lait, Dairy science technology, Vol.84, n° 3, pp. 221-241, May-June 2004.

[16] B. Martin, I. Verdier-Metz, S. Buchin, C. Hurtaud, and J. -B. Coulon, "How do the nature of forages and pasture diversity influence the sensory quality of dairy livestock products?", in Animal Science, Vol. 81, Issue 02, pp 205-212, October 2005.

[17] S. Bovolenta, "Il pascolo alpino come strumento di valorizzazione del territorio e di qualificazione dei prodotti caseari", in Agribusiness Paesaggio & Ambiente – Vol. 4, n°3, December 2000.

[18] S. Dovier, R. Valusso, M. Morgante, A. Sepulcri, and S. Bovolenta, "Quality differences in cheeses produced by lowland and highland units of the Alpine transhumant system", in Ital. J. Anim. Sci. Vol.4 (Suppl. 2), January 2010.

[19] M. Collomb, U. Butikofer, R. Siebera, B. Jeangrosb, and J.-O. Bosseta, "Correlation between fatty acids in cows' milk fat produced in the Lowlands, Mountains and Highlands of Switzerland and botanical composition of the fodder", in International Dairy Journal, Vol. 12, pp. 661–666, December 2002.

[20] G. Tornambè, A. Lucas, I. Verdier-Metz, S. Hulin, C. Agabriel, and B. Martin, "Effect of production systems on sensory characteristics of PDO Cantal cheese", in Ital. J. Anim. Sci., 4 (Suppl. 2) pp. 248–250, 2005.

[21] G. Bertoni, L. Calamari, and M. G. Maianti, "Producing specific milks for specialty cheeses", in Proceedings of the Nutrition Society Animal Nutrition and Metabolism Group Symposium on "Quality inputs for quality foods", Vol. 60 (2), pp. 231-46, May 2001.

[22] C. Agabriel, B. Martin, C. Sibra, J. C. Bonnefoy, M. C. Montel, R. Didienne, and S. Hulin, "Effect of dairy production systems on the sensory characteristics of Cantal cheeses: A plant scale study". Anim. Res., Vol. 53, pp. 221–234, January 2004.

[23] G. Chiopris, D. Pasut, E. Pittino, and M. Sanna, "Il monitoraggio degli alpeggi per lo sviluppo dell'alpicoltura in Friuli Venezia Giulia, Linee guida per la gestione delle malghe". Ersa Eds., 2014.

[24] E. Franciosi, I. Carafa, T. Nardin, S. Schiavon, E. Poznanski, A. Cavazza, R. Larcher, and K. M. Tuohy, "Biodiversity and γ-Aminobutyric Acid Production by Lactic Acid Bacteria Isolated from Traditional Alpine Raw Cow's Milk Cheeses", in BioMed Research International, Vol. 2015, 2015.

[25] E. Schornsteiner, E. Mann, O. Bereuter, M. Wagner, and S. Schmitz-Esser, "Cultivation-independent analysis of microbial communities on Austrian raw milk hard cheese rinds", in International Journal of Food Microbiology Volume 180, pp. 88–97, June 2014.

[26] L. Ponce De Leon-Gonzalez, W.L. Wendorff, B.H. Ingham, J. J. Jaeggi, and K. B. Houck, "Influence of Salting Procedure on the Composition of Muenster-Type Cheese", in Journal of dairy science, Vol. 83, Issue 6, pp. 1396–1401, June 2000.

[27] R. C. Lawrence, L. K. Creamer, and J. Gilles, J., "Texture development during cheese ripening", in J. Dairy Sci. - Symposium: Cheese Ripening Technology, Vol. 70, n°8, pp. 1748–1760, August 1987.

[28] R. A. Wilbey, J. E. Scott, and R. K. Robinson, "Cheesemaking Practice". Springer US, 3rd ed., 1998.

[29] E. Sorrentino, G. Tipaldi, G. Pannella, M. La Fianza, P. Succi, and P. Tremonte, "Influence of ripening conditions on Scamorza cheese quality", in Int. J. Agric. & Biol. Eng.. Vol. 6 n°.3/71, 2013 available online at: http://www.ijabe.org.

[30] G. Hough, M. L. Puglieso, R. Sanchez, and O. Mendes da Silva, "Sensory and Microbiological Shelf-Life of a Commercial Ricotta Cheese", in Journal of Dairy Science, Vol. 82, Issue 3, pp. 454–459, March 1999.

[31] S. Rainis, F. Sulli, S. R. S. Cividino, and E. Cossio, "The impact on landscape, environmental and society of new productive chains in a mountain area: strategies, analysis and possibilities of development", in Journal of Agricultural Engineering, vol. 43, n°1:e2, May 2012.

[32] M. Mrad, E. Sturaro, G. Cocca, and M. Ramanzin, "The alpine summer pastures in the Veneto Region: management systems", in Proc. 18th Nat. Congr. ASPA , Palermo (Italy), Ital. J. Anim. Sci. vol. 8 (Suppl. 2), pp. 313-315, 2009, available online at: http://www.aspajournal.it/index.php/ijas/article/view/ijas.2009.s2.313/427.

[33] S. Bovolenta, A. Romanzin, M. Corazzin, M. Spanghero, E. Aprea, F. Gasperi, and E. Piasentier, "Volatile compounds and sensory properties of Montasio cheese made from the milk of Simmental cows grazing on alpine pastures", in J Dairy Sci., Vol. 97(12), pp. 7373-85, December 2014.

[34] A. M. Hjalager, "Agricultural diversification into tourism: evidence of European Community development programme", in Tourism management, Vol. 17 (2), pp. 103-111, March 1996.

[35] M. Corti, "Le valenze turistiche ed educative del sistema delle alpi pascolive: indagine sugli eventi turistici sul tema dell'alpeggio"in Alpeggi – Quaderni Sozooalp. 1, pp. 53-89, May 2004.

[36] L. Berozzi, and G. Panari, "Cheeses with Appellation d'Origine Controlee (AOC): factors that affect quality", in Proceedings of the International Dairy Federation Seminar on Cheese Ripening, International Dairy journal, Vol. 3, pp. 297-312, 1993.

7

Effect of Selected Tree Species on Maximizing Soil Organic Carbon Sequestration in Imo State, Nigeria

Umeojiakor A. O.*, Egbuche C. T., Ubaekwe R. E., Nwaihu E. C.

Department of Forestry and Wildlife Technology, Federal University of Technology, Owerri, Nigeria

Email address:

heavenlytony@yahoo.com (Umeojiakor A. O.)

Abstract: The world is currently experiencing a period of warming and the role of soil carbon pools for mitigation of greenhouse gases has encouraged the need for more knowledge on the tree species effects on soil organic carbon. The study was conducted to evaluate the effect of tree species on maximizing soil organic carbon sequestration in Imo State, Nigeria. Four tree species (Teak, *Tectona grandis*, linn, Gmelina, *Gmelina arborea* Roxb, Rubber plant, *Hevea bransiliensis* Mull. Arg. and Black velvet, *Dialium guineense* Wild) were chosen for the study. Random soil sampling was used in field studies. Soil samples were collected at the depth of 0-15cm and 15-30cm. these soil samples were prepared and subjected to routine laboratory analysis. Soil organic carbon sequestration was calculated and relationships between soil organic carbon sequestration and soil properties were obtained by simple correlation. Results showed that *Tectona grandis* of sequestration value 154.1 and 116.8 at top soil and subsoil respectively provides the best option for maximizing carbon sequestration in the soil, followed by *Hevea bransiliensis* (147.4 and 91.1), *Gmelina arborea* (134.1 and 81.1) and least was in *Dialium guineese* (108.1 and 60.1) at all depth. There was significant ($P = 0.01$) positive correlation between base saturation, calcium, total nitrogen with soil organic carbon sequestration at r –values of 0.77, 0.74 and 0.97 respectively. Hence, negative correlation existed between soil pH, clay fraction potassium with soil organic carbon sequestration with r-values of – 0.37, -0.68 and -0.54 respectively. It can be concluded that soil organic carbon sequestration decreases with decreasing depths and were greatly affected by tree species, soil properties and management practices.

Keywords: Tree Species, Soil Organic Carbon Sequestration, Soil Properties, Management Practices

1. Introduction

The world is currently experiencing a period of warning and it is widely accepted that the cause of this warming is a direct result of the increased levels of carbon dioxide (CO_2) in the atmosphere, caused by both natural and anthropogenic factors, such as industrial development, deforestation, agricultural improvements, increased use of cars and release of green house gases (CH_4, CO_2, N_2O, etc) are the major contributing factors to the depletion of the ozone layer and its associated global warming and climate change. Carbon exists in many forms, predominantly as plant biomass, soil organic matter and as the carbon dioxide (CO_2) in the atmosphere. Soil especially the forest soil is one of the main sinks of carbon on earth because these soils normally contain higher soil organic matter. It has been propounded that one method to reduce atmospheric carbon dioxide (CO_2) is to increase the global storage of carbon in soils. Soils are the largest carbon reservoirs of the terrestrial carbon cycle. It contains twice the amount of carbon in the atmosphere as carbon dioxide and thrice the amount in global vegetation [1].

Interest in the ability of forest soils to sequester atmospheric carbon dioxide has increased because of the threat of projected climate change. Thus, understanding the mechanisms and factors of soil organic carbon dynamics in forest soils is important to identify and enhance natural sinks for carbon sequestration to mitigate the climate change. Soil carbon sequestration is the process of transferring carbon dioxide from the atmosphere into the long lived pools (soil) through plant residues and other organic solids, and in a form that is not immediately remitted [2]. Put simply, soil carbon sequestration involves the transfer of atmospheric carbon dioxide into soil organic carbon pool [3]. This transfer or "sequestering" of carbon helps off-set emissions from fossil fuel combustion and other carbon-emitting activities while enhancing soil quality and long-term agronomic productivity.

The role of soil carbon pools for mitigation of greenhouse gases has encouraged the need for more knowledge on the tree species effects on soil organic carbon. Forest management, including a change in tree species, has been accepted as a measure for mitigation of atmospheric carbon dioxide in national green house budgets [4]. Hence, carbon sequestration by forest plantations is being proposed as one method of positively affecting the balance of atmospheric levels of carbon dioxide. [5]; [6]. One of the most effective activities to improve soil carbon sequestration is choosing suitable forest tree species; unfortunately, there is limited knowledge of it [7]. Therefore, this study will aid us to understand the forest plantation that provides the best option for maximizing carbon accumulation in the soil in order to enhance the natural sink of green house gases, reduce the rate of soil organic carbon depletion there by reducing atmospheric carbon dioxide that contributes to global warming and climate change.

2. Materials and Methods

2.1. Description of the Study Area

The study was conducted in three different locations of four different forest plantations in Imo State, Imo state lies between latitude $4^0 45^1$ and $5^0 50^1$N and longitude $6^0 35^1$ and $7^0 30^1$E. The three locations of the plantations are Ohuba, Umudike – 11, both in Ohaji/Egbema Local Government Area, and Federal University of Technology Owerri (FUTO) in Owerri West Local Government Area in Imo State, southeastern Nigeria. It has a humid tropical climate, characterized by bimodal rainfall pattern with mean annual rainfall ranging from 1800 to 2500mm and mean annual temperature ranging from 26^0C to 31^0C [8]. [9] showed that the underlying geological material are coastal plan sand (Benin Formation) and the Bende-Ameki formation. It consists of mainly friable sands with minor intercalations of clay.

2.2. Selection of the Study Forest Plantations

The study area consists of four forest plantations and they are as follows:

1) *Gmelina arborea* forest plantation site was established in the year 2004. It covers about five (5) hectares of land. The site was originally a fallow land, which was cleared through manual bush clearing followed by burning. After which, the seedlings of *Gmelina arborea* gotten from Ohaji/Egbema forest reserve was introduced to the site with 11 x 17 spacing. There is no special management practice such as beaten up, fertilization, pruning etc given to the growing seedlings except clearing of the weeds/bushes which is mainly *Panicum maximum* and few of *chromoleana odorata*. Fuel-wood harvesting and hunting were observed in the plantation. Presently, the trees have attained appreciable height and have a lot of plant debris on the floor.

2) *Tectona grands* forest plantation was established on the same land area of *Gmelina arborea* in the year 2004. It has the same management practices with *Gmelina arborea*. The only different is in their growth sizes.

3) *Hevea brasilinensis* plantation site of study was established in the year 1965 and was registered with the government in 1968. The site was originally a farmland under yam cultivation, before it was cleared through manual cuttings followed by bush burning. Then, *hevea brasiliensis* seedlings were collected from Federal Government and planted using 11 x 22 spacing in an area of about 6 hectares of land. Planted seedlings were raised with NPK fertilizer and other locally made compost, weeding, beaten-up were also observed. Presently the stands are well matured with fruits/seeds which falls and germinate thus alters the original spacing. Pruning, thinning, and bush clearing are always carried out in the site for maintenance purposes, also tapping of latex from about 15 years old plants and fetching of fuel-wood were also observed in the site.

4) *Dialium guineense* is a dominant tree species in a secondary forest beside River Otamiri in Federal University of Technology Owerri (FUTO). The area is characterized by mostly plant species, arranged in storey, with close canopies, highly depleted by anthropogenic activities such as fuel-wood harvesting, deforestation, hunting and gathering from wildlife.

2.3. Soil Sampling Techniques

Soil samples were randomly collected from each plantation at varying depths of 0 – 15cm and 15 – 30cm with an aid of soil auger. The soil samples were randomly collected at five different points in each plantation (five replicates). The collected soil samples were bagged in a polyethene bags and carefully labeled according to the plantation types, replicate number and depth. A total of forty (40) samples were collected from four forest plantations at two (2) varying depth. Core samples were also used to collect soils at each plantation in two (2) varying depths for bulk density determination. A total of eight (8) undisturbed core soil samples were carefully packaged and labeled accordingly.

2.4. Laboratory Analysis

The soil samples were air-dried at room temperature, crushed and sieved using 2mm sieve. The composite soil samples were taken to the laboratory for determination of the physical and chemical properties. Particle size distribution was determined by hydrometer method [10]. Bulk density was determined by the core method using [11]. Textural class was determined by [12]. Soil pH was measured using pH meter [13]. Soil organic carbon was determined by the Walkley-Black method described by [14]. Soil organic matter was determined by [15]. Total nitrogen was determined by the Kjeldahl method [16]. Available phosphorus was determined using Bray 1 method [17] Cation exchange capacity was obtained using the procedure described by [18]. Base saturation was calculated as total exchangeable bases divided by effective cation exchange capacity multiply by 100 percent.

2.5. Data Analysis

Data collected were statistically analyzed using Genstat Discovery (3rd Edition). The relationship between soil carbon sequestration and soil properties were determined using simple linear correlation ($P \leq 0.05$) and ($P \leq 0.01$). Soil organic carbon sequestration was calculated according to [20].

Soil organic carbon sequestration (SOCS) = BD (g/cm^3) x SOC (g/kg) x soil depth (cm)

Where; BD = Bulk density

SOC = Soil organic

Carbon.

3. Results

Table 1 and 2 show the physical and chemical properties of the topsoil and subsoil in the four forest plantations of the study. The soils are generally sandy sand at 0 – 15cm depth except for Dialium guineense which is sandy loam. At 15 – 30cm depth, soils of Tectona grandis and Dialium guineese are sandy loam with sand value of 784gkg^{-1} and 824gkg^{-1} respectively, and clay value of 156gkg^{-1}. Gmelina arborea is sandy clay loam with sand value of 704gkg^{-1}, clay value 276gkg^{-1}. Hevea bransiliensis is loamy sand with sand value of 844gkg^{-1}, and clay value 136gkg^{-1}. However, [20], states that soil texture affects soil carbon sequestration due to its influence on soil microbial community and soil respiration.

Table 1. Soil Physical Properties of the Study Locations

Plantation Site	Depth (cm)	Sand (g/kg)	Silt (g/kg)	Clay (g/kg)	Textural Class	Bulk Density (g/cm^3)
Gmelina arborea	0–15	844	60	96	Sandy Sand	0.868
Hevea branisiliensis	0–15	884	20	96	Sandy Sand	0.936
Tectona grandis	0–15	864	40	96	Sandy Sand	0.849
Dialium guineense	0–15	804	40	156	Sandy Loam	0.775
	Mean	849	40	11	Sandy Sand	0.857
Gmelina arborea	15-30	740	20	276	Sandy Clay Loam	0.858
Hevea branisiliensis	15-30	844	20	136	Loamy Sand	0.832
Tectona grandis	15-30	784	60	156	Sandy Loam	0.916
Dialium guineense	15-30	824	20	156	Sandy Loam	0.761
	Mean	789	30	181	Sandy Clay Loam	0.84175

Table 2. Soil Chemical Properties of the Study Locations

Plantation	Depth (cm)	pH (H$_2$O)	pH (kcl)	Org. Carbon (g/kg)	Org. Matter (g/kg)	Total N (g/kg)	C/N	Avail. P (mg/kg)	Ca	mg	K Cmol/kg	Exc Na	H	Al	CEC	BS %
Gmelina arborea	0-15	5.03	4.05	10.3	17	0.8	12.9	6.36	3.2	1.6	0.19	0.11	0.1	0.4	5.6	91
Hevea branisiliensis	0-15	4.47	3.64	10.5	18.2	0.9	11.7	3.61	2.4	1.6	0.24	0.14	0.3	0.4	5.08	86.2
Tectona grandis	0-15	5.05	4.05	12.1	21	1	12.1	2.75	3.2	1.6	0.09	0.26	0.4	0.2	5.75	89.5
Dialium guineense	0-15	5	4.1	9.3	16.2	0.8	11.6	0.95	2	1.2	0.16	0.07	0.3	0.8	4.53	75.7
	Mean	4.89	3.96	10.55	18.1	0.88	12.08	3.42	2.7	1.5	0.17	0.15	0.28	0.45	5.24	85.6
Gmelina arborea	15-30	5.58	3.97	6.3	11.0	0.55	11.5	7.61	2.6	2	0.2	0.09	0.1	1	6	81.6
Hevea branisiliensis	15-30	4.36	3.76	7.3	12.7	0.6	12.2	6.82	2	1.2	0.29	0.11	0.2	0.7	4.5	80
Tectona grandis	15-30	4.63	3.8	8.5	14.8	0.7	12.1	5.42	2	1.2	0.12	0.23	0.2	0.8	4.55	78
Dialium guineense	15-30	5.25	4.38	5.3	9.3	0.4	13.3	0.86	1.2	0.8	0.26	0.12	0.5	0.3	3.18	74.8
	Mean	4.96	3.98	6.85	9.48	0.56	12.28	5.18	1.9	1.3	0.28	0.14	0.25	0.7	4.56	78.6

The bulk density of the soils differed among the forest plantations of study. The highest value of bulk density at 0 – 15cm depth was recorded on soils of Hevea bransiliensis (0.936g/cm^3) and the least value was on Dialium guineense (0.775 g/cm^3). At 15 – 30cm depth, the highest value of bulk density was recorded on soils of Tectona grandis (0.916 g/cm^3) while the least value was also on Dialium guineense (0.761 g/cm^3). The soils were acidic with pH value ranging from 4.36 – 5.58. It was observed that Hevea bransiliensis plantation is more acidic than others both in the topsoil (4.47) and subsoil (4.36). This may be attributed to the management

practice in the Hevea bransiliensis plantation such as application of fertilizer. [21] observed that continuous application of fertilizers such as ammonium sulphate (NH$_4$)$_2$SO$_4$ and urea to improve soil fertility without lime decrease the pH of the soil. The organic matter content of the soils ranging from (21 – 9.3g/kg), Tectona grandis plantation recorded the highest values in the topsoil (21g/kg) as well as subsoil (14.8g/kg). Dialium guineense has the least values both in the topsoil (16.2g/kg) and subsoil (9.3g/kg). However, the higher organic matter level is very important in tropical countries because it is the benchmark upon which forest soil properties depends. It also plays important role in soil quality and enhances agricultural productivity and sustainability. The percentage base saturation of the soils were between (91 – 74.8%). *Gmelina arborea* plantation recorded the highest values both in the topsoil (91%) and subsoil (81.6%).

Dialium guineense has the lowest values both in the topsoil (75.7%) and subsoil (74.8%). The higher percentage base saturation recorded in *Gmelina arborea* plantation may be attributed to higher clay percentage composition of the *Gmelina Arborea* plantation.

Table 3. Soil Organic carbon sequestrations among the Plantations at Different Depths

Plantation Site	Age of Plantation	Depth	Soil Organic Carbon Sequestration ($g^2/cm^2/kg$)
Gmelina arborea	9	0–15	134.1
Hevea branisiliensis	48	0–15	147.4
Tectona grandis	9	0–15	154.1
Dialium guineense	31	0–15	108.1
		Mean	135.93
Gmelina arborea	9	15-30	81.1
Hevea branisiliensis	48	15-30	91.1
Tectona grandis	9	15-30	116.8
Dialium guineense	31	15-30	60.5
		Mean	87.375

Table 4. Correlation between Soil Organic Carbon Sequestration and Soil Properties

Soil properties	R	Level of Significance
Available Phosphorus	-0.00576	NS
Bulk Density	0.62044	*
Cation exchange Capacity	0.57173	*
Base Saturation	0.76891	**
C/N ration	-0.27130	NS
Exch. Calcium	0.73675	**
Clay Fraction	-0.67718	*
Exch. Hydrogen	-0.0870	NS
Exch. Potassium	-0.54498	*
Exch. Magnesium	0.42487	NS
Organic Carbon	0.97556	**
pH (H_2O)	-0.37404	NS
Sand Fraction	0.57081	*
Total Nitrogen	0.96684	**

* and ** = significant at 0.05 and 0.01 probability levels respectively, NS = non significant

4. Discussion

Soil organic carbon sequestrations among the plantations at different depths are shown in table 3 above. The result shows that the mean value of soil organic carbon sequestration was highest (135.93) within the topsoil (0-15) with *Tectona grandis* plantation exhibiting the highest value (154.1) followed by *Hevea bransiliensis* plantation (147.4), *Gmelina arborea* plantation (134.1) and least was in *Dialium guineense* of secondary forest (108.1). Within the subsoil (15 - 30), the mean value of soil organic carbon sequestration was (87.38) with *Tectona grandis* value reduces to (116.8), followed by *Hevea bransiliensis* plantation (91.1), *Gmelina arborea* plantation (81.1) and least was also in *Dialium guineense* of secondary forest (60.5). Hence, the soil organic carbon sequestration decreased with depth at all study sites. This is in agreement with earlier findings of [21]. Also, the results indicated that the age of the plantation is not related to its soil organic carbon sequestrations. Since *Tectona grandis* plantation of 9 years, which is one of the youngest plantation has the highest value (154.1) of soil organic carbon

sequestration, *Hevea bransiliensis* plantation which is 48 years has the value of (147.4), *Gmelina arborea* plantation of 9 years has a value of (134.1) where as *Dialium guineense* in secondary forest of 31 years has the least value of (108.1). This is in accordance with similar findings by [22], where carbon change of the soil sampled below a depth of 10cm in an afforested area, had no significant relationship with stand age.

Furthermore, the results indicated that tree species affects soil properties as well as soil organic carbon sequestration that is not only natural and anthropogenic factors as documented by [23]. It was observed in the plantations of *Gmelina arborea* and *Tectona grandis* that was established on the same land area, with the same climate factors, given the same management practices, in the same year, yet they sequester carbon differently, *Tectona grandis* (154.1) and *Gmelina arborea* (134.1). This supports the earlier findings by various authors as reported below: [24] states that one the most effective activities to improve soil carbon sequestration are choosing suitable forest tree species. The selection of tree species is one factor to consider if we want to mitigate carbon dioxide emissions to the atmosphere through forest management. [25], [26], [27]. Forest management including a change in tree species, has been accepted as a measure for mitigation of atmospheric carbon dioxide in national green house gas budgets. Moreover, the results of simple correlation analysis are shown in table 4, indicated strong positive correlation between soil organic carbon sequestration and Base saturation (r = 0.76891, P = 0.01). This implies that soil organic carbon sequestration increases with increasing Base saturation. In the chemical properties of the soils shown in table 2, the least value of the base saturation was found in *Dialium guineense* of secondary forest (75.7% and 74.8%) at 0 – 15cm and 15 – 30cm depth respectively. *Dialium guineense* also recorded the least value of soil organic carbon sequestration both in topsoil and subsoil. Again, the results indicated strong positive correlation between soil organic carbon sequestration and total Nitrogen (r = 0.96684, p = 0.01) implying that soil organic carbon sequestration increases with increasing total Nitrogen. This is seen in the chemical properties of the soils, were *Tectona grandis* plantation has highest value of total Nitrogen (1.0g/kg) as well as soil organic carbon sequestration of (154.1) at 0 – 15cm depth. *Dialium guineense* in secondary forest with least value of soil organic carbon sequestration (108.1) and also has least value of total Nitrogen (0.08g/kg) at the same depth. At 15 – 30cm depth, *Tectona grandis* plantation has also the highest value of soil organic carbon sequestration (116.8) and total Nitrogen (0.07g/kg), while *Dialium guineense* in secondary forest with least value of soil organic carbon sequestration (60.5) and also has least value of total nitrogen (0.04g/kg). However, the results indicated negative correlation between soil organic carbon sequestration and exchangeable potassium (r = -0.54498, p = 0.05) implying that soil organic carbon sequestration decreases with increasing exchangeable potassium. That is the higher the value of soil organic carbon sequestration the lower its exchangeable potassium. This is seen in the results of the chemical properties in table 2 above, while *Tectona grandis* plantation with the highest value of soil organic carbon sequestration (154.1 and 116.8), has the least value of exchangeable potassium (0.09 and 0.12cmol/kg) at 0 – 15cm and 15 – 30cm depth respectively.

5. Conclusion

Soil organic carbon sequestration varied among the four plantations of study with maximum concentration in the *Tectona grandis* plantation, followed by *Hevea bransiliensis*, *Gmelina arborea* plantation and least concentration was found in *Dialium guineense* of secondary forest. It was ascertained that individual tree species had influence on the rates of soil organic carbon sequestration. Hence, *Tectona grandis* tree species provides the best option for maximizing carbon sequestration in the soil. Also it was observed that soil organic carbon sequestration decreased with decreasing soil depths. The highest concentration was found within the topsoil as compared to subsoil in all the plantations, due to

high concentration of humus at the topsoil layers, as a result of large amount of plant litter deposit at the topsoil surfaces etc. Furthermore, it was discovered that age of the individual tree species/plantations does not affect its soil organic carbon sequestration concentration. *Hevea bransiliensis* plantation of 48 years old should have the highest concentration of soil organic carbon sequestration, followed by *Dialium guineense* in secondary forest of 31 years old when compared with *Tectona grandis* and *Gmelina arborea* plantations of 9 years old. But reverse was almost the case. *Tectona grandis* plantation was highest in soil organic carbon sequestration concentration followed by *Hevea bransiliensis*, *Gmelina arborea* and least was *Dialium giuneense*. In addition, some activities in the plantations such as latex tapping as performed in *Hevea bransiliensis* plantation affects soil organic carbon sequestration according to earlier findings by [28]. The carbon sequestration decreased significantly at early stages of latex tapping which stabilizing during the continuous harvesting and finally increased when latex harvest ceased. Soil disturbances, deforestation, erosion, fuel-wood harvesting, land degradation, leads to reduction in soil quality, decrease in soil organic matter which result in a decrease in soil organic carbon and sequestration.

References

[1] Peng, C. H., M. J. Apps, D. T. Price, I. A. Nalder, andD. H. Halliwell, (this volume),Simulating carbon dynamics along the Boreal Forest Transect Case Study (BFTCS) in Central Canada, 1, Model testing,Global Biogeochem. Cycles,12, this issue.

[2] Perry, M. Rick, E., Bricklemyer R., (2004). Soil Carbon sequestration in Agriculture. Farm management practices can affect greenhouse Gas Emissions. Montana State University. Ext Services.

[3] Powlson, D.S., A.P. Whitmore and K.W.T. Goulding, 2011. Soil Carbon Sequestration to Mitigate Climate Change. A critical re-examination to identify the true and the false. Eur. J. Soil Sci., 62:42 – 55.

[4] Larsen, J.B., Nielsen, A.B., (2007). Nature-based Forest management – where are we going? Elaborating Forest Development types in and with practice. Forest Ecology and Management 238:107 – 117.

[5] Barson. M.M., Giford, R.M., (1989). Carbon dioxide Sinks the Potential Role of tree Planting in Austraria. In Swain, D.J. (Ed) Greenhouse and Energy, CSIRO, Mebbourne, PP 433 – 443.

[6] Adger, W.N., Brown, K., (1994). Land Use and the Causes of Global warming, Wiley, Chichester, UK, 271 pp.

[7] Vesterdal L, Schmidt IK, Callesen I, Nilsson L.O. Gundersen P. (2008) Carbon and Nitrogen in Forest Floor and Mineral Soil Under Six Common European Tree Species. Forest Ecology and Management 255 (1): 35 – 48.

[8] Onweremadu, E.U., Okon, M.A., Ihem, E.E., Okuwa, J., Udoh, B.T. and Imadojemu, P. (2011). Soil Exchangeable Calcium Mapping in Central Southeastern Nigeria using Geographic Information System. Nigeria Journal of Agriculture, Food and Environment. June 7(2): 24 – 29.

[9] Ofomata, G.E.K. 1975. Relief. In: Ofomata, G.E.K. (ed.) Nigeria in maps: Eastern States, Ethiope Publishing House, Benin City. pp.25-26

[10] Gee, G.W. and Or D. (2002). Particle Size Analysis. In Dane, J.H. and G.C. Topps (eds). Methods of Soil Analysis. Part 4, Physical Methods. Soil Sci. Soc. An. Book Series No. 5, ASA and SSSA, Madison, WI. Pp. 255 – 293.

[11] Grossman, R.B. and T.G. Reinsch. (2002). Bulk Density. In Dane, J.H. and G.C. Topps (eds). Methods of Soil Analysis. Part 4, Physical Methods. Soil Sci. Soc. An. Book Series No. 5, ASA and SSSA, Madison, WI. Pp. 201 – 228.

[12] Soil Survey Staff. 2006. Keys to Soil Taxonomy. 10th ed. Washington (DC): Natural Resources Conservation Services, US Department of Agriculture. P 332.

[13] Hendershot, W.H., Lalande, H. and Duguette, M. (1993). Soil Reaction and Exchangeable acidity. In Carter, M.R. (ed). Soil Sampling and Methods of Soil Analysis. Can Soc Soil Sci, Lewis pub. London. Pp. 141 – 145.

[14] Nelson, D. W. and Sommers L.E. (1996). Total Carbon, Organic Carbon and Organic Matter. Methods of Soil Analysis Part 3. Chemical Methods (ed. By D.L. Sparks), pp. 961 – 1010. American Society of Agronomy, Madison, W.L.

[15] Baldock, J.A. and Skyemstad, J.O. (1999). Soil Organic Carbon/Soil organic Matter. In 'Soil Analysis': an Interpretation Manual (Eds K.I. Peverill, L.A. Sparrow, and D.J. Reuter) Pp 159 – 170 (CSIRO Publishing: Colling Wood).

[16] Udo, E.J, Ibia, T.O., Ogunwale, J.A., Ano, A.O. and Esu, I.E. (2009). Manual of Soil, Plant and Water Analysis. Sibon Books Ltd, Flat 15 blk 6. Fourth Avenue Festac, Lagos.

[17] Bray, R.H and Kurkz I.T. (1945). Determination of Total Organic and Available forms of Phosphorus in Soil. Soil Science. J. 59:39-45.

[18] Soil Survey Staff. 2003. Keys to soil Taxonomy Ninth edition United State Department of Agriculture 332 pp.

[19] Davidson E.A., Belk, B. and Boone, R.D. (1998). Soil Water Content and Temperature as Independent or Confounded factors Controlling Soil Respiration in Temperate Mixed Hardwood Forest, Global Change boil, 4:217 – 227.

[20] Batjes, N.H., (1996). Total Carbon and Nitrogen in the Soils of the World Eur. J. Soil Sci. 47:151 – 163.

[21] Vesterdal L, Schmidt IK, Callesen I, Nilsson L.O. Gundersen P. (2008) Carbon and Nitrogen in Forest Floor and Mineral Soil Under Six Common European Tree Species. Forest Ecology and Management 255 (1): 35 – 48.

[22] Johnson, D.W. (1992). Effect of Forest Management on Soil Carbon Storage. Water, Air and Soil Pollution 64:83-120.

[23] De Wit, H.A. and Kvindesland, S. (1999). Carbon Stocks in Norwegian Forest Soils and Effect of Forest Management on Carbon Storage. Rapport Fra Skogforskningen, Supplement 14, norsk Institute for Skogforskining (NISK). 52p.

[24] Eviksson, E., Gillespie, A., Guestavesson, L., langvall, O., Olsson, M., Sathre, R. and Stendahl, J. (2007). Integrated Carbon Analysis of Forest Management Practices and Wood Substitution. Canadian Journal of Forest Research 37(3): 671 – 681.

[25] Yang, Y.S., Guo, J.F., Chen, G.S., Xie, J.S., Zhen, L., Zhao, J., 5005. Carbon and Nitrohen Pools in Chinese Fir and Evergreen Broadleaved Forests and Changes Associated with Delling and Burning in Mid-subtropical China. Forest Ecol. Manage. 216 (1-3), 216 – 226.

8

Batch Studies for the Investigation of the Mechanism of Pb Sorption in Selected Acid Soils of China

Nkwopara U. N.[1,2,*], Emenyonu-Chris C. M.[1], Ihem E. E.[1], Ndukwu B. N.[1], Onweremadu E. U.[1], Ahukaemere C. M.[1], Egbuche C. T.[3], Hu H.[2]

[1]Department of Soil Science and Technology, Federal University of Technology, Owerri, Nigeria
[2]College of Resources and Environment, Key Laboratory of Subtropical Agricultural Resources and Environment, MOA, Huazhong Agricultural University, Wuhan, China
[3]Department of Forestry and Wildlife, Federal University of Technology, Owerri, Nigeria

Email address:
ugoiken2003@yahoo.com (Nkwopara U. N.)

Abstract: The experiment focuses on mechanisms of Pb retention on acid soils. A batch experiment was conducted to investigate the effect of solution pH and ionic strength of electrolytes which will show the mechanisms of Pb retention on the soils. Result show that sorption of lead was affected strongly by solution pH and ionic strength of electrolytes. Retention of lead increased with increase in solution pH and decreased with increase in ionic strength of electrolytes. This suggests that surface complexation and ion exchange are the mechanisms of Pb retention on these acid soils. At pH above 6 there was precipitation of lead. SEM studies visualized the formation of white layers of Pb over the soil surface. Scanning electron microscopy (SEM) revealed that the adsorption of lead ions made the surface of the soil particles rougher than those without lead. This morphological change points to the formation of a surface coating on the soil particles.

Keywords: Sorption of Lead, Mechanism, Acid Soils, Solution pH, Ionic Strength of Electrolytes

1. Introduction

Heavy metals (such as lead) adsorption is usually described in terms of two basic mechanisms, specific adsorption e.g. surface complexation, and non specific adsorption such as ion exchange[1]. The chemistry of Pb in soils is affected by three main factors: specific adsorption to various solid phases, precipitation of sparingly soluble or highly stable compounds and formation of relatively stable complexes or chelates with soil components [2][3][4] [5] investigated the retention of Pb^{2+} ion by montmorillonites and found that cation exchange predominated in pH range 2 – 4, precipitation was the main mechanism at pH > 6, and a combination of both took place at intermediate pH 4 – 6. A strong dependence on ionic strength is typical for an outer- sphere complex, whereas insensibility of ionic strength is an indication for inner sphere surface complexation [6].

Scanning electron microscopy is widely used to study the morphological features and surface characteristics of the adsorbent materials [7][8].Scanning electron microscopy (SEM) has often been used to visually show the morphological changes of the adsorbents after adsorption with an adsorbate [9][10]. Scanning electron microscopy was used to assess morphological changes in the soils surface following adsorption of lead. This technique provides a deeper insight into the nature and the environment of the adsorbed species and lead to a sharper description of the surface involved [11].

The objective of this present work was to determine the mechanisms of Pb retention by the acid soils of eastern China. This study can give important information concerning the nature of selected samples of Chinese soil. The properties of these soils are usually not reported in the literature.

2. Materials and Methods

2.1. Soil Sample

Four representative variable charge soils were used in this study: Yellow brown soil (YBS), Red soil (RS), Latosol soil

(LS) and Latosolic red soil. They are Alfisol, Ultisol, and Oxisols in the American classification system. These soils were collected from Hubei, Hainan, and Guangxi provinces in China (Fig. 1). These soils with contrasting properties were

sampled at 0 – 20 cm depth. Samples of these soils were air-dried, ground, sieved through 2-mm mesh and homogenized prior to soil characterization and adsorption studies.

*= studied areas.

Fig. 1. *Map of China showing the studied areas.*

2.2. Effect of Solution pH on Sorption

A 100 mg L^{-1} solution of Pb $(NO_3)_2$ was prepared from a stock solution of 1000 mg L^{-1} using 0.01 mol L^{-1} KCl as background electrolyte. Metal ion solutions (25 mL of 100 mg L^{-1}) were added to 1g of soils weighed into 50 mL polyethylene bottles. The solutions were adjusted to pH 3 -7 using either 0.1 M NaOH or HCl. The suspensions were agitated for 2 h. The pH of the solutions was checked and readjusted to the initial pH. The suspensions were equilibrated for 20 h and readjusted to the pH after equilibration. The supernatants obtained after centrifugations were analyzed for Pb^{2+} using an atomic absorption spectrophotometer (AAS) (computed-aided Varian AA 240FS). The amount adsorbed was calculated by the difference between the total applied Pb^{2+} and the amount of Pb^{2+} remaining in the equilibrium solution.

2.3. Effect of Inorganic Anions on Sorption

25 mL of the various background electrolytes (KCl, KNO_3, K_2SO_4) in three concentrations of (0.1, 0.01, and 0.001 mol L^{-1}) containing 150 mg L^{-1} Pb^{2+} were separately added to 1 g soil. The pH of these solutions was not adjusted to avoid negating the pH effects of the electrolytes. These solutions were agitated in a shaker for 2 h, equilibrated for 22 h and centrifuged at 5000 rpm for 10 min. The supernatants obtained after centrifugation were analyzed for Pb^{2+} using an AAS. The amount adsorbed was calculated by the difference between the total applied Pb^{2+} and the amount of Pb^{2+} remaining in the equilibrium solution.

2.4. Effect of Inorganic Cations on Sorption

25 mL of the various background electrolytes (KCl and $CaCl_2$) in three concentrations of (0.1, 0.01, and 0.001 mol L^{-1}) containing 150 mg L^{-1} Pb^{2+} were separately added to 1 g soil. The pH of these solutions was adjusted to 5.5±0.1. These solutions were agitated in a shaker for 2 h, equilibrated for 22 h and centrifuged at 5000 rpm for 10 min. The supernatants obtained after centrifugations were analyzed for Pb^{2+} using an AAS. The amount adsorbed was calculated by the difference between the total applied Pb^{2+} and the amount of Pb^{2+} remaining in the equilibrium solution.

2.5. Scanning Electron Micrograph Studies

After adsorption, the soil residues were air-dried for 7 days and gold coated under vacuum in a JFC – 1600 sputter coater (JEOL, Japan) for 20 min. The SEM images obtained by a JSM 6390 scanning electron microscope (JEOL, Japan)

3. Results and Discussion

3.1. Physicochemical Properties and Mineralogical Properties of Soils

The results of the physicochemical properties showed that YBS had higher CEC and pH (27.4 cmol kg^{-1} and 5.2), while LS had higher organic carbon and crystalline Fe_2O_3 and Al_2O_3 (13.1 g kg^{-1}, 84 g kg^{-1} and 5.2 g kg^{-1}) respectively than the other soils (Table 1). The particle size analysis showed that LS was clay, YBS was silt loam while RS and LRS were clay loam.

Mineralogical composition showed that LS mainly consisted of kaolinite 75 %, hydroxyinterlayed vermiculite 15 % and goethite characterized by diffraction peaks at 0.71nm, 0.48nm and 0.27nm respectively. YBS consisted of illite 45 %, vermiculite 25 % and kaolinite 30 % characterized by diffraction peaks at 1nm, 1.396nm and 0.717nm respectively. RS consisted of vermiculite 35 %, kaolinite 40 %, and illite 25 % characterized by diffraction peaks at 1.404nm, 0.723nm, and 0.493nm respectively. LRS consisted of kaolinite 60%, and illite 40% characterized by diffraction peaks at 0.719nm and 1nm respectively (Fig .2). [12] reported that oxisols with high Fe oxide content are rich in clay;

however, the soils exhibit characteristics of sands. Considering the clay and organic matter content, CEC is very low in all soils, showing that they are low activity clay (LAC) soils because of strong weathering [13][14].This is confirmed by the clay mineralogy, showing that kaolinite is the dominant clay mineral. The clay mineralogy is also in accordance with extractable Al and Fe. The high contents of crystal Fe and relatively low contents of amorphous Fe in the LS, RS and LRS are consistent with the weathering and dominance of geothite in these Oxisols and Ultisol [13][14]. On the other hand, the amount of amorphous Al exceeds crystal Al, i.e Al_{ox}/Al_{DCB} ratio > 1 (Table 1) showing substantial amounts of poorly ordered aluminum oxides in all soils. These observations agree with reports by [14] on variable tropical soils from Tanzania.

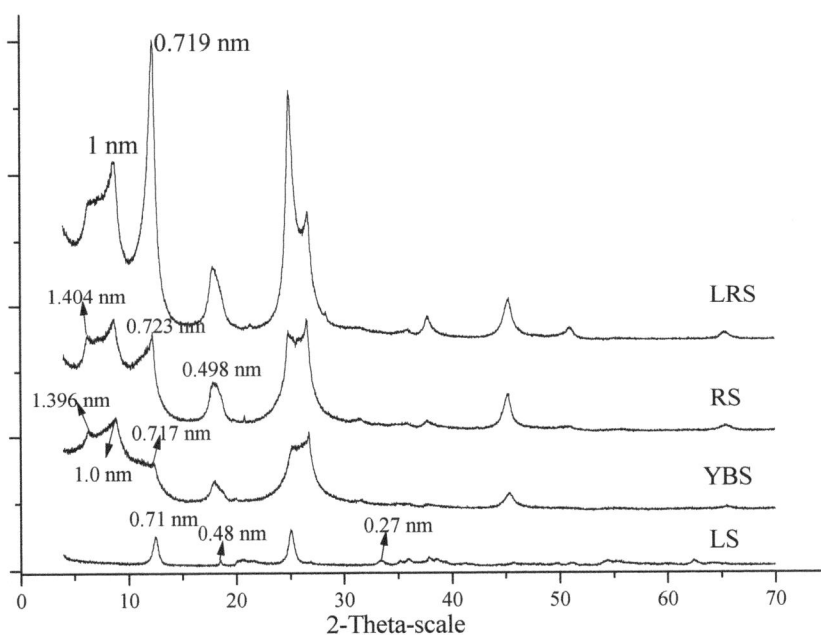

Fig. 2. *X-ray diffraction patterns of studied soils*

Table 1. *Basic properties of the tested soils*

Items	YBS soil	LS soil	RS soil	LRS soil
Sample site	Hubei	Hainan	Hubei	Guangxi
Order	Alfisol	Oxisol	Ultisol	Oxisol
pH (H₂O)	5.2	4.5	4.2	4.1
O.M (g/kg)	10.8	13.1	6.2	9.9
CEC (cmol/kg)	27.4	15.9	20.9	21.9
2-0.2mm Sand (g/kg)	64.8	270.7	304.4	279.2
0.2-0.02mm Silt (g/kg)	678.7	313.4	405.7	359.6
< 0.002mm clay (g/kg)	256.5	415.9	359.6	361.2
Crystal Fe (g/kg)	16.0	84.0	21.3	21.0
Amorphous Fe (g/kg)	13.8	8.7	5.9	3.5
Crystal Al (g/kg)	1.3	5.2	2.3	2.1
Amorphous Al (g/kg)	4.2	9.4	7.6	6.2
Clay minerals	I, V, K	GE, HIV,K	K, I, V	K, I
Background Pb (mg/kg)	Bdl	Bdl	Bdl	Bdl

YBS: Yellow brown soil, LS: Latosol soil, RS: Red soil, LRS: Latosolic red soil, O.M: Organic matter, CEC: Cation exchange capacity I: illite, K: kaolinite, V: vermiculite, GE: geothite, HIV: hydroxyinterlayed vermiculite, Bdl = below detection limit.

3.2. Effect of Solution pH on Sorption

The effect of pH on the adsorption of Pb^{2+} by soils is shown in Fig. 3. There was an increase in adsorption as the pH of metal ion solution increased. This result is consistent with previous findings that increasing pH generally causes an increase in the metal adsorption [4][15][16]. At pH 6.0, percent adsorptions of Pb^{2+} onto soils were 88.23 %, 31.17 %, 89 % and 88.88 % for YBS, RS, LS and LRS respectively. At pH 3.0, percent adsorptions of Pb^{2+} onto soils were 75.84 %, 7.71 %, 13.95 %, and 17.43 % for YBS, RS, LS and LRS respectively. [17] has reported 94 % adsorption of Pb^{2+} onto tripolyphosphate-impregnated kaolinite clay using 500 mg L^{-1}

solution of Pb^{2+} at pH of 7.0. Unlike the effect of the initial metal ion concentration, when the pH of the aqueous solution of metal ion increased, there was an increase in the equilibrium adsorption capacity as well as in the percentage of metal ion adsorbed (Table 2 and Fig. 3). There was significant difference in amount of Pb adsorbed at all pH and in all the soils (Table 2). At pH above 6 the pseudoequilibrium constant for adsorption (K) was 4.83 x10^{-20} while the solubility product constant (Ksp) was 1.42 x10^{-20}. The pseudoequilibrium constant for adsorption (K) was greater than solubility product constant (Ksp) which implies that there was precipitation of Pb at pH above 6 in this study.

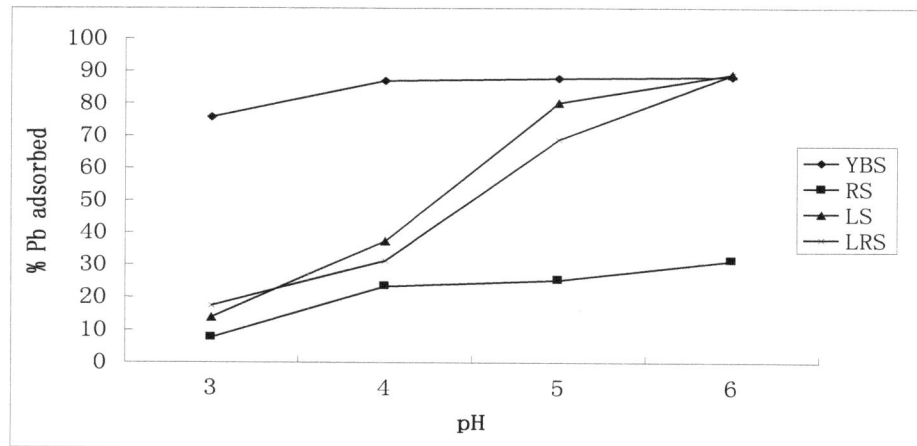

Fig. 3. *Effect of pH on adsorption of 100 mg/L of Pb ion by soils*

Table 2. *Amount of Pb adsorbed at different pH (mg/kg)*

pH	soils			
	YBS	**RS**	**LS**	**LRS**
3	1896d	193d	348d	436d
4	2173c	584c	933c	779c
5	2192b	631b	2003b	1720b
6	2206a	780a	2225a	2222a

All the values are means of three replicates. Within the same column, different letters (a-d) indicate significant difference (p ≤ 0.05). YBS = Yellow brown soil, LS = Latosol soil, RS = Red soil, LRS = Latosolic red soil.

As expected, the adsorption of Pb increased with increasing solution pH. Several reasons may contribute to the increased adsorption of metal ion relative to soil solution pH. First, the soil surface which contains a large number of active sites may become positively charged at very low pH, thus increasing the competition between H^+ and the metal ions for available adsorption sites because of increased amount of H^+ in solution. However, as pH increases this competition decreases as these surface active sites become more negatively charged, which enhances the adsorption of the positively charged metal ions through electrostatic force of attraction. Second, increasing pH has been shown to decrease the solubility of Pb^{2+}. Third, increasing pH encourages metal ion precipitation from solution in the form of hydroxides. Fourth, It is known that Pb^{2+} is the dominant species at lower pH but as pH increases

the concentration of $PbOH^+$ species that has high affinity for oxides become significant. Finally, increase in adsorption can be attributed to the favourable change in surface charge and to the extent of hydrolysis of the adsorbing Pb^{2+} change with varying pH. At low pH, Pb^{2+} are present in the form of M^{2+} and $M(OH)^+$. This leads to an effective competition between H^+ and H_3O^+ [1].

As the surface charge becomes more negative with increasing pH, the surface attracts bivalent metal cations for adsorption. Furthermore, the proportion of hydrated ions increases with pH and these may be more strongly adsorbed than unhydrated ions. Therefore, both of these effects are synergistically enhancing the amount of adsorption at higher pH [18].

The four tested soils behaved similarly, although at lower pH (3 -5), YBS adsorbed more Pb while at higher pH 6, LS adsorbed more Pb. This could be attributed to the fact that at lower pH the nature and properties of the soil surface control the adsorption behavior of these soils, but with an increase in solution pH, other factors such as precipitation, may become dominant, causing the difference in Pb^{2+} adsorption between the four soils. [19] reported that above pH 5.5 both adsorption and precipitation are the effective mechanisms to remove Pb^{2+} in aqueous solution. This result supports the previous finding in another aspect that increasing pH caused a decrease in desorption of lead [20]. When the pH of the aqueous solution of lead ion increased, there was an increase in the equilibrium adsorption capacity as well as in the percentage of lead ion adsorbed. Adsorption of Pb^{2+} ion onto soils was influenced by pH strongly. This suggests that the adsorption of Pb^{2+} ion onto soils is mainly dominated by surface complexation [21]. In general, surface complexation is mainly influenced by pH.

3.3. Effect of Inorganic Anion on Sorption

The effect of electrolyte anion on the adsorption of Pb^{2+} onto soils is shown in Table 3. During the equilibration period, no significant difference in electrolyte effect was observed (Table 4). The highest reduction occurred in LRS in the presence of NO_3^- (0.27), the lowest occurred in RS in the presence of Cl^- (0.11). No obvious pH changes were observed after equilibration (Table 4). The adsorption of Pb ion made no obvious difference to the final pH values. In the presence of K_2SO_4, adsorption of Pb onto soils was more than 90%, with the highest rate being 98.7% in YBS and the lowest rate being 98% in LRS. While in the presence of KNO_3 and KCl, the highest adsorption rate of 51.5% and 48.3% were recorded in YBS and the lowest rate of 15.6% and 15.5% were recorded in RS (Table 3). It was observed that more Pb^{2+} was adsorbed in the presence of K_2SO_4 while KNO_3 and KCl gave almost equal adsorption. However, the adsorption rate of soils was reduced in the presence of Cl^- and NO_3^- anions. There was stronger attraction between Pb^{2+} and the soils in the presence of K_2SO_4 than the other anions which show almost equal attraction.

The distribution coefficient values (K_d) which indicate attraction to soil were higher in the presence of K_2SO_4 than in KNO_3 and KCl. In the presence of K_2SO_4, at 0.1 mol L^{-1} YBS had the highest value of 1827 mL/g and LRS the lowest of 1215 mL/g, while in the presence of KNO_3 and KCl, the values were low, with the highest values of 27 mL/g and 23 mL/g occurring in YBS and the lowest of 5 mL/g in RS and LRS (Fig. 4). Higher K_d value indicates stronger attraction to the soil solids and lower susceptibility to leaching loss.

The solution pH after equilibration decreased by less than one pH unit. This observation is contrary to [22] on adsorption of Cu by clinoptilolite that solution pH after equilibration increased by more than one pH unit in the presence of KCl and KNO_3 but less than one unit in the presence of K_2SO_4. The lack of pH dependence in the soil that has highest permanent charge clay indicates that the silicate clay may form the predominant sites for Pb adsorption in this soil. For other soils it seems that the Fe and Al oxyhydroxides and organic matter are providing most sites for adsorption. It is suggested that Pb sorption enhancement due to sulfate is as a result of reduction of oxide surface charge caused by anion adsorption and could not be attributed to formation of ternary complexes. The finding agrees with the assumption of [23] that chloride and nitrate metal ion species are not significantly adsorbed. Thus, when these anions form charged complexes with metal ions, they reduce the amount of metal ions adsorbed and hence the adsorption equilibrium capacity of the adsorbent, especially Cl^- which has the ability to form charged and uncharged species (e.g. MCl_2^0, MCl_3^- and MCl_4^{2-}) with metal ions [17]. Similar results were reported by [24][4] for adsorption of Pb^{2+}

and Cd^{2+} onto phosphate – modified and unmodified kaolinite clay and adsorption of Cd^{2+} onto goethite respectively. Besides, Cl^- adsorption on the surface of soils changes the surface state of soils and decreased the availability of binding sites [25].

Heavy metals have more affinity for sulphur group and sulphide than other anions as can be seen from the natural occurrence of most of the metals in their sulphide form [26]. Chemical affinity of lead towards sulphur groups is higher. Since, lead chemical species show a high affinity toward sulphur, adsorption capacity of the soils could be enhanced by introduction of sulphur materials into the soils. Sulfate ions are attracted by the positive charges that characterize acid soils containing iron and aluminum oxide and 1:1 type silicate clay. They also react directly with hydroxyl groups exposed on the surfaces of these clays. However much sulfate may be held by the iron and aluminum oxide and 1:1 type silicate clays that tends to accumulate in the subsoils horizons of the Ultisols and Oxisols of the warm, humid regions [27].There is extensive ion pairing of SO_4^{2-} with multivalent cations, no ion pairing of cations with Cl^- and small amount with $NO3^-$ [22].

Also there could be precipitation of lead in the presence of potassium sulfate. The greater distribution coefficient values (K_d) in YBS than the other soils indicate that Pb^{2+} has a greater affinity for YBS than the other soils. The interpretations mentioned above are reasonable in principle.

3.3.1. Effect of Nature and Ionic Strength of Different Anions on Adsorption

The effect of inorganic anion on the adsorption of Pb^{2+} onto soils is shown in Table 3. It was observed that more Pb was adsorbed in the presence of SO_4^{2-} while NO_3^- and Cl^- gave almost equal Pb adsorption. However, the adsorption rate of soils was reduced in the presences of Cl^- and NO_3^- anions compared to SO_4^{2-} anion. There was stronger attraction between Pb^{2+} and the soils in the presence of SO_4^{2-} than the other anions which show almost equal attraction.

Increasing the ionic strength of the anions (from 0.001 – 0.1 mol L^{-1}) affected the adsorption of lead. In the presence of SO_4^{2-}, Pb adsorption increased with increasing ionic strength but the reverse was the case in the presence of Cl^- and NO_3^- (Table 3). In the presence of anions, the K_d values decreased with increase in ionic strength with the exception of SO_4^{2-}. At ionic strength of 0.001 mol L^{-1} SO_4^{2-}, Cl^- and NO_3^- had 954 mL/g, 587 and 489 mL/g in YBS respectively, while at 0.1 mol L^{-1}, SO_4^{2-}, Cl^- and NO_3^- had 1827 mL/g, 23 and 27 mL/g in YBS respectively (Fig.4). It was observed that increasing the ionic strength of SO_4^{2-} increased the affinity of Pb^{2+} for soils while increasing the ionic strength of Cl^- and NO_3^- decreased the affinity of Pb^{2+} for soils. There was no significant difference in percent Pb^{2+} adsorbed in the presence of SO_4^{2-} in all the soils and at all concentrations, while there was significant difference in the presence of Cl^- and NO_3^- (Table 3).

Ionic strength of anions effects can be attributed to changes in adsorbent-suspension pH through its effect on the diffuse double layer. However, the negative effect of increasing ionic strength of anion on adsorption of metal ion operates through its effect on electrostatic potential in the plane of adsorption, rather than through its effect on surface charge. It is also suggested that increasing anion concentration can cause screening of surface negative charge by the anion leading to a drop in the adsorption of the Pb ion [28]. More adsorption of Pb ion in the presence of SO_4^{2-} than the other anions can be attributed to increase in the negative charge in the presence of SO_4^{2-} than the other anions. Among anionic ligands, sulfate is believed to form inner-sphere complexes with surface active sites [29] and to increase the net negative surface chage,

increasing metal adsorption. On the contrary, it was stated that Cl^- and NO_3^- form outer-sphere complexes and enhance metal adsorption indirectly [29]. It should be noted that higher adsorption in the presence of potassium sulfate was not entirely due to the adsorption on the soils but, likely, involved precipitation of lead. It is not possible to distinguish adsorption from precipitation by conventional techniques. However, we may make a reasonable assumption that adsorption was the dominant process.

Table 3. Percent Pb^{2+} adsorption in the presence of anions

	0.001 (mol L⁻¹)			0.01 (mol L⁻¹)			0.1 (mol L⁻¹)		
soils	SO_4^{2-}	NO_3^-	Cl^-	SO_4^{2-}	NO_3^-	Cl^-	SO_4^{2-}	NO_3^-	Cl^-
YBS	98.5a	95.2a	96a	98.5a	82.2a	83.2a	98.7a	51.5a	48.3a
LS	91.5a	49.4b	52.5 b	94.7a	32.7b	30.7b	98.5a	31b	30.1b
RS	92.5a	59.6c	60.9c	96.6a	27.6c	27.7c	98.4a	15.6c	15.5c
LRS	89.3a	43.6b	45b	91.2a	20.4c	18.4c	98a	17c	15.7c

All the values are means of three replicates. Within the same column, different letters (a- c) indicate significant difference (p ≤ 0.05). YBS = Yellow brown soil, LS = Latosol soil, RS = Red soil, LRS = Lateritic red soil, SO_4^2 = sulphate, NO_3 = nitrate, Cl^- = chloride

Table 4. Changes in solution pH at initial and equilibrium solution

Electrolyte	Ionic strength (mol/L)	YBS		LS		RS		LRS	
		Initial	Final	Initial	Final	Initial	Final	Initial	Final
K₂SO₄	0.001	4.81	4.78	4.39	4.26	4.37	4.29	4.16	4.01
	0.01	5.02	4.94	4.74	4.68	4.60	4.47	4.30	4.22
	0.1	5.11	4.90	4.89	4.68	4.65	4.43	4.53	4.26
KNO₃	0.001	4.58	4.54	4.13	4.02	4.27	4.22	3.96	3.88
	0.01	4.46	4,42	4.12	4.06	4.23	4.17	4.03	3.94
	0.1	4.38	4.12	4.20	4.03	4.17	3.98	4.27	3.91
KCl	0.001	4.64	4.53	4.15	4.00	4.21	4.14	4.05	3.88
	0.01	4.48	4.37	4.15	4.09	4.21	4.18	4.06	3.89
	0.1	4.40	4.19	4.18	4.06	4.21	3.98	4.04	3.85

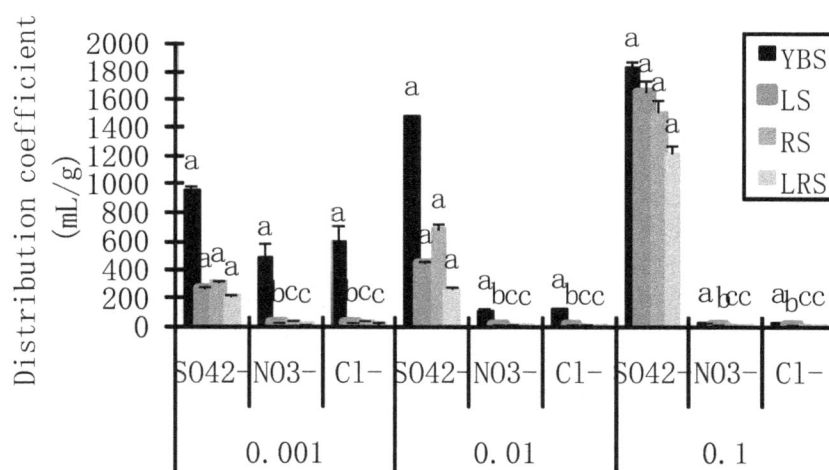

Fig. 4. Distribution coefficient of Pb^{2+} at different ionic strength of anions

3.3.2. Effect of Nature and Ionic Strength of Different Cations on Adsorption

Sorption of Pb^{2+} in the presence of K^+ was more than that observed in the presence of Ca^{2+} solutions at all concentrations and in all the soils (Table 5). Increasing the concentration of

inorganic cation from 0.001- 0.1 mol L⁻¹ caused a drop in adsorption capacity of soils, which suggests that the cations greatly affects Pb^{2+} adsorption. Ca^{2+} had a more negative effect than K^+. It reduces the adsorption capacity of the soils by 19.2 – 21.9 %. Fig. 5 shows the distribution coefficient value (K_d) of Pb in the presence of inorganic cation. The values were higher in K^+ than Ca^{2+} and decreases as the ionic strength increase. At ionic strength of 0.001 mol L^{-1} K^+ had 1688 mL/g in YBS and Ca^{2+} had 747 mL/g while at 0.1 mol L^{-1}, K^+ had 190 mL/g in YBS and Ca^{2+} had 87 mL/g. This indicates that increasing ionic strength decreased the affinity of Pb^{2+} for the soils and that affinity of Pb^{2+} for soil was greater in the presence of K^+ than Ca^{2+} cation. There was no significant difference in percent Pb^{2+} adsorbed in the presence of Ca^{2+} in all the soils and at all concentrations, while there was significant difference in the presence of K^+. However, in the presence of K^+ there was no significant difference in percent Pb^{2+} adsorbed in LS, RS and LRS (Table 5).

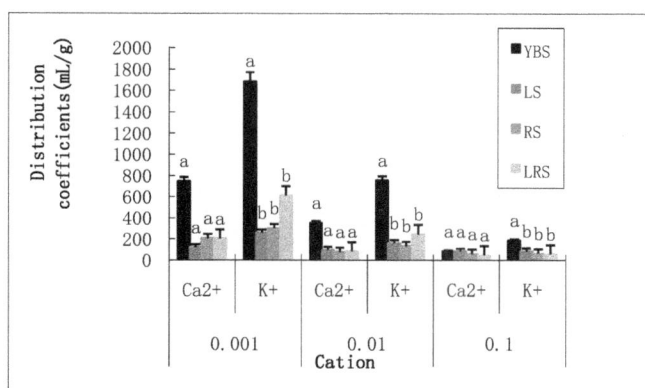

Fig. 5. *Distribution coefficient of Pb^{2+} at different ionic strength of cations*

Table 5. *Percent Pb^{2+} adsorption in the presence of cations*

soils	0.001 mol L^{-1}		0.01 mol L^{-1}		0.1 mol L^{-1}	
	Ca^{2+}	K^+	Ca^{2+}	K^+	Ca^{2+}	K^+
YBS	96.8a	98.5a	93.5a	96.8a	77.6a	88.4a
LS	83.2a	91.3b	79.8a	86.8b	77.1a	77.6b
RS	89.4a	92.4b	76.5a	84.6b	71.3a	72.3b
LRS	89.1a	96b	77.3a	90.9b	66.2a	70.1b

All the values are means of two replicates. Within the same column, different letters (a- c) indicate significant difference (p ≤ 0.05). YBS = Yellow brown soil, LS = Latosol soil, RS = Red soil, LRS = Lateritic red soil, Ca^{2+} = calcium, K^+ = potassium

The effect of cation may be attributed to increased competition between Pb^{2+} and Ca^{2+} for adsorption sites on the surface of soils and to the greater specificity of the divalent Ca^{2+} and its effect on the thickness of the double layer [17][30]. It has been shown by [17] that the potential in the plane of adsorption is related to the valence of the ion through its effect on surface charge density. Increasing the valence of the cation makes the potential in the plane of adsorption less negative thereby reducing Pb^{2+} adsorption. The valence of the cation has a dominant influence on the case of displacement of ions from a charged surface. The higher the valence of the cation, the higher the displacing ability of the ion [31].Therefore, Pb^{2+} ions displace H_3O^+ and other monovalent ions from soil surface in the presence of K^+. This explains why Ca^{2+} had greater negative impact on the equilibrium adsorption capacity of soils than K^+. The difference in the adsorption capacities of the soils for lead ion in the presence of K- and

Ca- electrolytes is due to the difference between K^+ and Ca^{2+} with respect to competition with Pb ions sorption sites. This indicates that the soils show stronger preference for Ca^{2+} (a divalent cation) than for K^+ (a monovalent cation) with the resultant effect of a further drop in monolayer adsorption capacities of the soils for the lead ion when Ca-electrolyte was used. Similar trends have been observed by [30] for adsorption of Cd^{2+} on soils; [24] for the adsorption of Pb^{2+} and Cd^{2+} on orthophosphate-modified and unmodified kaolinite and [32] for the adsorption of Cu^{2+}, Zn^{2+}, Co^{2+} and Cd^{2+} on kaolinite.

The effect of ionic strength on metal ion adsorption is often attributed to changes in adsorbent-suspension pH effect on the diffuse double layer [33]. [30] found that pH of soil suspension decreased by 0.4 – 1 unit on the average, with increase in ionic strength of cation. This may explain in part the decrease in metal ion adsorption. However, the negative effect of increasing ionic strength of inorganic cation on the adsorption of metal ion operates through its effect on electrostatic potential in the plane of adsorption, rather than through its effect on surface charge. According to[34], if increasing ionic strength of inorganic cation decreases metal ion adsorption, then this implies that increasing ionic strength is making the potential of the adsorbent surface less negative (the potential in the plane of adsorption is getting increasingly positive) and this would decrease metal ion adsorption. In addition, the effect of ionic strength could be ascribed to: (1) the competition of high concentrations of cations in electrolyte with lead for the adsorption sites on soil surface. An increase of ionic strength will supply more K^+ or Ca^{2+} ions which compete with lead for the adsorbing sites on soil surface (2) the decrease of activities of lead due to the increase of ionic strength; (3) the formation of ionic pairs or chelating compounds [35][36]. An increase in ionic strength will supply more K^+ or Ca^{2+} ions which compete with lead for the adsorption sites on soil surface. This resulted in a decrease of exchangeable adsorption. Additionally, the presence of salts may compress the electric double layer surrounding negative charged surfaces which contributed to the release of adsorbed lead[37][38] . Beside, the cations in solution may influence the double layer thickness and interface potential, and thereby affect the binding of the adsorbed species. The observed variation in Pb sorption with ionic strength may be explain by the formation of outer-sphere complexes since K^+ and Ca^{2+} in the background electrolyte could compete with the Pb^{2+} ion adsorbed on the outer-sphere sorption sites and reduced the adsorption. Whereas K^+ and Ca^{2+} would not have competed for the inner-sphere sites, as independence of sorption with background electrolyte concentration has been interpreted to indicate that the sorption process is primarily non-electrostatic in nature [39]. Ion exchange and outer-sphere complexes are affected by the variations of ionic strength more easily than the inner-sphere complexes, since the background electrolyte ions are placed in the same plane as outer-sphere complexes [25]. [40] studied the adsorption of Pb(II) onto montmorillonite and found that increasing ionic strength from 10 to 100 mM sensibly decreased lead adsorption at the pH below the Pb hydrolysis point. [41] investigated the adsorption of Pb(II) onto oxidized multiwalled carbon nanotubes (MWCNTs) and found that the adsorption was independent of ionic strength. The present results showed that the adsorption of Pb^{2+} ion onto soils was influenced by ionic strength strongly. This suggests that the adsorption of Pb^{2+} ion is mainly dominated by ion exchange [21]. Adsorption was dependent on ionic strength, suggesting that they were mainly adsorbed on the outer-sphere sites [42].

3.4. Scanning Electron Microscopy Analysis

The morphology of the soils with Pb^{2+} ion showed some important observations. Typical SEM photographs are shown in Fig. 6(a–h). Coverage of the surface of the soil due to adsorption of Pb^{2+} ion presumably leading to formation of layer of lead molecules over the soil surface. It is evident from the formation of white layer (molecular cloud). There was higher surface coverage of lead in YBS, RS and LRS than LS. It is noticed that soils have bigger pore structures, $1 - 5\mu m$ and after adsorption, the pore size have been reduced to $0.4 - 0.6\ \mu m$. There was spherical like structures attached in bundles. This could be due to adsorption of Pb^{2+} ion on soils. Surface adsorbed spherical shaped Pb particles on the soil could be found. There was increase in the roughness of the soil surface after adsorption of lead ions, indicating that the adsorption reaction occurred on the surface of the soil particles (Fig.6 (a–h). The increase of roughness may be attributed to the formation of lead ion complexes on the surfaces [9].

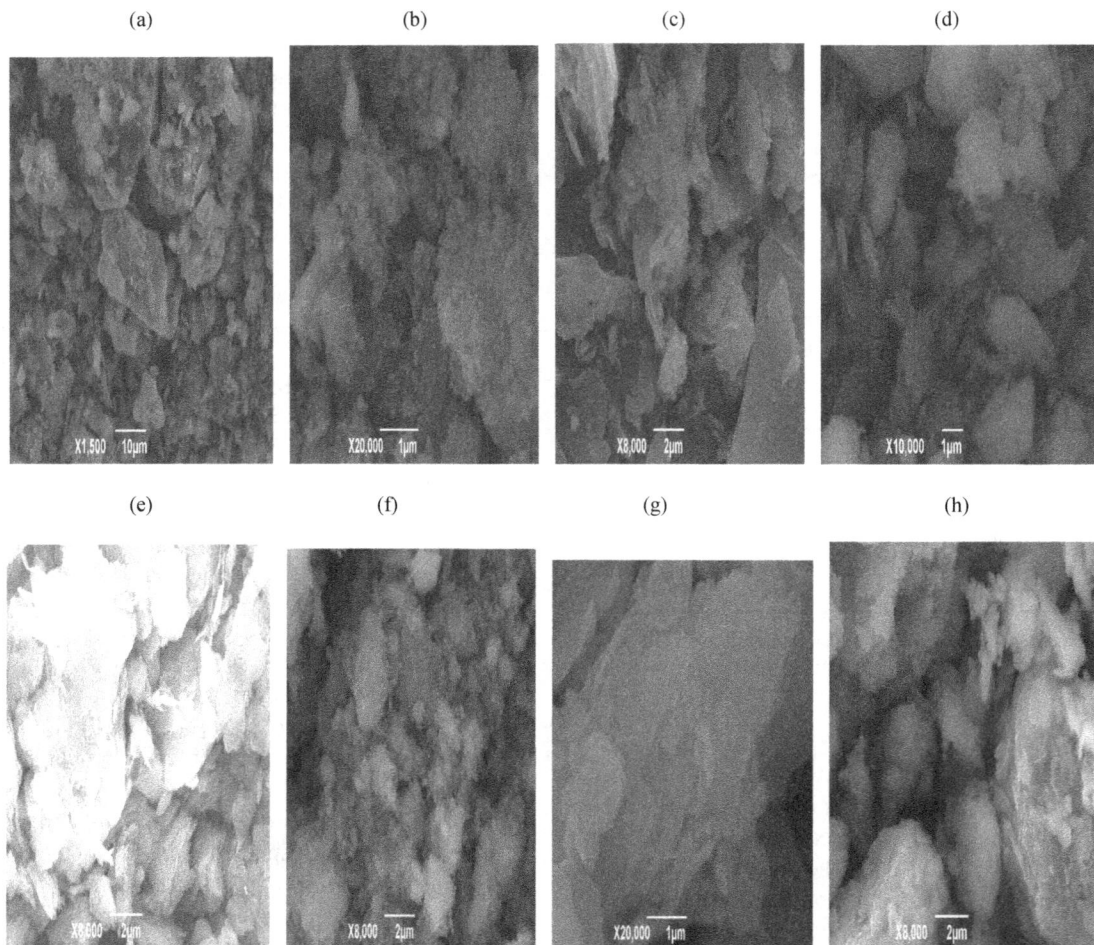

Fig. 6. Scanning electron microgram of soils without and with Pb adsorption a. LRS, b. LS, c YBS, d. RS without Pb, e. LRS, f. LS, g. YBS, h. RS with Pb

4. Conclusion

The greater content of illite and the moderate content of vermiculite provide YBS with greater exchange capacity than the other soils, as this minerals have larger net surface than kaolinite. In YBS cations will be commonly adsorbed than anions compared to other soils. In general, the soils are acid, of low fertility and poor in organic matter. Low 2:1 clay mineral is attributable to the high weathering of the soils. Adsorption of Pb^{2+} ion onto soils was influenced by pH strongly. This suggests that the adsorption of Pb^{2+} ion onto soils is mainly dominated by surface complexation. Scanning electron microscopy (SEM) revealed that the adsorption of lead ions made the surface of the soil particles rougher than those without lead. The present results showed that the adsorption of Pb^{2+} ion onto soils was influenced by ionic strength strongly. This suggests that the adsorption of Pb^{2+} ion is mainly dominated by ion exchange and was mainly on the outer-sphere sites. In general, ion exchange is mainly influenced by ionic strength.

References

[1] Mouni, L., Merabet, D., Robert, D., Bouzaza, A. Batch studies for the investigation of the sorption of the heavy metals Pb2+ and Zn2+ onto Amizour soil (Algeria). Georderma, 2009. 154, pp 30-35.

[2] Strawn, D.G, Sparks, D.L. The use of XAFS to distinguish between inner- and outer-sphere lead adsorption complexes on montmorillonite. Journal of Colloid Interface Science, 1999. 216, pp 257 – 269

[3] Appel, C., Ma, L.Q. Concentration, pH, and surface charge effects on cadmium and lead sorption in three tropical soils. Journal Environmental Quality, 2002. 31, pp581-589.

[4] Bradl, H.B. (2004). Adsorption of heavy metal ions on soils and soil constituents. Journal Colloid Interface Science, 1999. 277, pp 1 – 18.

[5] Taylor, R.W. Kinetics and mechanism of metal retention/release in geochemical processes in soil. Final report. U.S. Department of Energy.

[6] Lutzenkirchen, J. Ionic strength effects on cation sorption to oxides: macroscopic observations and theur significance in microscopic interpretation. Journal of Colloid and Interface Science, 1997. 195, pp149 – 155.

[7] Nelly, J.W., Isacoff, E.G. Carbonaceous Adsorbents for the Treatment of Ground and Surface water, Marcel Dekker, New York. 1982.

[8] Gupta, S. Pal, A., Ghosh, P.K. Bandyopadhyay, M. Performance of waste activated carbon as a low-cost adsorbent for the removal of anionic surfactant from aquatic environment. Journal Environmental Science Health, 2003. A38, pp 381-397.

[9] Deng, S., Renbi, B., Chen, J.P. Behaviour and mechanisms of copper adsorption on hydrolyzed polyacrylonitrile fiber. Journal of Colloid and Interface Science, 2003. 260,pp 265-272.

[10] Zhang, X., Bai, R.B. Deposition/adsorption of colloids to surface-modified granules: effect of surface interactions. Langmuir, 2002. 18, pp3459-3465.

[11] Namasivayam, C., Kavitha, D. IR, XRD and SEM studies on the mechanism of adsorption of dyes and phenols by coir pith carbon from aqueous phase. Microchemical Journal, 2006. 82 (1), pp 43 – 48.

[12] Foth, H.D. Fundamentals of Soil Science. 7th Ed. John Wiley and Sons, New York, 1984.

[13] Borggaard, O.K., Elberling, B. Pedological Biogeochemistry. Paritas Grafik, Brondby, Denmark. 2004

[14] Gimsing, A.L., Szilas, C., Borggaard, O.K. Sorption of glyphosate and phosphate by variable-charge tropical soils from Tanzania. Geoderma, 2007, 138, pp127-132.

[15] Elzahabi M, Yong R. N. pH influence on sorption characteristics of heavy metal in the vadose zone. Engineering Geology, 2001, 60, pp 61-68.

[16] Pagnanelli F, Esposito A, Toro L, Veglio' F. Metal speciation and pH effect on Pb, Cu, Zn, and Cd biosorption onto sphaevotilu natan: Langmuir –type emperical model. Water Resources, 2003, 37, pp 627-633.

[17] Unuabonah E I, Olu–Owolabi, B I, Adebowale K O, Ofomaja A E. Adsorption of lead and cadmium ions from aqueous solutions by triphosphate – impregnated kaolinite clay. Colloids and surface A: Physicochemical and Engineering Aspects, 2007, 292, pp 202 – 211.

[18] Chang, J., Law, R., Chang, C. Biosorption of lead, copper and cadmium by biomass of Pseudomonas aeruginosa. Water Resources, 1997, 31, pp1651- 1658.

[19] Erenturka S, Malkocb E. Removal of lead (II) by adsorption onto Viscum album L.: Effect of temperature and equilibrium isotherm analyses. Applied Surface Science, 2007, 253 (10), pp 4727 -4733.

[20] Yang J Y, Yang X E, He Z L, Li T Q, Shentu J L, Stofella P J. Effects of pH, organic acids and inorganic ions on lead desorption from soils. Environmental Pollution, 2006, 143, pp 9 – 15.

[21] Tan, X.L., Chang, P.P., Fan, Q.H., Zhou, X., Yu, S.M., Wu, W.S., Wang, X.K. Sorption of Pb(II) on Na-rectorite: effects of pH, ionic strength, temperatuire, soil humic acid and fulvic acid. Colloids Surface, 2008, A328, pp 8-14.

[22] Doula, M., Ioannou, A. The effect of electrolyte anion on Cu adsorption-desorption by clinoptilolite. Microporous and Mesoporous Materials, 2003, 58 (2), pp 115 – 130.

[23] Neal R H, Sposito G. Effects of soluble organic matter and sewage sludge amendments on cadmium sorption by soils at low cadmium concentrations. Soil Science, 1986, 142, pp 164-172.

[24] Adebowale, K.O., Unuabonah, I.E., Olu–Owolabi, B.I. (2006). The effect of some operating variables on the adsorption of lead and cadmium ions on kaolinite clay. Journal of Hazardous materials B, 134 : 130 – 139.

[25] Wang, S. Jun, H. Li, J. and Dong, Y. Influence of pH, soil humic/fulvic acid, ionic strength, foreign ions and addition sequence on adsorption of Pb(II) onto GMZ bentonite. Journal Hazardous Material, 2009, 167, pp 44 -51.

[26] Goel J, Kadirvelu K, Rajagopai C, Garg V.K. Removal of lead (II) by adsorption using treated granular activated carbon: Batch and Column studies. Journal of Hazardous Material, 2005, 125 (1 -3), pp 211- 220.

[27] Qin F, Shan X, Wei B. Effects of low – molecular –weight organic acids and residence time on desorption of Cu, Cd, and Pb from soils. Chemosphere, 2004, 57, pp 253–263.

[28] Schofield, R.K. and Samson, H.R. Flocculation on kaolinite due to the attraction of oppositely charged crystal faces, Discuss. Faraday Society, 1954, 18, pp 135- 145.

[29] Sposito, G. The chemistry of soils. Oxford press, London.1989.

[30] Naidu R., Bolan N. S., Kookana R. S., Tiller K. G. Ionic strength and pH effects on surface charge and Cd sorption characteristics of soils. Journal of Soil Science, 1994, 45, pp 419 -429.

[31] Li, L.Y. Li, R.S.. The role of clay minerals and the effect of H+ ions on removal of heavy metals (Pb2+) from contaminated soils. Canadian Geotechnical Journal, 2000, 37, pp296- 307.

[32] Spark, K.M. Johnson, B.B. and Wells, J.D. Characterizing trace metal adsorption on kaolinite. European Journal of Soil Science, 1995, 46(4), pp 633-640.

[33] Barrow, N.J. and Ellis, A.S. [1986] Testing a mechanistic model V. the points of zero salt effect for phosphate retention, for zinc retention and acid/ alkali titration of a soil. Journal Soil Science, 37, pp303 -310.

[34] Barrow, N.J. (1985) Reactions of anions and cations with variable charge soils. Advances in Agronomy, 38, pp 183 – 230.

[35] Chen, H.M. Heavy metal pollution in soil-plant system. Science Press Beijing, 1996.

[36] Songhu, Y., Zhimin, X., Yi, J., Jinzhong,W., Chan, W., Zhonghua, Z., Xiaohua, L. Desorption of copper and cadmium from soils enhanced by organic acids. Chemosphere, 2007, 68, pp1289-1297.

[37] Philips, I.R. Lamb, D.T. Walker, D.W. Burton, E.D. Effect of pH and salinity on copper, lead, and zinc sorption rates in sediments from Moreton bay, Australia. Bulletin Environmental Contamination Toxicology, 2004, 73, pp 104-108.

[38] Xu, Y.H., Zhao, D.Y. Removal of copper from contaminated soil by use of poly (amidoamine) dendrimers. Journal Environmental Science Technology, 2005, 39, pp2369 – 2375.

[39] Stumn, W., Morgan, J. Aquatic Chemistry. Wiley, New York, 1996.

[40] Businelli, M., Casciari, F., Businelli, D., Gigliotti, G. Mechanisms of Pb(II) sorption and desorption at some clays and goethite-water interface. Agronomie, 2003, 23, pp 219-225

[41] Xu, D., Tan, X.L., Chen, C.L., Wang, X.K. Removal of Pb(II) from aqueous solution by oxidized multiwalled carbon nanotubes. J. Hazard. Mater., 2008. 154, pp407-416.

[42] Li, J., Xu, R.K., Tiwari, D., Ji, G.L. Effect of low- molecular-weight organic anions on exchangeable aluminum capacity of variable charge soils. Applied Geochemistry, 2006, 21, pp1750 – 1759.

9

Characterization and Classification of Selected Rice Soils of Tropical Rainforest Region, Southeastern Nigeria

S. N. Obasi[1], E. U. Onweremadu[2], Egbuche C. T.[3], U. P. Iwuanyanwu[1]

[1]Department of Agricultural Technology, Imo State Polytechnic, Umuagwo Imo State, Nigeria
[2]Soil Science Department, Federal University of Technology, Owerri, Imo State Nigeria
[3]Department of Forestry and Wildlife Technology, Federal University of Technology, Owerri, Imo State Nigeria

Email address:

nathanielobasi2000@yahoo.com (S. N. Obasi)

Abstract: The study carried out in Amachi Izzi, Abakaliki, tropical rainforest region, Southeastern Nigeria was aimed at characterizing and classifying the lowland rice soils of the area. Three pedons were sunk, at the region. Results showed that soils were strongly acidic (pH <5.0), profiles 1 and 3 had low organic matter while profile 2 was moderate. Total nitrogen was also low to medium while available P was medium. Sand exhibited low variation (CV ≤15%) at Profile 2 while it indicated moderate variation (CV>15≤35%) at Profile 1 and Profile 3. Silt varied lowly (CV≤15%) at Profile 1 and moderately (CV>15≤35%) at Profile 2 and Profile 3. Clay varied highly (CV>35%) in all pedons of Abakaliki lowland soils. Base saturation varied highly (CV>35%) at Profile 1 and Profile 2 and lowly (CV≤ 15%) at Profile 3. Al saturation varied highly (CV>35%) in all pedons at Ebonyi north lowlands. Ochric epipedons were observed in pedons 1, 2 and 3 and diagnostic subsurface horizons were kandic. The temperature regime and percentage base saturation were considered at the subgroup level in the soil taxonomy. Pedons 1 and 2 were classified as Fluvaquentic Eutrudepts and Eutric Cambisol (FAO – WRB) while Pedon 3 was classified as Fluvaquentic Dystrudepts and Dystric Cambisol (FAO – WRB).

Keywords: Soil Classification, Rice Soils, Soil Variability and Southeastern Nigeria

1. Introduction

It is still challenging to meet the specific target of the new millennium development goals (MDGs) on food security and sustainability in Nigeria. At an assumed growth rate of 3.2%, Nigeria's population was expected to have exceeded 160 million by the year 2011. This population density ranked among the highest in the sub-Saharan Africa [1]. The population however depends on crop agriculture to provide food and support their livelihood in an agro-ecology reputed to be sensitive and fragile [2]. In order to make adequate and strategic plan towards sustaining the food needs of the country's growing population, serious attention is required in the area of the country's productive potential for major crops that the agro-ecology supports. Rice among other cereal crops such as sorghum, millet, is expected to contribute substantially in reducing food insecurity in Nigeria [3].

Hydromorphic (phreatic) and wetland (fluxial) rice cultivation requires specific characteristics of the soils of the colluvial footslopes and the valley bottoms. Rice being a semi-aquatic plant, poor to moderately-well drained conditions are best suited for hydromorphic and wetland rice cultivation. Well-drained and excessively drained soils carry a drought risk. [4] Stated that coarse-textured soils generally are less productive than soils with a fine texture. This is due to the lower inherent fertility of the former, but also to their higher percolation rates, by which nutrients (including fertilizers) are easily leached beyond the root zone. The inherent fertility of sandy soils in general is very low. In such soils, the cation exchange capacity (CEC) depends largely on the organic matter. This serves as storage for plant nutrients, but it is a source of nutrients too, mainly of nitrogen. In clayey soils, the fertility is generally higher [5]. In the valley bottoms, however, the fertility of the soils is governed by the hydrology of the valley [6].

Although Ebonyi State has been described to have the potential for upland and lowland rice production [7]; [8], there is very limited information on the classification of rice soils of northern Ebonyi State. Therefore, the objective of this research was to characterize and classify the soils of Ebonyi State using the USDA Soils taxonomy and correlated with the World Reference Base (WRB) legend.

2. Materials and Methods

The study was conducted in Amachi-Izzi near Abakaliki, tropical rainforest region, Southeastern Nigeria which lies between latitudes 6°20' and 6°30' N and longitudes 8°05' and 8°40' E. Parent material has been identified as a mixture of shale, Alluvium and sandstone within the tropical rainforest zone of the southeastern Nigeria. It experiences rainfall most of the months of the year and having highest intensities occurring between June-August. Mean annual rainfall is between 1800-2300 mm.

2.1. Field Work

Lowland rice land use was considered in the experiment, three profile pits were sunk in the investigated lowland rice soils through which soils were characterized and classified.

2.2. Laboratory Analyses

Particle size distribution was determined by hydrometer method according the procedure of [9]. Bulk Density was measured by core method [10]. Soil pH was measured electrometrically in 1:2.5 soil/ water ratio [11]. Organic Carbon was determined using method described by [12]. Total nitrogen was determined using modified micro Kjeldahl method [13]. Total available phosphorus was determined using Bray II method [14]. Exchangeable cations were extracted with 0.1M BaCl2 method [11] and analyzed with atomic absorption spectrometer. Exchangeable acidity was determined by the KCl extraction method and the

extractant was titrated against 0.05N NaOH. Effective cation exchangeable capacity (ECEC) was calculated as the summation of the exchangeable bases and KCl extractable Al and H. The percentage base saturation (BS) was the summation of total exchangeable bases expressed as a percentage of ECEC.

2.3. Data Analyses

Soil data was subjected to Coefficient of variation (C.V.) ranked as follow; Low variation (≤15 %), Moderate variation (>15≤35 %), High variation (>35 %) was used to estimate the degree of variability of soil properties [15].

3. Results and Discussion

The profiles were deep enough for rice cultivation and drainage indicated imperfectly drained due to high water table (Table 1). The color of the soils ranged from very pale brown (10YR 7/4) to yellow (10 YR 7/8) in Profile 1, Very pale brown (10 YR 8/3) to yellow (10 YR 8/6) in Profile 2 and Very pale brown in Profile 3, all colors were tested under wet condition. The imperfectly drained condition may have induced this matrix color. Textural class is from silty loam (SiL) to clay loam (CL) in all pedons in this mapping unit. There were common and medium distinct to many coarse and distinct mottles at the Btg and BCg of Profile 1 respectively. At Profile 2 mottles ranged from few fine and faint to very coarse and distinct mottles. Also at Profile 3 mottles ranged from few fine and faint to common coarse and distinct. Soil structure indicated weak crumby and medium to moderate crumby and coarse at the Profile 1. At the Profile 2 and Profile 3 there were weak crumby and medium to strong blocky and very coarse structure.

Table 1. Soils Morphological and Physical Properties

Horizon	Depth (cm)	Colour (moist)	Mottles	Text. Class	Struct.	Soil Texture			BD	Porosity
						Sand	Silt	Clay		
	Profile 1	(P1)								
Ap	0-13	10YR7/4	-	SiL	1,cr,m	428.0	505.6	66.4	1.31	50.0
AB	13-34	10YR8/2	-	SiL	1,cr,m	368.0	485.6	126.4	1.33	49.8
Btg	34-66	10YR8/4	c,2,d	L	3,cr,c	275.2	445.6	279.2	1.47	44.5
BCg	66-98	10YR7/8	m,3,d	CL	2,cr,c	287.9	385.7	326.4	1.57	40.8
Mean						339.0	455.6	199.6	1.42	46.4
	Profile 2	(P2)								
Ap	0-9	10YR8/3	f,1,f	SiL	1,cr,m	380.8	552.8	66.4	1.33	49.8
AB	9-26	10YR7/4	-	L	1,cr,m	384.5	465.6	146.4	1.37	48.3
Btg	26-54	10YR8/4	f,2,d	L	2,bk,m	348.0	405.6	246.4	1.45	45.3
BCg	54-103	10YR8/6	v,3,d	CL	3,bk,vc	348.0	332.8	319.2	1.57	40.8
Mean						365.3	439.2	194.6	1.43	46.05
	Profile 3	(P3)								
Ap	0-15	10YR8/4	f,1,f	SiL	1,cr,m	346.1	535.6	116.4	1.33	49.8
AB	15-35	10YR7/4	-	L	1,cr,m	485.2	345.6	169.2	1.36	48.6
Btg	35-71	10YR8/2	f,3,f	L	3,bk,c	308.0	445.6	246.4	1.45	45.3
BCg	71-112	10YR8/4	c,3,d	CL	3,bk,vc	255.2	456.4	288.4	1.48	44.2
Mean						348.6	445.8	205.1	1.41	46.97

Mottles: 1= fine, 2= medium, 3= coarse, fe= few, f= faint, c= common, d= distinct, p = prominent, v = very many, Texture: L = loam, SL Silty Loam, Structure: 1= weak, 2= moderate, 3= strong, cr= crumb, f=fine, m= massive, bk= blocky, vc = very coarse, m= medium.

Table 2. Ebonyi North Lowland Soils Chemical Properties

Depth (cm)	pH (H₂0)	OM g/kg	TN g/kg	Av.P Mg/kg	Ca	Mg	K	Na	TEB Cmol/kg	Al	H	TEA	ECEC	B. Sat. (%)	Al Sat (%)
P 1.															
0-13	3.77	29.6	1.5	7.39	1.33	0.80	0.03	0.05	2.21	1.43	0.98	2.41	4.62	47.83	30.95
13-34	4.66	22.7	1.1	13.9	1.34	0.40	0.07	0.07	1.88	0.80	1.00	1.80	3.68	51.08	21.74
34-66	4.88	13.1	0.7	7.25	1.28	0.73	0.04	0.07	2.08	1.32	0.96	2.28	4.36	47.71	30.27
66-98	4.36	13.1	0.6	8.34	8.24	1.02	0.09	0.05	9.40	0.01	0.40	0.41	9.81	95.82	0.10
Mean	4.42	19.6	1.0	9.25	3.05	0.74	0.06	0.06	3.89	0.89	0.84	1.73	5.62	60.61	20.77
CV	10.89	40.98	42.2	34.24	113.6	34.8	48.0	19.3	94.4	72.8	34.8	53.1	50.3	38.81	69.35
Rank	*	***	***	**	***	**	***	**	***	***	**	***	***	***	***
P 2.															
0-9	4.88	39.2	2.0	7.32	1.20	0.34	0.02	0.05	1.61	0.72	0.72	1.44	3.05	52.79	23.60
9-26	3.55	17.2	0.9	12.67	1.23	0.38	0.09	0.02	1.72	0.52	1.56	2.08	3.80	45.26	13.68
26-54	4.85	15.8	0.8	12.59	1.48	0.45	0.09	0.07	2.09	2.00	1.12	3.12	5.21	40.11	38.39
54-103	5.55	11.0	0.6	16.77	4.00	0.73	0.08	0.34	5.15	0.01	0.16	0.17	5.32	96.80	0.18
Mean	4.70	20.8	1.1	12.34	1.98	0.48	0.07	0.12	2.64	0.81	0.89	1.70	4.35	58.74	18.96
CV	17.77	60.0	58.6	31.39	68.48	37.1	48.1	123.	63.74	104.	66.9	72.49	25.5	44.09	85.01
Rank	**	***	***	**	***	***	***	***	***	***	***	***	**	***	***
P 3.															
0-15	4.52	36.5	1.8	9.91	1.12	0.60	0.06	0.07	1.85	1.16	1.20	2.36	4.21	43.94	27.55
15-35	4.67	24.1	1.2	8.26	1.44	0.54	0.05	0.08	2.11	0.76	1.12	1.88	3.99	52.88	19.05
35-71	3.84	0.70	0.1	7.55	1.44	0.64	0.05	0.10	2.23	1.60	0.40	2.00	4.23	52.71	37.82
71-112	4.67	12.4	0.4	6.99	1.51	0.72	0.09	0.21	2.53	0.80	1.67	2.47	5.00	50.60	16.00
Mean	4.43	18.4	0.9	8.18	1.38	0.63	0.06	0.12	2.18	1.08	1.10	2.18	4.36	50.03	25.11
CV	8.96	83.5	88.2	15.49	12.69	12.1	30.3	56.1	12.94	36.2	47.8	12.96	10.1	8.38	38.97
Rank	*	***	***	**	*	*	**	***	*	***	***	***	*	*	***

OM = organic matter, TN = total nitrogen, Av.P = available phosphorus, TEB = total exchangeable bases, TEA = total exchangeable acidity, ECEC = effective cation exchange capacity, B. Sat = base saturation, Al Sat. = aluminum saturation, CV= Coefficient of Variation, *= Low variability, **= Moderate variability, ***= High Variability.

The sand content of profile 1 decreased down the profiles in all pedons with the means as follow; 339, 365.3 and 348.6 g/kg in Profiles 1, 2 and 3 respectively. Silt consistently decreased in Profile 1 and Profile 2; in their four pedogenic horizons respectively. There was an irregular silt distribution in Profile 3. Mean silt content was 456.6, 439.2 and 445.8 g/kg in Profile 1, Profile 2 and Profile 3 respectively. Clay content increased down the profile in all pedons investigated at the Ebonyi North lowland mapping unit. In Profile 1, clay ranged as follow; 66.4, 126.4, 279.2, and 326.4 g/kg at the Ap, AB, Btg and BCg respectively. There was a similar trend in Profile 2; at their pedogenic horizons. The same trend continued in Profile 3 where clay ranged as; 116.4, 169.2, 246.4 and 288.4 g/kg in four pedogenic horizons respectively. The mean clay content in Abakaliki lowland mapping unit was 199.6, 194.6 and 205.1 g/kg in Profile 1, Profile 2 and Profile 3 respectively. The high silt content of soils of Abakaliki lowland soils could be attributed to low intensity of weathering of soils [16]. Also the consistent clay increase with depth was an indication of clay migration by lessivage to produce the process of illuviation. This has been observed by [17]. Sand exhibited low variation (CV ≤15%) at Profile 2 while it indicated moderate variation (CV>15≤35%) at Profile 1 and Profile 2. Silt varied lowly (CV≤15%) at Profile 1 and moderately (CV>15≤35%) at Profile 2 and Profile 3. Clay

varied highly (CV>35%) in all pedons of Abakaliki lowland soils.

Silt/clay ration (SCR) in Ebonyi North lowland mapping unit indicated a decrease from top to bottom of three pedons in this location. SCRs were high (1.04 – 7.62) suggesting that soils were of young parent materials as values exceeded 0.15 which has been reported [18] as benchmark for soils of old parent materials. SCR had high variation (CV>35%). Bulk density in Abakaliki lowland mapping unit also took an increasing trend from top to bottom of the pedons. Bulk density increased with depth in all the farms studied. This finding corroborates with those of [19] and [20]. Low bulk density reported in the soil may be a consequence of organic matter content of the soil. [21] Reported that low soil organic matter was responsible for increased bulk density in cultivated soils of Southeastern Nigeria. Within the profiles, possibility of migrating clay filling up the pore spaces in the horizons of illuviation may be account for the high bulk density values in the sub-surface horizons. Moreover, frequent cultivation of land made the soil loose and ultimately contributed for the low density in these layers. Results on bulk density were less than the critical limits for root restriction (1.75 – 1.85 g/cm3) [22] indicating the potential of the soil to support rice production. The percentage total porosity of the soil ranged between 46 - 50%, with the surface horizons containing higher pore spaces. [23] Reported similar trend in selected paddy soils under long-term intensive fertilization. High pore spaces recorded in the surface horizons may be attributed to loosening of soil materials during puddling/cultivation of soil Bulk density and porosity varied lowly (CV≤15%) at Profiles 1 and 2 while they varied moderately (CV>15≤35%) at profile 3.

The Chemical properties of soils of Ebonyi North Lowland mapping unit as investigated in the study were shown in Table 2. The pH distribution indicated very acidic in all pedons of Abakaliki lowlands. Mean pH in water for profiles 1, 2 and 3 were 4.42, 4.70 and 4.43 respectively. This pH range is strongly acidic (<5) and could be due to Al saturation in soil solution [24]. The problem of strongly acid soils may also be caused by strong leaching from high rainfall, and mainly from oxidation of sulfidic material [25]. Soil pH exhibited low variation (CV ≤15%) in all Abakaliki lowland pedons except Profile 2 where it moderately varied (CV>15≤35%). OM and TN all decreased down the profile in all the pedons. Organic matter had means of 19.6, 20.8 and 18.4 g/kg while mean total nitrogen were 1.0, 1.1 and 0.9 g/kg all in Profiles 1, 2 and 3 respectively. Organic matter was low (< 20 g/kg) at Profiles 1 and 3 while moderate (20 – 42 g/kg) [25] at Profile 2. TN was moderate or medium (0.5 – 1.5 g/kg) [25] in all pedons investigated at Abakaliki lowland soils. Organic matter and total nitrogen all had high variation (CV>35%) in all pedons at Abakaliki lowland profiles. The distribution of available phosphorus took an increasing trend in Profile 2, decreasing trend in Profile 3 and no particular trend in Profile 1. Mean available P was highest in Profile 2 (12.34 mg/kg), followed by Profile 1 (9.25 mg/kg) while the lowest occurred at Profile 3 (8.18 mg/kg). Available P was medium in Abakaliki lowland soils. [25] Reported the critical P values as <5, 5-15 and >15 mg/kg for low, medium and high respectively. Available P varied moderately (CV>15≤35%) in all pedons at Abakaliki lowland soils.

Total exchangeable bases (TEB) in Abakaliki lowland mapping unit increased down the profile in all the pedons in this unit. The mean TEB distribution was highest in Profile 1 (3.89 cmol/kg) followed by Profile 2 (2.64 cmol/kg) whereas the lowest occurred at ALP3 (2.18 cmol/kg). TEB varied highly (CV>35%) at ALP1 and P2 while it exhibited low variation (CV≤15%) at Profile 3. Effective cation exchange capacity (ECEC) increased down the profiles in all the pedons. Mean ECEC were 5.62, 4.35 and 4.36 cmol/kg. ECEC varied highly (CV>35%) at P1, moderately (CV>15≤35%) at P2 and lowly (CV≤15%) at ALP3. Base saturation also increased down the profile in all pedons. Mean base saturation distribution was highest in P1 (606.1 g/kg), followed by P2 (587.4 g/kg) while the lowest occurred at P3 (500.3 g/kg). Base saturation varied highly (CV>35%) at P1 and P2 and lowly (CV≤ 15%) at P3. Mean Al saturation was highest at P3 (251.1 g/kg) followed by P1 (207.7 g/kg)

and the lowest at P2 (189.6 g/kg). Al saturation varied highly (CV>35%) in all pedons at Ebonyi north lowlands.

4. Taxonomic Classification

Mean annual soil temperatures were over 22^{0} C and nearly constant within the study area [26]. Ochric epipedons were observed in pedons 1, 2 and 3 and diagnostic subsurface horizons were kandic. Isohyperthermic soil temperature regime placed pedons 1, 2 and 3 on the suborder Tropepts, organic matter contents and stratification qualified pedons as Fluvents, there was an irregular decrease in organic-carbon content (Holocene age) between a depth of 25 cm and either a depth of 125 cm below the mineral soil surface. The temperature regime and percentage base saturation were considered at the subgroup level in the soil taxonomy. A base saturation (by NH_4OAc) of more than 60 percent or more at a depth between 25 and 75 cm from the mineral soil surface at Pedons 1 and 2 while a base saturation (by NH4OAc) of less than 60 percent or more in one or more horizons at a depth between 25 and 75 cm from the mineral soil surface or directly above a root-limiting layer in Pedon 3. Therefore Pedons 1 and 2 were classified as Fluvaquentic Eutrudepts and Eutric Cambisol (FAO – WRB) while Pedon 3 was classified as Fluvaquentic Dystrudepts and Dystric Cambisol (FAO – WRB).

5. Conclusion

There was very high variability in the physical and chemical properties investigated. Agricultural intensification on the lowland rice soils requires the application of optimum rates of N and P fertilizers on soils derived from Shale and Recent Alluvium. The classification of the soils will also enhance the transfer of information about the soils within and between communities of soil users.

References

[1]   IITA (1990) Annual Report 1989/90.International Institute of Tropical Agriculture, Ibadan Nigeria.

[2]   IITA (1994) Annual Report 1993/94.International Institute of Tropical Agriculture, Ibadan Nigeria.

[3]   Akpokodje, G; Lancon, F and Erenstein, O (2001) Nigeria\s rice economy: State of the Art. *USAID Project report on the Nigerian rice economy in a competitive world:Constraints, opportunities and strategic choices.* West Africa rice development Association (WARDA), Bouake, Cote d'Ivoire.

[4]   Windmeijer, P.N., and W. Andriesse, (1993). Inland Valleys in West Africa: An Agro- Ecological Characterization of Rice Growing Environments. International Institute for Land Reclamation and Imporvement.P.O. Box 45, 6700 AA Wageningen.The Netherlands.

[5]   IRRI (1978).Soils and Rice. IRRI, Los Barios, The Philippines Kern, J. S., 1994. Spatial patterns of soil organic carbon in the contiguous United States, *Soil Science Society of America Journal*, 58: 439-55.

[6] Moorman, F. R (1981). Representative of toposequence of soils in southern Nigeria and their pedology. In: *Characterization of soils in relation to their classification and management for crop production* D. J Greenland (ed). Clarendo Press Oxford. Pp10-29.

[7] Mbah, C.N., Anikwe, M.A.N. Onweremadu, E.U. and Mbagwu, J.S.C. (2007). Soil organic matter and carbohydrate contents of a Dystric Leptosol under organic waste management and their role in structural stability of soil aggregates. Int. J. Soil Sci. 2(4)L 268-277.

[8] Edeh, H.O., Eboh E.C., and Mbah, B.N., (2011). Analysis of Environmental Risk Factors Affecting Rice Farming in Ebonyi State, Southeastern Nigeria. World Journal of Agricultural Sci. 7(1): 100-103)

[9] Gee, G.W., and Or D. 2002. Particle size distribution: In Dane J.H., Topp G.C. (eds). Methods of soil analysis Part 4 Physical methods. Soil Sci. Soc. Am Book series No. 5 ASA and SSSA, Madison WI; 225-293.

[10] Grossman, R.B., Reinsch, T.G.(2002). Bulk density and linear extensibility, In: Dane, J.H., Topp, G.C. (eds). Methods of soil analysis, part 4 physical methods. Soil Sci. Soc. Am. Book Series No.5 ASA and SSSA, Madison WI 2002; 201-228

[11] Hendershot, H.W., H.L. Lalande and Duquette, 1993, Soil Reaction and Exchangeable Acidity in: Soil Sampling and Methods of Analysis. Cartar, M.R. (Ed.), Can. Soc. Soil Sci. Lewis Publishers, London, pp: 141-145.

[12] Nelson, D. W and L. E. Sommers. (1982). Total Carbon, Organic Carbon and Organic Matter in: Sparks, D. L. (Eds.), Methods of Soil Analysis. Part 3. SSSA Books Series No 5 SSSA Madison,

[13] Bremner, J.M. and Mulvany, C.S. 1982. Nitrogentotal. In: Methods of Soil Analysis, Part 2, Ed. A.L. P.G Miller and D.R. Keeney (eds.) American Society of Agron. Madison WI, pp 593-624

[14] Olson, S. R. and L. E. Sommers. (1982). Phosphorus in: Methods of Soil Analysis. A. L. Page, R. HMiller and D. R. Keeney (eds). Maidson, W. I AmericalSciety of Agronomy: 1572pp

[15] Wilding, L.P. (1985). Spatial variability, its documentation, accommodation and implication to soil surveys, pp. 166-194. In D.R. Nielsen and J. Bouma (eds.) Soil Spatial Variability: Produc, Wageningen, Netherlands.

[16] Moberg, J. P. and Esu I. E., (1991) Characteristics and composition of some savanna soils of Nigeria *Geoderma 48:113-129.*

[17] Essoka, A.W and Esu, I.E., (2001) Physical properties of inland valley soils of central Cross- River State of Nigeria. *In Proceedings of the 27th annual conference of Soil Science Society of Nigeria 5-9th Nov. 2001 Calabar, Nigeria.*

[18] Ayolagha, G.A (2001) Survey and Classification of Yenagoa meander belt soils in the Niger- Delta. In Proceedings of the 27th annual conference of Soil Science Society of Nigeria 5-9th Nov. 2001 Calabar, Nigeria.

[19] Onweremadu, E.U., Ndukwu, B.N., Opara, C.C and Onyia, V.N (2007).Characterization of wetland soils of Zarama, Bayelsa, State, Nigeria in relation to Fe and Mn distribution. Int. J. of Agriculture and Food Systems 1(1):80-86.

[20] Chikezie, A., Eswaran, H., Asawalam, D.O., and Ano, A.O. (2009). Characterization of the benchmark soils of contrasting parent materials in Abia State, South-eastern Nigeria. Global J. of Pure and Applied Sciences 16(1) 23-29.

[21] Akamigbo, F.O.R., (1999). Influence of land use on soil properties of the humid tropical agro-ecology of Southeastern Nigeria. Niger Agric J. 30: 59-76.

[22] Soil Survey Staff (1996), Soil quality information sheet; soil quality indicators Aggregate stability National Soil Survey Center in collaboration with NRCS, USDA and the national Soil Tilth.Laboratory, ARS, USDA.

[23] Khan, M.S.H, Abedin, M.M.J., Afrin A and Khosruzzaman, M. (2006). Physicochemical changes of paddy soils under long term intensive fertilization. Asian J. of plant Sciences 5(1) 105-110.

[24] Soil Survey Laboratory Staff, (1992). Soil Survey Laboratory methods manual. USDA-SCS Soil Survey Investigations Report No. 42 Version 20. US Government Print Office Washington DC.

[25] Tabi, F.O., M. Omoko, A. Boukong, Ze.A.D. Mvondo, D. Bitondo and C. Fuh Che (2012), Evaluation of lowland rice (Oryza sativa) production system and management recommemdations for Logone and Chari flood plain- Republic of Cameroon. Agr Sc. Research J. Vol. 2(5): pp261 – 273.

[26] Lekwa, G., (1986). Soils of tidal marshes in the Kono-Imo River estuary, Rivers State, Nigeria. *Niger J. Soil Sci.* 6:47-56.

10

Calculation of FCR and RBC with Varied Effect of Iron in Broiler

Barkat Ali Kalwar[1], Hakim Ali Sahito[2, *], Mehmood Ahmed Kalwar[1], Zaibun Nisa Memon[2], Madan Lal[1]

[1]Department of Nutrition and Animal Product Technology, Faculty of AHV, Science, SAU, Tandojam- Sindh
[2]Department of Zoology, Faculty of Natural Sciences, SALU- Khairpur- Sindh

Email address:

hakimsahito@gmail.com (H. A. Sahito)

Abstract: One hundred and fifty hubbard broiler were studied to examine their response to various levels of iron in relation to FCR and blood parameters. The experiment was conducted at poultry experimental station, Faculty of Animal Husbandry and Veterinary Sciences, Sindh Agriculture University Tandojam during, 2013. Commercial feed was supplemented with iron concentration of 0 (Control), 40, 80, 120, 160mg/kg in groups A, B, C, D and E, respectively. Result revealed lowest feed (3780g) and water (8160ml) consumed by group E. Better (P<0.05) live weight (1939g), FCR (1.94), dressing percentage (64.93%), RBC (3.33×10^6/µl), HB (9.30g/dL), PCV (31.1%) and Rs. 47.35 per bird net profit was also recorded in group E where, 160mg/ kg iron was supplemented in broiler ration. Lowest mortality (6.66%) was also observed in group E, while non-significant differences in edible parts were observed among the groups. Increasing level of iron showed better performance in the groups. It is concluded that 160mg/kg iron level can be supplemented in broiler ration for better FCR, dressing % and per bird net profit along with better performance in blood parameters.

Keywords: Ration, Red Blood Cells, Mortality, Profitand Poultry

1. Introduction

A 100 g edible portion of broiler meat contains 74.6 g moisture, 12.1 g proteins, 11.1 g lipid, 1.0 g minerals and 158-175mg cholesterol. To overcome this gap, poultry industry can play its role by providing the best source of palatable, nutritious and high quality animal protein in a comparatively short duration, at an appropriate and affordable cost, because broiler meat is a high source of nutrients and is easily and completely digested. It is also a good source of protein and vitamins. The growth promoting value of the protein in chicken compares favorably with that of fish protein mainly meat diets, in response to gain weight (Bhatti, 2001).

The economic analysis shows that the feed is a major expense in poultry production amounting to about 70 to 75% of the total cost. Hence, the most important factor is the ratio between the feed/meat. Different feed give different results in terms of growth production. Thus, chemical composition of a feed is of great economic importance, i.e. preparation of feed by formulation of balanced amounts of feed sources to adequate required quantities of vitamins, minerals and trace elements. Of trace elements, iron is one of the most important elements for smooth function of body muscles. Within the body iron exist in two oxidation states: ferrous (Fe^{2+}) or ferric (Fe^{3+}). Because iron has an affinity to electronegative atoms such as oxygen, nitrogen and sulfur, these atoms are found at the heart of the iron-binding centers of macromolecules. Further the body temperature and blood cholesterol level of broiler are also some time effect the fat supplementation on the growth (Sahito et al., 2012). Thus, the partial mango pulp mixing in ration on can change its behaviour and production of broiler (Soomro et al., 2013). The poultry production also depends upon its well management to acquire the production through advance techniques being applied in Sindh- Pakistan (Abbasi et al., 2013).

Iron is a trace element whose deficiency or an excess can

compromise the immune system. It has been well documented that serum iron falls early in response to bacterial and viral infections and rebounds quickly with recovery(Larry, 1995). If iron is in great excess, then the key proteins required to initiate recognition and antibody production may be masked. Contrary, anemic animals are much more susceptible than those with adequate iron (Larry, 1995). Iron is the fourth most abundant element in the earth crust and is only a trace element in biologic systems, which makes up only 0.004% of the body mass(Uthman, 1998). Yet it is an essential component or co-factor of numerous metabolic reactions. By weight, the great proportion of the body's iron is dedicated to its essential role as a structural component of hemoglobin. On the other hand, iron is a toxic substance. Too much iron accumulating in vital structures, especially the heart, pancreas and liver produces a potentially fatal condition, hemochromatosis (Uthman, 1998). Considering the importance of iron, the present study was carried out to determine the calculation of FCR and RBC with varied effect of iron in broiler.

2. Materials and Methods

Iron is a trace element that has linear association with immune system of broiler and its deficiency or excess can compromise the immune system. In order to ascertain the effect of different concentration of iron on the growth and blood parameters of broiler, the present study was conducted during, 2013 at Poultry Experiment Station, Faculty of Animal Husbandry and Veterinary Sciences, Sindh Agriculture University, Tandojam.

2.1. Housing

The house was initially cleaned and washed with pressure water. White wash was done by using limestone later-on sanitizer (Snucop) (1gm powder in 1 liter of water) by spraying on the walls, floors and equipment etc. the shed door was closed for 30 minutes for disinfections prior to starting the research. To make the house comfortable, wooden dust was used as a litter. Before spreading the wooden dust on the floor, it was first dried under sun light. Calcium carbonate 500 gm over 100 square feet of litter was used to prevent bacterial and other parasitic growths and also prevented for moisture absorbent.

2.2. Brooding

The day-old chicks were brooded under the electric blub which were about (40 to 60 watts) used to maintain the light. Temperature ranged from 90 to 95°F during the first week and it was reduced by 5°F each week and maintained at 70°F. Humidity was maintained around 55 to 60 % by providing external water spray and light provision was ensured 24 hours. Floor space area of 1 square feet per broiler was provided.

2.3. Vaccination

The chicks were protected against common viral diseases through the inoculation i.e. new castle (ND), infectious bronchitis (IB), and infectious bursal disease (IBD) and hydro-pericardium syndrome throughout the study period (Table-1).

Table 1. Vaccination schedule for experimental broilers

S. No.	Vaccination	Days				
		6th	11th	14th	21st	26th
1	Vaccine	N.D+IB	Gumboro	Hydro pericardium sydrome	N.D La Sota	Gumboro
2	Route of administration	Intraocular	Intraocular	Subcutaneous	Drinking water	Drinking Water
3	Dose/b.wt	One drop	One drop	0.5 cc	2 ml	2ml

2.4. Grouping of Broilers

One hundred fifty (150) day old hubbard broiler were purchased, initially weighted and randomly divided into five groups A, B, C, D and E having 30 chicks in each group. All groups were provided ration that contains different iron concentration.Group A, kept as control (0.0%) , group B, C, D and E were given iron concentration of 40, 80, 120, and 160 mg/ kg feed, respectively.

2.5. Feeding

The chicks were fed *ad libitum,* the commercial starter ration was provided during first 0-4 weeks and finisher ration during 5-6th weeks of age. Three types of feeder were used such as plate type, rod/tube and adult feeder made up of plastic and aluminum. The feed refusal from each group was collected and weighted daily twice a day.

2.6. Watering

Underground fresh water was provided 24 hrs. Initially two chick drinkers (2 liter) were provided, later on adult drinker was provided to each group. The system was manual and water was provided twice daily at morning and evening. Refusal water was measured and replaced with fresh water twice a day. When, 42 days completed, the experimental broilers of each group were marketed on the individual weight basis and the final live weight was recorded. For other parameters, 10 broilers in each group were randomly selected and slaughtered to record dressed weight and percentage and weight of liver, heart and gizzard.

2.7. Blood Analysis

For recording RBC, hemoglobin and PCV, the blood

samples were collected from the wing vein of broiler before slaughtering. The samples were kept in a sterilized bottle contain sodium citrate 3%. The collected samples were analyzed for RBC count (by Hemocytometer), Haemoglobin (by Sahlis method). Further, under given parameters were kept under study period:

1. Live weight of broilers, 2. Feed intake (g), 3. Water intake (ml), 4.Mortality (%), 5.Dressing weight (g/b), 6.Weight of edible parts (g), 7. RBC count (x 10^6 ul), 8. Haemoglobin (g/dL), 9.PCV (%), 10.Economics (Rs).

2.8. Data Analysis

The data on various related parameters were collected and arranged in the tables, analyzed statistically to discriminate the superiority of treatment means. The collected data were fed in the computer in Mstat-C, Microsoft program. Duncans Multiple Range (DMR) test was employed to compare the individual means as suggested by (Gomez and Gomez, 1984).

3. Results

In order to examine the effect of different iron concentration on FCR and RBC count in broilers the study was carried out at Sindh Agriculture University during, 2013. The data thus obtained along with the outcome interpretation are presented in the following tables.

3.1. Feed Intake

Average feed intake of broiler in group A, B, C, D and E was 3880.0, 3850.0, 3830.0, 3810.0 and 3780.0 g/b, respectively (p<0.001). This indicated that additional iron supplementation resulted in reduced feed intake and it was lowest (3780.0g/b) in group E, where the broilers were fed on feed with highest iron concentration (160mg/kg feed) (Table-2).

Table 2. Average feed intake of broiler (g/b)

Weeks	Groups				
	A	B	C	D	E
W_1	143.3	142.1	139.1	136.5	134.5
W_2	353.5	351.1	338.5	339.5	341.5
W_3	485.3	459.8	448.6	442.0	432.0
W_4	665.6	959.5	666.5	660.0	650.8
W_5	989.8	986.3	973.7	975.0	971.2
W_6	1269.5	1251.2	1263.6	1257.0	1250.0
Total intake	3880.0^a	3850.0^{ab}	3830.0^{ab}	3810.0^{ab}	3780.0^b

Note: Group probability = 0.001 and LSD 0.05 = 35.827

3.2. Water Intake

Broiler consumed less water in the first week and later with the increase in their age, the water consumption was increased simultaneously (Table-3). Average water intake of the broiler in group A, B, C, D and E was 8905.2, 8710.0, 8315.0, 8183.0 and 8160.0 ml/b, respectively (p<0.001). Water intake was higher in group A (Control) then the rest of groups. It was observed that additional iron supplementation caused a reduction in water intake by the broilers and inversely with

increased iron concentration in feed the broiler water consumption was decreased considerably. Hence, the minimum water intake (8160.0 ml) was recorded in broilers fed on ration containing highest iron concentration in group E.

Table 3. Average water intake of broiler (ml/b)

Weeks	Groups				
	A	B	C	D	E
W_1	111.2	111.5	108.0	106.8	105.5
W_2	586.3	586.5	535.5	499.0	529.5
W_3	1090.9	1085.3	1020.3	999.5	980.0
W_4	1538.0	1437.0	1360.2	1294.5	1284.0
W_5	2495.0	2400.0	2341.2	2390.0	2280.6
W_6	3083.8	3089.7	2949.8	2893.0	2980.4
Total intake	8905.2^a	8710.0^b	8315.0^c	8183.0^d	8160.0^d

Note: Group probability = 0.001 and LSD 0.05 = 60.75

3.3. Live Body Weight Gain

Average live body weight gain of broilers in group A, B, C, D and E was 1650.0, 1730.0, 1780.0, 1820.0 and 1939.0 g/b, respectively (p<0.001). Live body weight was significantly high (p<0.001) in broilers group E, then rest of groups. The results further showed that live body weight of broilers was significantly improved with iron supplementation (Table-4). The statistical analysis illustrated that the live body weight was affected under different iron concentrations.

Table 4. Average live body weight of broiler (g/b)

Weeks	Groups				
	A	B	C	D	E
W_0	41.7	41.5	41.5	41.0	41.5
W_1	145.0	142.0	146.0	147.5	148.5
W_2	345.0	365.5	370.5	372.5	380.0
W_3	550.0	580.4	591.2	593.5	600.0
W_4	870.0	885.5	900.0	898.2	905.5
W_5	1230.0	1245.2	1250.0	1265.0	1272.0
W_6	1650.0^d	1730.0	1780.0^{bc}	1820.0^c	1939.0^a

Note: Group probability = 0.001 and LSD 0.05 = 47.17

3.4. Liver Weight

Different iron concentrations did not affect the liver weight which is known as edible part of broiler (p<0.091). Average liver weight of broiler in groups A, B, C, D and E was 41.5, 41.6, 42.0, 41.8, g/b.Increasing iron concentration in broiler ration did not have significant effect on heart weight (p<0.061). Average heart weight in groups A, B, C, D and E was 10.5. 10.1, 10.7, 10.8 and 11.5 g/b.Average gizzard weight in groups A, B, C, D and E was 48.6, 49.3, 48.7, 48.5 and 47.8 g/b, respectively. The results shows that there is no significant difference (p<0.093) in gizzard weight of all groups, respectively.

3.5. Dressing Percentage

Average dressing weight of broilers in group A, B, C, D and E was 910.0, 987.0, 1055.0, 1112.0 and 1259.0 g/b. This indicated that average dressing percentage was 55.15, 57.05, 59.26, 61.09 and 64.93%, respectively. Dressing weight was remarkably higher 1259.0 g (64.93.0), in broiler of group E, while the lowest dressed weight of 910.0 g/b(55.15%) was recorded in group A (control). The result indicates that the most efficient dressing weight and dressing percentage were obtained from the group E (p<0.001). The quantity of dressed weight in broiler improved with increased iron concentration. However, dressing percentage was slightly increased when iron concentration was increased over 120 0r 160 mg/kg feed.

3.6. Feed Conversion Ratio (FCR)

Average feed conversion ratio of broiler observed in group A, B, C, D and E was 2.35, 2.22, 2.15, 2.09 and 1.94, respectively (Table- 5). The results indicated that the most efficient feed conversion ratio (1.94) was recorded in broilers of group E (p<0.001), respectively. While, highly significant (p<0.01) when compared with group B and control A. Results indicate that it was significantly improved with supplementation of iron to various proportions. However, this improvement in FCR was marginal when iron concentration increased beyond 160 mg/kg feed.

Table 5. *Average liver, heart and gizzard weight of broiler (g/b) and dressing weight*

| Parameters | Groups | | | | | Probability |
	A	B	C	D	E	
Liver weight	41.5	41.6	42.0	41.8	41.8	0.091
Heart weight	10.5	10.1	10,7	10.8	11.5	0.061
Gizzard weight	48.6	49.3	48.5	48.5	47.8	0.093
Dressing weight (g/b)	910.0[b]	987.0[c]	1055.0[a]	1112.0[b]	1259.0[b]	
Dressing (%)	55.15[b]	57.05[b]	59.26[a]	61.09[b]	64.93[b]	
FCR	2.35[c]	2.22[b]	2.15[a]	2.09[a]	1.94[a]	

Note: Group probability = 0.001; LSD 0.05 = 36.65 and FCR: LSD 0.05 = 0.096

3.7. Mortality Rate

Average mortality in groups A, B, C, D and E was 5, 5, 4, 3 and 2 b/g which is the rate of mortality 16.6, 16.6, 13.3, 10.0 and 6.66 percent respectively (Table-6). Mortality was relatively higher (16.6%) in group A and B, where lowest iron concentration (0-40mg/kg feed) was given, while the lowest mortality of 6.66 % was seen in group E, where highest iron concentration (160mg/kg feed). The average PCV count in groups A, B, C, D and E was 28.1, 29.3, 30.0, 30.9 and 31.1 % (p<0.001), respectively. PCV count was comparatively higher in group E (31.1%), where higher iron concentration (160 mg/ kg feed) was given, while the lowest PCV (28.1%) was determined in group A (Control). It is obvious from the results that there was a highly significant (p<0.001) association of PCV count with increasing iron concentration. However, the PCV count in all the groups remained with the normal recommended ranges.The hemoglobin level in blood on average in groups A, B, C, D and E was 8.10, 8.50, 8.90, 8.99 and 9.30 g/dL, respectively. Hemoglobin level in blood was significantly higher (9.30 g/dL) in group E, where higher iron concentration (160 mg/ kg feed) was given, while the lowest hemoglobin level in blood (8.10 g/dL) was determined in group A (Control). The determinations of hemoglobin indicate that with supplementation and increasing iron concentration of broiler ration, the blood hemoglobin level was considerably improved. The differences in hemoglobin level in blood under different iron concentrations were statistically significant (p<0.003).The RBC (Red Blood Cell) level in blood on average in groups A, B, C, D and E was 2.73, 2.90, 2.99, 2.99, and 3.33 (x 10^6μl), respectively (Table-6). RBC level in blood was significantly higher i.e., 3.33 (x 10^6μl) in group E, where higher iron concentration (160 mg/kg feed) was given, while the lowest RBC level in blood count 2.73 (x 10^6μl) was determined in group A (Control). The determination of RBC indicated that with supplementation and increasing iron concentration of broiler ration, the blood RBC level was significantly improved. The differences in RBC level in blood under different iron concentrations were statistically significant (p<0.001).

Table 6. *Average mortality of broiler (%) and packed cell volume in blood of broiler*

| Parameters | Groups | | | | |
	A	B	C	D	E
Dead broiler (#)	5	5	4	3	2
Mortality (%)	16.6	16.6	13.3	10.0	6.66
PCV (%)	28.1	29.3	30.0	30.90	31.1
Hb (g/dL)	8.10	8.50	8.90	8.99	9.30
RBC (x 10^6μl)	2.73	2.90	2.99	2.99	3.33

Note: group probability = 0.032, PCV (%), LSD = 0.063, Hb, LSD = 0.0063 and RBC, LSD = 0.1825

3.8. Economics

The economic analysis of the broiler ration supplemented with different iron concentrations indicated the total cost of production of broilers in groups A, B, C, D and E was Rs. 105.60, 106.55, 107.55, 108.85 and 109.70, respectively (Table- 7). After marketing of broiler at the rate of 81.00 per kg live weight, the total revenue received was Rs. 133.65, 140.13, 144.18, 147.42 and 157.05 in groups A, B, C, D and E, showing a net profit of Rs 28.05, 33.58, 36.63,38.57, 47.35 per broiler, respectively. Group E, with iron concentration of 160 mg/ kg feed proved to be most profitable concentration, followed by group D (120 mg/ kg feed).

Table 7. *Economic analysis of the broiler ration supplemented with different iron concentrations*

S. No.	Economic parameter	Groups				
		A	B	C	D	E
1	Day old chick cost (Rs./b)	32.5	32.5	32.5	32.5	32.5
2	Feed cost (Rs./b)	85.20	57.75	57.45	57.15	56.70
3	Supplement cost (Rs./b)	0.0	1.4	2.7	4.3	5.6
4	Medication and vaccination cost (Rs./b)	4.3	4.8	4.8	4.8	4.8
5	Liter and limestone cost (Rs./b)	3.6	3.6	3.6	3.6	3.6
6	Labor cost (Rs./b)	1.5	1.5	1.5	1.5	1.5
7	Miscellaneous expenditure (Rs./b)	5.0	5.0	5.0	5.0	5.0
8	Total expenditure (Rs./b)	105.60	106.55	107.55	108.85	109.70
9	Final live body weight (kg/b)	1650.0	1730.0	1780.0	1820.0	1939.0
10	Marketing (Rs/kg)	81.0	81.0	81.0	81.0	81.0
11	Income (Rs/b)	133.65	140.13	144.18	147.42	157.05
12	Net profit / loss (Rs/b)	28.05	33.58	36.63	38.57	47.35

4. Discussion

In the present investigation, the feed and water intakes were decreased with increasing iron concentration in the range of 160mg/kg feed, while live body weight was increased considerably which resulted an improved feed conversion ratio. Moreover, the decreased weight 1259 g/band dressing percentage of 64.93 percent was superior when broiler was fed on ration containing 160 mg/kg iron concentration, followed by 120 mg/kg feed. Ruiz et al. (2000) examined broiler feed supplemented with iron concentration 50, 100 and 150 mg/ kg feed and found that iron concentration at rate of 100 mg/ kg feed proved optimally positive results for all the studied parameters. The effect of increasing iron concentration on liver, heart and gizzard weight was not significant, but this increased level was within the normal ranges. Apparently, the low mortality of broiler had some association with supplementation or increasing iron concentration in feed.

Packed cell volume (PCV), hemoglobin level and red blood cell (RBC) level in blood of the broiler increased considerably with increasing Fe concentration in feed. However, this increase remained within the normal recommended levels and hence there was no negative effect on these blood parameters due to iron concentration upto 160 mg/kg feed. In a similar study, Huff et al., (1997) reported that broiler diets containing lower Fe levels significantly decreased packed blood cell volume and hemoglobin concentration without altering the number of circulating erythrocytes. Similarly, Vahl et al. (1997) found that broiler feed was supplemented with Fe concentrations and concluded that apparently Fe requirement was 100 mg/kg diet (80 mg from dietary components and 20 mg Fe from supplement and in blood, the hemoglobin concentration increased slightly with extra dietary Fe. Julio et al. (1998) concluded that chicks fed high levels of Fe consistently showed increased packed cell volumes. In a similar study, Pecelunas et al., (1999) reported that the effect of dietary Fe was significant (p<0.05) on growth of broiler, while PVC values remained unchanged; while in China, Jiang and Zhang, (2003) examined Fe concentrations, 0, 30, and 60 mg/kg feed

and reported that low Fe level in diets decreased feed conversion ratio and increasing Fe level increased weight gain. The findings of the present study are further supported by Pek et al., (2005) who determined the effects of different amounts of iron on body weight gain, feed consumption, feed efficiency and some blood parameters and reported that supplementation of 100 mg/ kg feed Fe significantly increased body weight gain, feed consumption and feed efficiency; while Skrivan et al., (2005) recommended 120 mg of Fe / kg for high broiler growth. The comparison of the findings of present study with the results reported by various researchers from different parts of the world, it can be concluded that iron concentration of 160 mg /kg feed could be most beneficial and economical range for quantitative meat and profitable broiler production may be due to iron produce connective tissue within the body, maintain immune system, carries oxygen within the body and iron metal complex that binds molecular oxygen in the lungs and carries it to all of other cells within the body (e.g. muscles) that needs oxygen to perform their activities.

Economic broiler production has always been the aspect under discussion with the producers and scientists, because feed industry experienced a tremendous development with rapid progress in poultry industry. The broiler farmer is attracted by the feed with growth promoting efficiency, while scientists favour the feed that results better feed conversion efficiency along with optimum level of hematological characteristics to ensure good broiler health and produce quality meat (Michael, 2006). Iron is an interesting trace element in that either a deficiency or an excess can compromise the immune system. It has been well documented that serum iron falls early in response to bacterial and viral infection and rebounds quickly with recovery. To the contrary, anemic animal are much more susceptible to infections than those with adequate iron (Larry, 1995). The study conducted to determine the effect of various level of iron on the FCR and RBC count in broilers. On the basis of findings of present investigation, it was concluded that iron supplementation in broiler feed improved FCR, dressing % and per bird net profit coupled with hemoglobin, RBC and PCV level in blood. Further, it is suggested that the iron supplementation at level of 160 mg/kg feed of broiler can be supplemented for better broiler performance. Studies on meat quality are recommended. Feed manufactures are suggested that the broiler ration may be formulated with iron supplementation for better net return of farmers.

References

[1] Abbasi IHR, HA. Sahito, F. Abbasi, MA. Kalwar, AA. Soomro, DR. Menghwar, M. Memon, MI. Sanjrani. 2013. Management and production of live stock and poultry through advance techniques in Sindh, Pakistan. International J. Innovative Agri. & Biology Res., 1(1): 31-42.

[2] Abro, MR, HA. Sahito, A. Memon, RN. Soomro, H. Soomro and NA. Ujjan. 2012. Effect of various protein source feed ingredients on the growth performance of broiler. International J. Medicinal Plant Res., 1(4): 038-044.

[3] Bhatti, MY. 2001. Emerging prospectus of poultry production in Pakistan at the dawn of 21st century monthly Agro- Vet. News (Oct-2001), Mehmoodcentre, bc-11, block9, Clifton, Karachi.20.

[4] Gomez, KA and AA. Gomez. 1984. Statistics for Agriculture research (2nd edition), John Wiley and sons, New York.

[5] Huff, WE., CF. Chang, MF. Warren and PB. Hamilton. 1997. Ochratoxin a- induced iron deficiency anemia. Appl. Environ. Microbiol. J. of An. Sci., 39(5):7-8.

[6] Julio, L. P., JL. Gerger, ME. Cook and LJ. Stahl.1998. Iron metabolism in chicks fed various levels of zinc and copper. Dept: of nutritional sciences and poultry sciences, university of Wisconsin, Madison, WI, USA. The J. Nutr. Bioch., 3(3): 140-145.

[7] Larry, L. B .1995. Trace minerals: keys to immunity. Animal science. University of Illinois. Pp: 55.

[8] Micheal, WK. 2006. Medical biochemistry. IU iron introduction, school of medicine. 15: 9-24.

[9] Pecelunas, KS., DP. Wages and JD. Helm. 1999. Botulism in chickens associated with elevated iron levels. Avian Dis. 43(4):783-787.

[10] Pek, H., M. Avci, M. Yertk and N. Aydlek. 2005. Effects of copper and iron addition to diets on growth performance and some blood parameters in quail. Veteriner. Bilimleri. Dergisi., 21(1/2):45-50.

[11] Ruiz, J A., AM. Perez-Vendrell and EE. Garcia. 2000. Effect of dietary iron and copper on performance and oxidative stability in broiler leg meat. 290 ASL, deptt of aimal sciences, 1207 West Gregory Drive, Urbana, IL 61801. Br. Poul. Sci., 4:163-167.

[12] Sahito HA, RN. Soomro, A. Memon, MR. Abro, NA. Ujjan and A. Rahman. 2012. Effect of fat supplementation on the growth, body temperature and blood cholesterol level of broiler. Glo. Adv. Res. J. Chem. and Mat. Sci., 1(2): 023-034.

[13] Skrivian, M., V. Skrivanova and M. Marounek. 2005. Effect of dietary zinc, iron and copper in layer feed on distribution of these elements in eggs, liver, excreta, soil and herbage. Research institute of animal production, Prague, CZ, Czech republic. Poult. Sci., 84(10) 1570-1575.

[14] Soomro H, MI. Rind, SN. Sanjrani, AS. Magsi, GS. Barham, SA. Pirzada and HA. Sahito. 2013. Effect of partial mango pulp mixing in ration on behaviour and production of broiler. International J. Plant and Animal Sci., 1(1): 030-036.

[15] Uthman, Ed. MD. 1998. Iron and its metabolism. Nutritional anemias, diplomate, American board of pathology. 24.

[16] Vahl, HA. and AT. Klooster. 1997. Dietary iron and broiler performance. Clo- institute for animal nutrition de schothorstmeerkoetenweg, lelystad, the Netherlands. Br. Poult. Sci., 28(4)567-576.

11

Carbon Storage and Climate Change Mitigation Potential of the Forests of the Simien Mountains National Park, Ethiopia

Habtamu Assaye[1], Zerihun Asrat[2], *

[1]College of Agriculture and Environmental Sciences, Bahir Dar University, Bahir Dar, Ethiopia
[2]School of Forestry, Hawassa University, Hawassa, Ethiopia

Email address:

zerasrat@yahoo.com (Z. Asrat)
*Corresponding author

Abstract: The study assessed land cover change, carbon stock and sequestration potential of Simien Mountains National Park (SMNP), Ethiopia. Landscape was stratified into four zones based on the vegetation ecology and land uses: Afro-alpine grassland (AAGL), Afro-alpine woodland (AAWL), Afro-montane forest (AMF) and Cultivated and overgrazed land (COL). 40 sample plots were taken randomly (10 from each zone). Nested plot design with size of 50m*50m and subplots of 20m*20m, 10m*10m, 5m*5m, 2m*2m and 1m*1m was used for the measurement of trees of different diameter classes. Soil sampling was done at the four corners of the 10m*10m subplots to a depth of 30cm and taken to laboratory for analysis along with litter and undergrowth. Allometric equation was used for determination of above ground biomass (AGB) carbon. Below ground biomass (BGB) carbon was taken as 24% of AGB carbon. Land cover change was analyzed comparing satellite images of different periods. It was found that the COL has increased from 20% in 1972 to 48% in 2013. As a result, the AMF and AAWL have shrunk by nearly 50%. A future projection with a simple linear model indicated 73ha and 251.3ha of annual deforestation rate in the AAWL and AMF zones respectively, implying that it will take only 71 and 49 years for the AMF and AAWL respectively to be completely lost. Above ground carbon (AGC), below ground carbon (BGC) and soil organic carbon (SOC) holds 34.4%, 8.3% and 55.2% of the total carbon stock respectively. Dead wood and Litter Biomass together contributed only to the 2.2%. From land cover point of view AMF, AAGL, AAWL and COL stored 47.5%, 22%, 20.9% and 9.6% of the total carbon stock in the area respectively. A linear regression of Shannon diversity index against total carbon and AGC was calculated for AMF zone and as such no strong relationship was found for the total C ($R^2 = 0.242$) and also AGC ($R^2 = 0.337$), but it appeared that the stored carbon tends to decrease as the Shannon diversity index increases.

Keywords: Land Cover Change, Carbon Sequestration Potential, Simien Mountains National Park, Tree Species Diversity

1. Introduction

Climate change has been proved by scientific evidences and unequivocally accepted by the global community as a common issue of interest. Since the industrial revolution, the burning of fossil fuels and the destruction of forests have caused the concentrations of heat-trapping Green House Gases (GHGs) to increase significantly in our atmosphere, at a speed and magnitude much greater than natural fluctuations would dictate. If concentrations of GHGs in the atmosphere continue to increase, the average temperature at the Earth's surface will increase by 1.8 to 4°C by the end of the century [24]. Thus, the rapid increase in global surface temperature is mainly due to the rise in the amount of carbon dioxide in the atmosphere primarily due to anthropogenic activities [39]. As a result of change in global climate there has been a widespread and growing concern that has led to extensive international discussions and negotiations. In seeking solutions for this, the overwhelming priority is to reduce emissions of GHGs and to increase rates of carbon sequestration. The concerns have led to efforts of reducing emissions of GHGs, especially CO_2, and measuring carbon absorbed by and stored in forests, soils, and oceans. To slow down the rise of GHGs concentrations in the atmosphere, and thus possible climate change, is to increase the amount of carbon removed by and stored in forests [23, 4, 25].

As a natural solution, the role of trees and forests in the process of carbon cycle is quite significant as it stores more carbon among the terrestrial ecosystems [26, 41, 35]. This will make forest ecosystems to be the largest terrestrial carbon pool. Protected areas, with their all and diverse ecosystems including forests are vital systems to capture and store carbon from the atmosphere and to help people and ecosystems adapt to the impacts of climate change [11]. Ethiopia, being party to the United Nations Environmental Program and signatory to its treaties and protocols, is striving to contribute to the international effort of climate change adaptation and mitigation. It has adjusted its development strategy aiming at meeting net zero emissions by 2030 and developed climate resilient green economy (CRGE) strategy. Conserving and enriching existing forests, establishing new forests, enhancing of the existing protected areas and establishing new ones are some of the measures undertaken by the government. The role of forests to capture and store carbon from the atmosphere has been studied by several researchers [43, 30, 2, 33, 1]. However, studies on carbon storage process at a landscape level for instance in a protected area with different land covers are lacking. Therefore, this study was undertaken to assess the carbon storage potential of Simien Mountains National Park (SMNP) in Ethiopia through its different land cover zones and the overall dynamics of land cover changes.

2. Methodology

2.1. Site Description

The study was conducted in SMNP found in Amhara National Regional State, north Gondar, Ethiopia located at about 846 km North of Addis Ababa. The park has an area of 412 km^2 and geographically situated around 13°11'N and 38°04'E. SMNP was established in 1966 and officially recognized in 1969 for its rich of rare and endemic wildlife species, diverse fauna and flora composition and for its spectacular landscape and unique scenery. The park was inscribed in the World Heritage List for fulfilling criterion III (exceptional beauty) and criterion IV (importance for biodiversity) in 1978. SMNP is the first natural World Heritage Site inscribed in Ethiopia. The climatic condition within the park ranges from woina-dega at lower altitude (1500 – 2400 m.a.s.l) to wurch zone at the upper elevations (above 3700 m.a.s.l). High-dega and temperate climate zones are found in between the two. Mount Ras-Dejen, with 4620 m.a.s.l, the highest peak in the area as well as in the country is also part of this park. Approximately 75% of precipitation in the area falls between June and September as predominantly hail, rain and mist resulting in a mean annual rainfall of 1550mm. Temperatures are relatively consistent

throughout the year, however there are large diurnal fluctuations ranging from a minimum of -2.4 - 4°C at night to a maximum of 11 - 18°C during the day [17, 16, 27, 7, 21]. The national park is important conservation site for rare and endemic animals that made the park as last habitat such as Walia Ibex (*Capra walie*), Chelada baboon (*Theropithecus gelada*), Ethiopian Wolf (*Canis simensis*), etc.

2.2. Stratification of the Area

Considering vegetation differences and land uses within the study area, SMNP was stratified into four zones. These zones include, the Afro-alpine grassland (AAGL) occupying the highest altitude ranges, Afro-alpine woodland (AAWL) that is dominated by the *Erica arborea* trees, the Afro-montane forest (AMF) surrounding the mountain base and steep slopes, and Cultivated and overgrazed land (COL). Image analysis for land cover change study and stratification of the area was done using satellite image of four different years (1972, 986, 2000 and 2013) with ERDAS Imagine and Arc GIS software.

2.3. Sampling Technique

A square grid of 1km*1km was drawn on the map of the park considering the outer gridlines as reference. 10% of the square grids were considered for the sampling and proportionally distributed to the different vegetation zones. Accordingly, 41 samples were needed for SMNP, which would have been distributed as 7, 9, 5 and 20 plots for the AAGL, AAWL, AMF and COL respectively. However, taking into consideration the fact that there is high variability and carbon stock in the AAWL and AMF zones as compared to the COL, and also in consideration of taking fairly equal minimum number of plots, 10 samples were taken randomly from each zone making it 40 total sample plots. Nested plot design, which is appropriate for inventories in natural forests where there is high variability in tree size, distribution and structure, was used. Forest carbon assessments in particular use nested plot designs that present variable size subplots for the different tree size classes and also for the different forest carbon pools. Accordingly, 50m*50m (the outer most) plot was used for trees above 30cm DBH, 20m*20m subplot was used for trees with DBH between 10cm and 30cm, 10m*10m subplot was used for trees between 5cm and 10cm DBH, 5m*5m subplot was used for small trees of DBH between 2cm and 5cm, 2m*2m subplot was used for regeneration and undergrowth and the inner most 1m*1m subplot was used for litter. Tree height was also measured along with DBH. Soil samples were taken at four corners of the 10m*10m subplot to a depth of 30cm, and one composite sample was taken for soil carbon determination.

2.3.1. Vegetation Data Collection and Identification

The estimations of above and below ground carbon depend on the above ground biomass of living tree species. To estimate the above ground biomass all tree species within selected sample plots were identified, measured and recorded as specified above. Plant identification was done according to

Flora of Ethiopia and Eritrea [12, 13, 14].

2.3.2. Carbon Stock Measurement

Carbon stock has been assessed in five forest carbon pools, which is in accordance with the IPCC 2006 guideline [15]. Hence, the major activities of carbon measurement during the field data collection were focused on above-ground tree biomass, below-ground biomass, dead wood, litter, and soil organic carbon as stated here below.

i. Aboveground vegetation biomass (AGB) carbon

Carbon in the AGB was assessed through measurement of standing trees and shrubs using proper mensuration techniques. DBH and height of trees were measured according to their size class in the respective subplots. Therefore, species type, diameter at breast height (DBH) and height of trees (H) were the interest of measurement for trees. GPS was used to identify exact location of plots. DBH was measured with caliper/diameter tape depending on the size of the tree. Tree height was measured using haga hypsometer, and slope was measured with suunto clinometer to adjust the size of the plots to proper size. Carbon stock assessments in Africa are highly variable and have high degree of uncertainty due to lack of consistency in techniques of inventory and lack of site and species specific allometric equations. There are few species specific allometric equations developed in Africa, and most of the carbon stock assessments used general allometric equations despite the high degree of variability in site growth conditions and growth characteristics of species [22]. Equations indicated in [5, 6, 9, 22] are some of the most used general allometric equations in Africa for the purpose of biomass and carbon stock assessments. The one in [9] is particularly used by many studies and has been the best general model for carbon stock assessment so far [22]. Therefore, allometric equation [9] was used for this study is:

$$Y(kg) = exp(-2.187 + (0.916 * ln(\rho D2H)))$$ (1)

Where, Y = tree biomass, H = tree height (m), D = DBH (cm) and ρ = Wood density (kg m^{-3}). While DBH and tree height are directly measured, wood density of species is obtained from other studies and databases. Average wood density value of the known species was used for species which wood density was not found.

Fresh weight of all the undergrowth had been measured in the 2m*2m subplot and a small sample of known weight were taken for dry matter analysis. Regeneration was counted in this subplot. Tree biomass and respective carbon stock were calculated using allometric equations, and dry matter content of the undergrowth was determined after oven drying the fresh undergrowth sample and converting that proportionally to the 2m*2m subplot and hectare levels. Therefore, the AGB is the sum of the two vegetation biomasses. Then the AGB carbon is calculated from the AGB using a biomass-carbon conversion factor of 0.5 [29].

ii. Belowground biomass (BGB) carbon

Below ground biomass carbon is directly derived from aboveground vegetation carbon using known conversion factors. Below ground root biomass is estimated using root to shoot ratio which varies 20 to 50% depending on species. However, for carbon accounting purposes conservative values are recommended. Accordingly, 24% was used as a conversion factor for belowground biomass from above ground biomass as also recommended by other authors [8].

$$BGB = AGB \times 0.24$$ (2)

iii. Dead wood carbon

Dead wood carbon was estimated by applying general log volume estimation techniques using Smalian formula, and converting estimated volume to biomass and then to carbon.

$$V = f(Ds2 + Dl2) * L/2$$ (3)

Where: V = volume of the wood (m^3), Ds = small diameter (cm), Dl = large diameter (cm), L = length (m), f = adjustment factor = 0.00007854.

iv. Carbon Stocks in the Litter Biomass

The litter layer is defined as all dead organic surface material on top of the mineral soil. Some of this material will still be recognizable (for example, dead leaves, twigs, dead grasses and small branches). The following formula was used to determine litter carbon stock of the study area. The total dry weight was determined in the laboratory after oven drying of the sample. Oven-dried samples were taken in pre-weighed crucibles. The samples were ignited at 550°C for one hour in furnace. After cooling, the crucibles with ash were weighed and percentage of organic carbon was calculated.

$$LB = \frac{Wfield}{A} * \frac{Wsub\_sample\ (dry)}{Wsub\_sample\ (fresh)} * \frac{1}{10,000}$$ (4)

Where: LB = Litter biomass (t ha^{-1}); W field = weight of wet field sample of litter sampled within an area of size 1 m^2 (g); A = size of the area in which litter were collected (ha); Wsub_sample (dry) = weight of the oven-dry sub-sample of litter taken to the laboratory to determine moisture content (g), and Wsub_sample (fresh) = weight of the fresh sub_sample of litter taken to the laboratory to determine moisture content (g).

Carbon stock in litter biomass was then determined using the following formula:

$$CL = LB \times \% C$$ (5)

Where: CL = total carbon stocks in the litter in t ha^{-1}, % C = carbon fraction determined in the laboratory [36].

v. Soil organic matter

Soil organic matter contributes to more than 50% of the forest carbon stock in some forest types [40]. In some conditions the soil carbon stock is less dynamic and hence is less interesting to carbon stock assessment although it is the largest forest carbon pool. However, when there is high anthropogenic impact on the soil, particularly when there is a land use change, it is important to address the soil carbon content change related with land use changes.

In the current study, soil organic carbon (SOC) was assessed as there is dynamic process of land use change, forest land being converted to agriculture or grazing field, and hence it was found important to assess SOC content. Soil samples were taken at four corners of the 10m*10m subplot using 10 cm diameter core sampler to a depth of 30cm. The four subsamples were then mixed together and weighed for the soil bulk density determination. Then a composite sample of 100g was taken. Soil bulk density has been determined by drying soil samples in oven at 103°C for 24 hours. Then, SOC was determined following the loss-on-ignition [36] method and the calculations were made as follows:

$$BD = \frac{Wav.dry}{V} \qquad (6)$$

Where: BD = bulk density of the soil sample per the quadrant, Wav. dry = average dry weight of soil sample per the quadrant, V = volume of the soil sample in the core sampler auger in cm^3

$$SOC = BD * d * \% C \qquad (7)$$

Where: SOC = soil organic carbon stock per unit area (t ha^{-1}), BD = soil bulk density (g cm^{-3}), d = the total depth at which the sample was taken (30 cm), and %C = Carbon concentration (%)

vi. Estimation of Total Carbon Stock

The total carbon stock is calculated by summing the carbon stock densities of the individual carbon pools. Accordingly, carbon stock density of a study area is:

$$CT = AGC + BGC + DWC + LC + SOC \qquad (8)$$

Where: CT = Total Carbon stock for all pools (t ha^{-1}), AGC = above ground carbon stock (t ha^{-1}), BGC = below ground carbon stock (t ha^{-1}), LC = litter carbon stock (t ha^{-1}) and SOC = soil organic carbon (t ha^{-1}). The total carbon stock was then converted to tons of CO_2 equivalent by multiplying it by 3.67 as stated by [37].

3. Results and Discussion

3.1. Land Cover Change

As the satellite image analysis of this study revealed, the cultivated and overgrazed land has increased from 20% in 1972 to 48% in 2013 (Figures 1 and 2). As a result, the AMF and AAWL have shrunk by nearly 50%. The AMF and AAWL land have been shrinking, on average, by 118.4 and 200.8 ha per annum, respectively. A future projection using a simple linear model (equation 9 and 10) developed from the general land cover change that has been prevailed from satellite image analysis indicated 73ha and 251.3ha of annual deforestation in the AAWL and AMF zones, respectively. If this rate of deforestation continues, it will take only 71 and 49 years for the AMF and AAWL, respectively, to be completely lost if the management is not improved.

$$AMF\ (in\ ha) = 8268 - 72.9x \qquad (9)$$

$$AAWL\ (in\ ha) = 22790.4 - 251.3x \qquad (10)$$

Where; x is the number of years starting from 1972.

However, with increasing population and diminishing resources, rate of deforestation will increase and it may not take that long unless swift management approaches are implemented on the ground.

According to [44] the analysis of land-use/land-cover change and knowing its dynamics is a fundamental tool for adoption of conservation strategies within hotspot sites like protected areas. The phenomenon of forest loss in the nation seems to be quite common as outlined by some other researchers in different parts of the country [28, 18, 19]. The finding of the current study is also in consistent with the outcome of those researchers, but what is surprising about this finding is that the situation has happened in a protected area. Therefore, this is an alarming sign for improved and more effective conservation measures to be taken soon. Agricultural land has been expanding at the expense of the natural vegetation with a pressure both from within inhabitants and adjacent communities. The conversion of forested areas to cropland has been also a serious issue globally particularly from 1950s onward [32]. In the current study area 80% of the total park is directly affected by human activities such as settlement, cultivation, grazing and extraction of wood for fuel and construction [21]. Such activities will be catastrophic to the grass resource base and grass species diversity in the park. This will also significantly affect not only the grazing field, but the regeneration capacity of forest and woodland zones. The side effect of grazing on the regeneration capacity of forests has already been observed in this study. In the high forest areas far from villages, grazing is common experience and regeneration has already been affected. Since other interests like agricultural land expansion will also increase, the potential grazing land will shrink adding more grazing pressure on grasslands and forests, which will ultimately lead to forest degradation and then to land degradation. Fuel wood collection and selective logging of construction wood, together with grazing are primary factors for the forest degradation in the park.

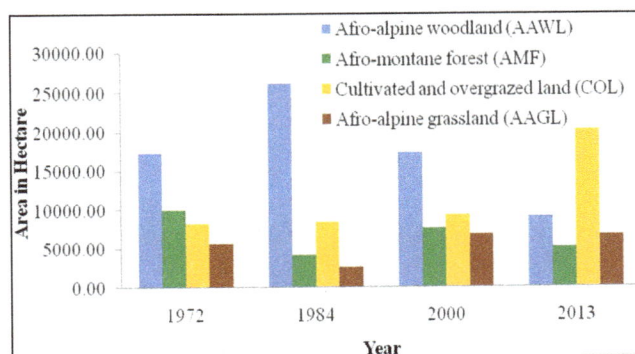

Figure 1. Land cover change trend of SMNP.

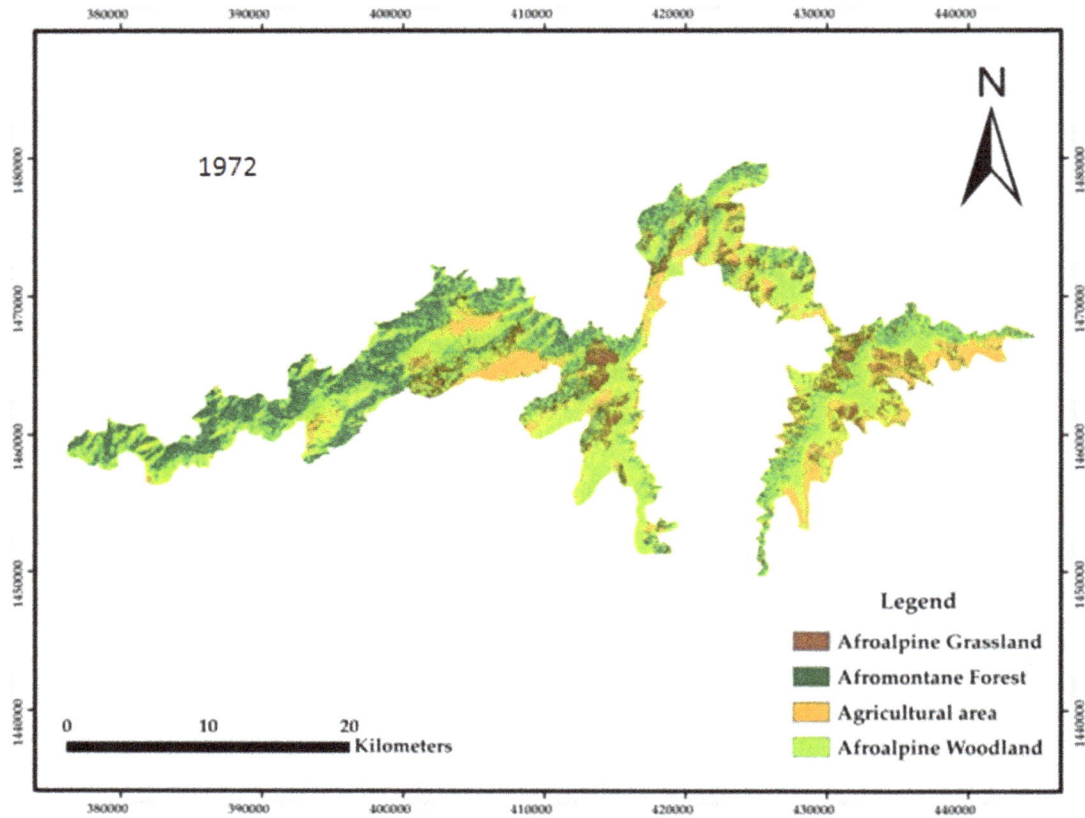

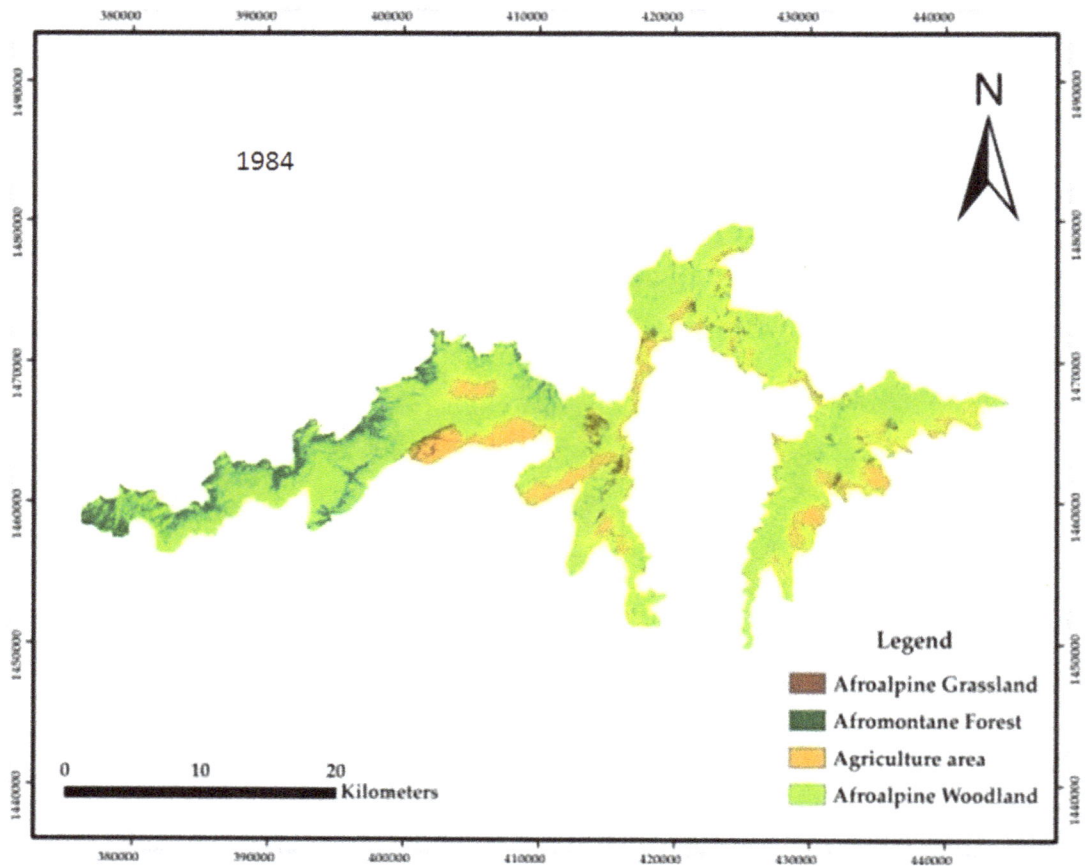

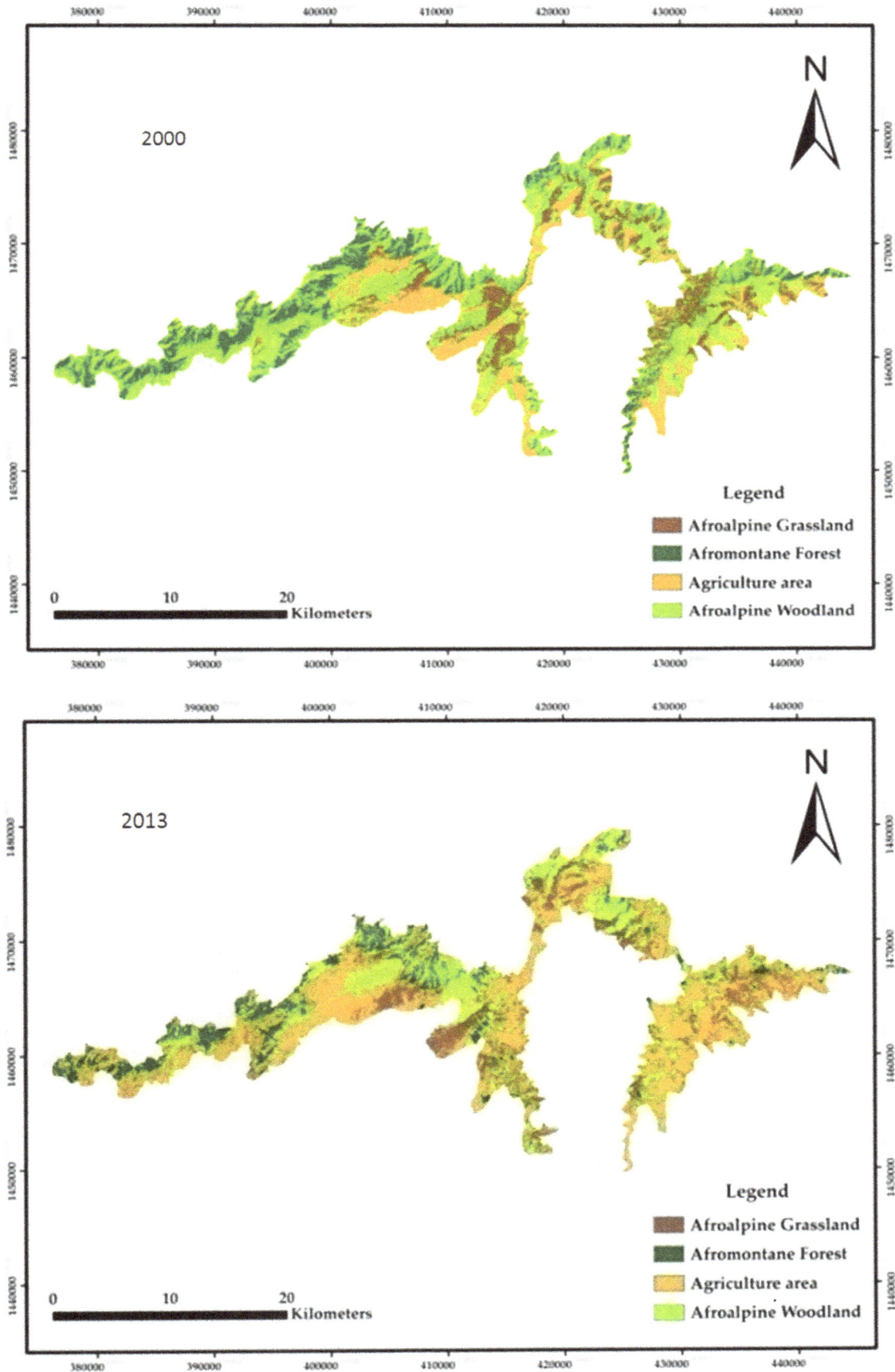

Figure 2. *Land cover change of SMNP from 1972 up to 2013.*

3.2. Carbon Stock of SMNP

One of the important points regarding carbon management is to identify the carbon pool that has high stock as well as one that is highly dynamic and sensitive. As seen in table 1 and figure 3 below, most of the carbon stock is concentrated in three carbon pools: Aboveground, belowground and soils. AGB, is a pool that holds about 34.4% of the forest carbon stock and hence it is the second largest pool next to the SOC which holds about 55.2%. The third largest pool is BGB, which holds 8.3% of the total carbon stock. AGB and BGB together constitute 42.7% of the total carbon stock.

Therefore, the tree component is the largest carbon pool next to soil which implies that removal of trees is apparently removal of the bulk of carbon stock. The other two carbon pools contained less amount of carbon where Deadwood Biomass (DWB) and Litter Biomass (LB) both together contributed only 2.2%. Conservation of forests for sustaining their existing carbon stock and future sequestration potential; assisting regeneration has to be the central focus of any carbon management project.

Table 1. Summary of mean carbon stock of aboveground, belowground, litter, deadwood and soil (ton/ha) of the study site.

| Total N | Different Carbon Pools | | | | | |
	AGC t/ha	BGC t/ha	DWC t/ha	LC t/ha	SOC t/ha	Total C t/ha
40	57.83 (14.13)	13.88 (3.39)	2.77 (1.11)	0.85 (0.34)	92.7 (8.61)	168.02 (21.79)

Numbers in bracket are standard error.

In this study on average soil carbon was found to be the main contributor of the overall carbon stored in the study area. This finding is consistent with the report of [20] that states soil is the largest pool of organic carbon in the terrestrial biosphere, and hence, minor changes in SOC storage can impact atmospheric carbon dioxide concentrations.

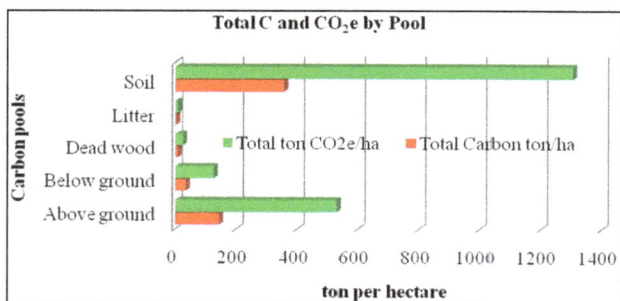

Figure 3. Total carbon and carbondioxide equivalent of the study site by pools.

When biomass accumulation and carbon storage was evaluated from land cover point of view, AMF was found to be the largest reservoir by storing nearly half (47.5%) of the total carbon stock in the area followed by the AAGL (22%) and AAWL (20.9%) as a second and third carbon reservoir respectively. The least carbon storage was observed in COL (9.6%) zone. Similarly, the large amount of carbon observed in AAGL zone was obtained from SOC may be as a result of high and rapid decomposition rate of the grass material and its incorporation to the soil as organic matter. From the table 2 and figure 4 it can be seen that land cover zones with trees and grass vegetations are found to be the most important depositors of biomass and carbon; hence it is evident that conversion of any form of natural vegetation to cultivated and overgrazing field will result in reduced AGC, BGC and SOC content which will affect the general holding capacity and sustainability of the area.

Table 2. Summary of mean carbon stock of aboveground, belowground, litter, deadwood and soil (ton/ha) of the four land covers of the study site.

| Zone (Land cover) | Total N | Different Carbon Pools | | | | | |
		AGC t/ha	BGC t/ha	DWC t/ha	LC t/ha	SOC t/ha	Total C t/ha
AAWL	10	20.98	5.04	0.52	0.00	87.58	114.12
AMF	10	123.35	29.60	6.40	2.13	97.90	259.38
AAGL	10	0.80	0.19	0.00	0.00	119.36	120.36
COL	10	0.00	0.00	0.00	0.00	52.21	52.21

The corresponding carbondioxide equivalent (CO_2e) was calculated for all land cover zones and all carbon pools (figure 3 and 4) by multiplying the amount of carbon by 3.66 value, and it followed the same trend for all land cover zones and pools like that of carbon storage since it was derived using a constant figure. Considering the AMF land cover only for comparison with others report the following discussions were made. The AGC of AMF of this study is comparable to those reported for the global above ground carbon stock in tropical dry and wet forests that ranges between 13.5 to 122.85 t ha^{-1} and 95 to 527.85 t ha^{-1},

respectively [34]. Yet this finding is similar to [43] who reported 122.85 t ha^{-1} AGC for selected church forests; and [30] obtained 133 t ha^{-1} AGC for Menagesha Suba state forest. However, it shows variation from the findings of [33] that obtained 306 t ha^{-1} of AGC for Tara Gedam forests and [1] found 237.75 t ha^{-1} of AGC for mount Zequalla Monastery forest. Also the AGB value of this study (246.7 t ha^{-1}) is a bit lower than the AGB of the Amazonian Brazil forests ranged between 290- 495 t ha^{-1} [3]. The BGB and BGC have similar trend with that of the aboveground values due to the fact that it is derived from the aboveground results using a constant conversion factor 0.24 [8]. The variation is perhaps due to methodological differences referring to the sample size and models or allometric equations used to calculate the biomass and also site variations. The result of carbon stock in litter layer for AMF is 2.13 t ha[-1] and this value is larger than the findings of [33] and [42] 0.9 and 0.017 t ha[-1] respectively; whereas it is smaller than the findings of [30] and [43] 5.26 and 4.95 t ha[-1] respectively. The result of carbon stock in litter layer for AMF is 2.13 t ha[-1] and this value is larger than the findings of [33] and [42] 0.9 and 0.017 t ha[-1] respectively; whereas it is smaller than the findings of [30] and [43] 5.26 and 4.95 t ha[-1] respectively.

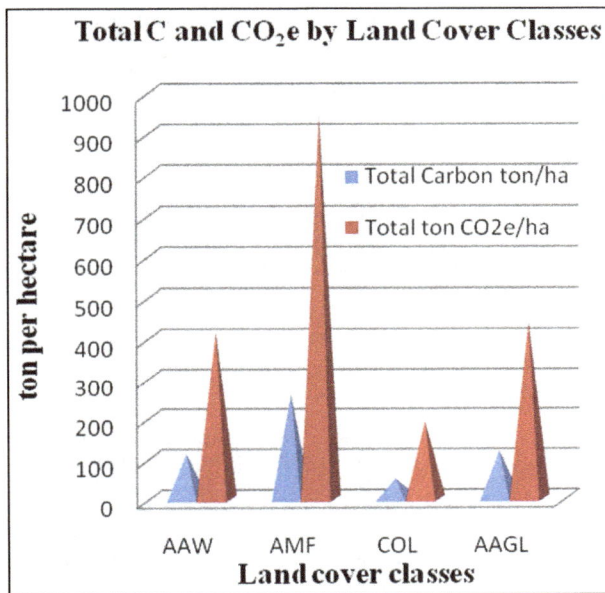

Figure 4. Total carbon and carbondioxide equivalent of different zones/ land cover

3.3. Carbon Stock and Tree Species Diversity

Biodiversity is one important issue in the management of forests for carbon dioxide sequestration and carbon stock purposes. It is generally required if there is direct relationship between diversity and carbon stock, so that the carbon stock management and biodiversity conservation can go hand in hand. However, there is no general conclusion reached at so far regarding biodiversity and carbon stock relation. As it is seen in figure 5 below, a linear regression between Shannon-wiener diversity index and total carbon and AGC was calculated for the observations within the AMF land cover and as such no clear and strong relationship was seen for the total C ($R^2 = 0.242$), but it appeared that the stored carbon tends to decrease as the Shannon diversity index increases. The relation seems to be stronger ($R^2 = 0.337$) for Shannon diversity index and AGC in AMF. The more diverse the forest implies less carbon stock. The AMF has the highest carbon stock per hectare. This could be due to the fact that in the AMF zone, there are big trees which occupy the upper canopy and discourage other trees not to grow. In addition, the larger tree sizes are the fewer in number so that reducing

the diversity index. In this context, it is important that some gap is created in the AMF zone either naturally or artificially so that biodiversity is promoted, which has direct side effect on the carbon stock. The diversity index was not calculated for the AAWL zone since in almost all the plots only one tree species, *Erica arborea*, was observed. In fact, the relation between tree species diversity and amount of carbon stored needs to be studied in detail and case by case using more robust data. So far some researchers studied the issue, but their findings vary and might be difficult to draw a general conclusion. For instance, [10] reported that forest carbon storage depends on species composition and on the way in which species are lost. [31] in their study of tree species diversity and AGB revealed that there is a complex and highly variable relationship between biomass and species diversity within Central African rainforests. Some plots with high diversity have relatively low biomass, and some plots with low diversity had high biomass.

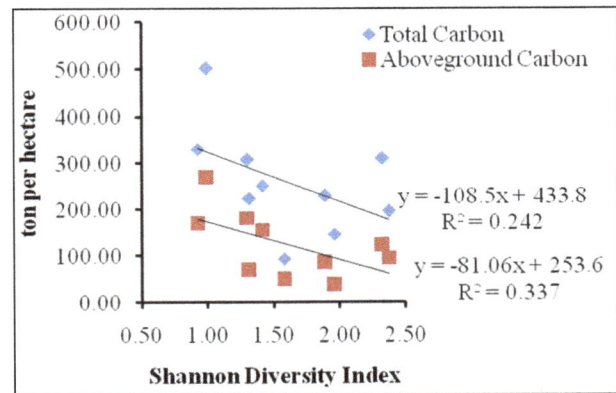

Figure 5. Relationship between carbon and tree species diversity for the AMF land cover.

4. Conclusion

Consensus over climate change phenomena is reached globally and a wide range of adaptation and mitigation measures have been taken. Enhancing carbon sequestration potential of forests and other similar land use or land cover zones is one of the feasible mitigation measures. Hence, carbon sequestration within protected areas is a valuable resource of GHG removals which would not be available if these protected areas were converted to other land use types. Even though, the management of the SMNP has strongly oriented towards the protection of wildlife and their habitat, there is high potential to address the climate change adaptation and mitigation issues as it still holds huge forest area that has ample carbon stock. In the study area land cover change dynamics is very much significant. It was found that there is an annual degradation of 73ha and 251.3ha in the AMF and AAWL zones respectively. AMF zone has high stock of carbon particularly in the AGB and the highest tree species diversity with different storey structure. The AAGL zone holds large carbon stock in the soil and grass species diversity. Therefore, conservation of these zones in particularly and protected areas in general is relevant in meeting double objectives of emission reduction from deforestation and also carbon sequestration as well as biodiversity conservation. Detailed investigation with robust data from other similar protected areas would reinforce these findings.

Acknowledgement

The authors would like to thank Population Health and Environment Ethiopia Consortium (PHEEC) for supporting the project financially through the Strategic Climate Initiative (SCIP) project.

References

[1] Abel G., Teshome S. and Tesfaye B. 2014. Forest Carbon Stocks in Woody Plants of Mount Zequalla Monastery and It's Variation along Altitudinal Gradient: Implication of Managing Forests for Climate Change Mitigation. *Science, Technology and Arts Research Journal.* 3(2): 132-140.

[2] Adugna F., Teshome S., and Mekuria A. 2013. Forest Carbon Stocks and Variations along Altitudinal Gradients in Egdu Forest: Implications of Managing Forests for Climate Change Mitigation. *Science, Technology and Arts Research Journal.* 2(4): 40-46.

[3] Alves L. F., Vieira S. A., Scaranello M. A., Camargo P. B., Santos F. A. M., Joly C. A. and Martinelli L. A. 2010. Forest structure and live aboveground biomass variation Along an elevational gradient of tropical Atlantic moist forest. *Forest Ecology Management* 260: 679-691.

[4] Broadmeadow M., and Robert M. 2003. Forests, Carbon and Climate Change: The UK contribution. Forestry Commission Bulletin 125. Forestry Commission, Edinburgh.

[5] Brown S. A. J., Gillespie J. R. and Lugo A. E. 1989. Biomass estimation methods for tropical forests with application to forest inventory data. *Forest Science* 35(4): 881-902.

[6] Brown S. 1997. Estimating Biomass and Biomass Change of Tropical Forests: A Primer. UN FAO Forestry Paper, Rome 134:55.

[7] Busby G. B. J., Busby J. S. E., Grant J., Hoolahan R. A. and Marsden C. D. 2006. The Lone Wolf Project Final Report: An expedition to the Simien Mountains, 29th June to 2005. Unpublished expedition report. 127 pages.

[8] Cairns M., S. Brown E. H. Helme and G. A. Baumgardner. 1997. Root biomass allocation in the world's upland forests. *Oeclogia* 111(1).

[9] Chave J., Andalo C., Brown S, Cairns M. A., Chambers J. Q., Eamus D, Folster H, Fromard F, Higuchi N., Kira T., Lescure J. P., Nelson B. W., Ogawa H., Puig H., Riera B., Yamakura T. 2005. Tree allometry and improved estimation of carbon stocks and balance in tropical forests. *Oecologia* 145: 87-99.

[10] Daniel E. B., Fabrice C., Jason C. B., Robert K. C., Ivette P., Oliver L. P., Mahesh S., and Shahid N. 2005. Species Loss and Aboveground Carbon Storage in a Tropical Forest. *SCIENCE VOL 310.* www.sciencemag.org

[11] Dudley N., S. Stolton, A. Belokurov, L. Krueger, N. Lopoukhine, K. MacKinnon, T. Sandwith and N. Sekhran [editors] 2010. *Natural Solutions: Protected areas helping people cope with climate change,* IUCNWCPA, TNC, UNDP, WCS, The World Bank and WWF, Gland, Switzerland, Washington DC and New York, USA.

[12] Edwards S., Mesfin T. and Hedberg I. [editors] 1995. Flora of Ethiopia and Eritrea, Vol. 2, Part 2: Canellaceae to Euphorbiaceae. The National Herbarium, Addis Ababa, Ethiopia and the Department of Systematic Botany, Uppsala, Sweden.

[13] Edwards S., Sebsebe D. and Hedberg I. [editors] 1997. Flora of Ethiopia and Eritrea, Vol. 6: Hydrocharitaceae to Arecaceae. The National Herbarium, Addis Ababa, Ethiopia and the Department of Systematic Botany, Uppsala, Sweden.

[14] Edwards S., Mesfin T. and Hedberg I. [editors] 2000. Flora of Ethiopia and Eritrea, Vol. 2, Part 1: Magnoliaceae to Flacourtiaceae. The National Herbarium, Addis Ababa, Ethiopia and the Department of Systematic Botany, Uppsala, Sweden.

[15] Estrada M. 2011. Standards and methods available for estimating project-level REDD+ carbon benefits: reference guide for project developers. Working Paper 52. CIFOR, Bogor, Indonesia.

[16] Falch F. and Keiner M. 2000. Simien Mountains National Park Management plan. Final Draft (unpublished) Amhara National Regional State, Bahir Dar.

[17] Friis L. and Vollesen K. 1984. Additions to the flora of Ethiopia. - Willdenowia 14: 355-371. ISSN 0511-9618.

[18] Gete Z. and Hurni H. 2001. Implications of land use and land cover dynamics for mountain resource degradation in the northwestern Ethiopian highlands. *Mountain Research and Development* 21(2): 184–191.

[19] Gessesse Dessie and Kleman, J. 2007. Pattern and Magnitude of Deforestation in the South Central Rift Valley Region of Ethiopia. *Mountain research and development* 27(2): 162-168.

[20] Girmay G., Singh B. R., Mitiku H., Borresen T., And Lal R. 2009. Carbon Stocks In Ethiopian Soils In Relation To Land Use And Soil Management. Land Degrad. Develop.19: 351–367.

[21] GMP. 2009. Simien Mountains National Park General Management Plan, 2009-2019.

[22] Henry M., Picard N., Trotta C., Manlay R. J., Valentini R., Bernoux M. and Saint-André L. 2011. Estimating tree biomass of sub-Saharan African forests: a review of available allometric equations. *Silva Fennica* 45(3B): 477–569.

[23] IPCC (Intergovernmental Panel on Climate Change). 2000. Land Use, Land-Use Change, and Forestry. Edited by Watson R. T, Nobel I. R, Bolin B, Ravindranath N. H, Verardo D. J, Dokken D. J, Cambridge University Press, Cambridge.

[24] IPCC. 2007a. Highlights from Climate Change 2007. The Physical Science Basis. Summary for Policy Makers. Contribution of Working Group I to the Fourth Assessment Report of the Intergovernmental Panel on Climate Change. Cambridge University Press, Institute of Terrestrial Ecology, Edinburgh. pp. 545-552.

[25] IPCC. 2007b. Facts on climatic change. A summary of the 2007 assessment report of IPCC. Cambridge University Press, Cambridge, UK.

[26] Jandl R., Rasmussen K., Tomé M., Johnson D. W. 2006. *The role of Forests in carbon Cycles, Sequestration and Storage.* Issue 4. Forest management and carbon sequestration. Federal Research and Training Centre for Forests, Natural Hazard and Landscape (BFW), Vienna, Austria.

[27] Julia Grunenfeld 2005. Livestock in the Simien Mountains, Ethiopia. Its role for the livelihoods and landuse of local small holders. Masthers Thesis, University of Berne.

[28] Kebrom T. and Hedlund L. 2000. Land cover changes between 1958 and 1986 in Kalu district, southern Wello, Ethiopia. *Mountain Research and Development* 20(1): 42–51.

[29] Liu X., Ekoungoulou R., Loumeto J. J., Ifo S. A., Bocko Y. E., Koula F. E., 2014. Evaluation of carbon stocks in above- and below-ground biomass in Central Africa: case study of Lesio-louna tropical rainforest of Congo. Biogeosciences Discuss., 11, 10703–10735.

[30] Mesfin S. 2011. Estimating and Mapping of Carbon Stocks based on Remote Sensing, GIS and Ground Survey in the Menagesha Suba State Forest. M. Sc. Thesis, Addis Ababa University, Addis Ababa.

[31] Michael D., Cristina B., Ervan R. and Terry C. H. S. 2013. Relationships between tree species diversity and above-ground biomass in Central African rainforests: implications for REDD. *Environmental Conservation* 41 (1): 64–72.

[32] Millennium Ecosystem Assessment. 2005. Ecosystems and human well-being: biodiversity synthesis. World Resources Institute, Washington, DC.

[33] Mohammed G. 2013. Forest carbon stocks in woody plants of Tara Gedam forest and its variations along environmental factors: Implication for climate change mitigation, in South Gondar, Ethiopia.

[34] Murphy P. G. and Lugo A. E. 1986. Structure and biomass production of a dry tropical forest in Puerto Rico. *Biotropica* 18: 89-96.

[35] Pan Y., Birdsey R. A., Fang J., Houghton R., Kauppi P. E., Kurz W. A., Philips O. L., Schivdenko A., Lewis S. L., Canadell J. G., Ciasis P., Jackson R. B., Pacala S. W., McGuire A. D., Piao S., Rautiainen A., Sitch S. and Hayes D. 2011 A large and persistent carbon sink in the world's forests. *Science* 333: 988–993.

[36] Pearson T., Walker S. and Brown S. 2005. Sourcebook for land-use, land-use change and forestry projects. Winrock International and the Bio-carbon fund of the World Bank. Arlington, USA, pp. 19-35.

[37] Pearson T. R., Brown S. L. and Birdsey R. A. 2007. Measurement guidelines for the sequestration of forest carbon: Northern research Station, Department of Agriculture, Washington, D. C, pp. 6-15.

[38] Per-Marten S., Felix H., Christoph L., Ann-Catrin F. and Hermann F. J. 2014. Higher subsoil carbon storage in species-rich than species-poor temperate forests. Environmental Research Letters 9 (2014).

[39] Petit J., Jouzel J., Raynauud D., Barkov N. I., Barnola J. M., Basile I., Bender M., Chappelaz J., Davis M., Delaygue G., Delmote M. 1999. Climate and atmospheric history of the past 420,000 years from the Vostok ice core in Antarctica. *Nature* 339: 429-436.

[40] Roshetko M., Rodel D. L. and Marian S. D. A. 2007. SMALLHOLDER AGROFORESTRY SYSTEMS FOR CARBON STORAGE. Mitigation and Adaptation Strategies for Global Change. Springer 12: 219–242.

[41] Sundquist E., Robert B., Stephen F., Robert G., Jennifer H., Yousif K., Larry T. and Mark W. 2008. Carbon Sequestration to Mitigate Climate Change. U. S. Geological Survey, New York, USA, pp. 1-4.

[42] Tibebu Y., Teshome S. and Eyale B. 2014. Forest Carbon Stocks in Lowland Area of Simien Mountains National Park: Implication for Climate Change Mitigation. *Science, Technology and Arts Research Journal.* 3(3): 29-36.

[43] Tulu T. 2011. Estimation of carbon stock in church forests: implications for managing Church forest for carbon emission reduction.

[44] Turner II B. L., Meyer W. B., Skole D. L. 1994. Global land-use/land-cover change: Towards an integrated program of study. Ambio 23 (1), 91–9.

12

Effect of Processing and Packaging Materials on the Storability and Microorganisms Associated with *Garcinia kola* (Bitter kola)

Ihejirika G. O.[1], **Nwufo M. I.**[1], **Ibeawuchi I. I.**[1], **Obilo O. P.**[1], **Ofor M. O.**[1], **Ogbedeh K. O.**[1],
Okoli N. A.[1], **Mbuka C. O.**[1], **Agu G. N.**[1], **Ojiako F. O.**[1], **Akalazu J. N.**[2], **Emenike H. I.**[3]

[1]Department of Crop Science Technology, Federal University of Technology, Owerri, Nigeria
[2]Department of Plant Science and Biotechnology, IMO State University, Owerri, Nigeria
[3]Cooperative Information Network, OBAFEMI AWOLOWO University, Ile-Ife, Osun State, Nigeria

Email address:
ihejirikagabriel@yahoo.co.uk (Ihejirika G. O.)

Abstract: The study on the effect of processing and packaging materials on the storability of *Garcinia kola*, Heckel, harvested from a local farm at Ngokpola was carried out in the green house of Federal University of Technology Owerri. It was laid out in a two factor factorial using Randomized Complete Block Design (RCBD) with 12 treatments and was replicated 5 times. The pods was processed by three different processing methods, which are , cut fresh pods immediately it was harvested, kept the pods outside on a shade and allowed to decay for one week and soaked the pods in water and allowed to ferment for 1 week. It was observed that the pods kept outside on the ground and those soaked in water was significantly different at 5% level of probability. The seeds extracted from the different processing methods was stored and packaged in polythene bag, dry plantain leaves and cocoyam leaves then control. It was found that polythene bag retained moisture than others. Alkaloid has high phyto-chemical content in the seeds that was processed by keeping the pods outside on a ground and allowed to decay and packaged in cocoyam leaves, which might be as a result of the processing method and packaging materials used. The seeds contain Saponin, Cyanide, Tannin and Ash which makes it to be an anti-oxidant and anti-nutrient. The pathogens isolated from the seeds are *Aspergillus sp*, *Penicillum sp* and *Diplodia sp*, affect stored seeds. The respondents from the 60 questionnaires administered to people in 3 different zones in Imo state, showed that 60% of the pods are harvested when fallen pods are picked, 30% harvest when someone climbed the tree and pluck it with hand, 10% when the fruits are plucked with sticks while the harvester is on the ground. 80% processed the fruits by keeping it outside on the ground and allowed to decay and 20% cut the fresh pods immediately it was harvested. Based on the findings, I recommend that farmers and marketers should engage in good processing method and packaging materials such as the ones used in this work for preservation.

Keywords: Processing, Packaging, Storability, Microorganisms, *Gracinia kola*

1. Introduction

The seed of *Garcinia kola* Heckel is commonly called bitter kola in Nigeria, Aki ilu in Ibo, Orogbo in Yoruba and Namijiu goro in Hausa. According to (14), it is an indigenous medicinal tree belonging to the family of Clusiaceae formerly Gultiferae. It is mostly found in Central and Western Africa. As a tropical fruit tree species, it is characterized by slow rate of growth. It grows as a medium sized tree up to 12m high and 1.5m wide. According to (3), it has been shown to posses a caloric value of 35.8kcal/g. The fruits are normally harvested from July - October. A mature fruit tree produces 85 to 1,717 fruits, with 208 to 6,112 annually, having mean values of 834 fruits and 2,627 nuts per tree. It produces 26 tones/ha/annum, with 278 trees/ha at 6m×6m spacing. The fruit is reddish-yellow and seeds contain 10% carbohydrates, 5% crude protein and >10% crude fats and sodium 215.10ppm (Seed Information Data Base, 2004). The fruit is about 6.25cm in diameter and each fruit contains two to four brown seeds embedded in an orange coloured pulp (18).

Factors that have discourage farmers from growing *Garcinia kola* include difficulties encountered in attempting to raise seedlings in nurseries and long gestation period before flowering and fruiting. However, many of the germination difficulties have been overcome by methods developed by (23) and (16). The extracts from seed and dry powered seeds have been made into various forms, such as tablets, cream and tooth paste. *Garcinia kola* was initially consumed as a stimulant before the remarkable bioactivities were explored. The stems and twigs of the plants are used as chewing sticks in many parts of Africa. It has been commercialized for years in major cities and has offered natural dental care to human (28).

Traditionally, bitter kola holds a high position among all the Nigerian tribes, particularly the Yoruba and Ibo communities. The Yoruba use bitter kola as an important component of the material used in traditional naming and marriage ceremonies while the Ibos use it in their traditional 'Fetish' recipes. Traditional herbalists use bitter kola in various pharmacopoeia preparations for various ailments. *Gacinia kola* biflavones used in traditional Africa system of medicine have been found to be active against a wide variety of micro-organisms. The seeds are used as stimulant and to cure cough (30).

(26) stated that *Gacinia kola* seeds are used as extractive in dietary food supplement while FDA, (13) reported that they are used as flavor enhancer in the beverage industry and also as hop substitute in several indigenous alcoholic drinks. Medicinally, the seeds are used as antidote for *Strophantus gratus* infection. The seeds are used for the treatment of bronchitis throat infections, antpurgative and antiparasitic (21). Other known uses include guinea worm remedy, anti-atherogenic effects and anti-lipoperoxative effects (1).

According to (15) *Gacinia kola* is known to exhibit a complex mixture of phonetic compounds including posses' anti-inflammatory, anti-microbial, anti-diabetic and anti-viral properties. The indigenous practices used by farmers to protect the species include selective clearing, during land preparation for cropping sustainable bark, harvesting of stands in wild populations and recognition of individual property ownership on certain wild of the tree.

Traditionally, *Gacinia kola* is harvested by post-harvest methods, which are a deliberate action used to separate food stuff from its medium, ripping cereals, picking fruits and all succeeding actions at the farm. Post harvest period begins at the separation of food stuff item) from the medium of immediate growth or production. Fruits become post harvest after it has been picked, but fruits that falls from the tree and allowed to decay or rot on the ground, is not a post harvest losses because, it was never harvested. Farm should be kept in a state of good sanitation by removing and destroying fallen fruits which are infested by fruit fly larvae.

In developing countries, processing is one of the main problems of agricultural produce. Unfortunately, much of agricultural resources are wasted because of short shelf life of the fruits after harvest. Since over 85% of fruits are harvested during fruiting season. (11), report that, there is a glut during the period of harvesting. The major consideration is how to ensure stable supply of agricultural produce all the year round.

Processing and sale of bitter kola is largely a family base home industry by which when the product is harvested at sustainable levels, has little negative impacts on the mother tree. Because of high demand of seeds in Nigeria, processing of bitter kola pulp should be effective so that, the seeds will be available in the market and for industry production.

According to some rural farmers, bitter kola are process faster by cutting it without ferment or soften which often results to physical damage of seeds and disease infestation. Soaked pulps leads to the improvement in seed quality but often result to micro-organisms that has well defined or potential commercial value.

When fruit ripe, the green pericarp turns a reddish yellow color and the fruit falls from the tree. The fallen fruits are collected and kept in an open, cool place till the pericarp and the pulpy mesocarp become soft. Once softened, the fruits are threshed to release the seeds which are thoroughly washed to remove the sticky mucilaginous materials that sheath the seed. According to (20), processing of bitter kola for international market is always required in two forms, which are fresh or dried. But most importers always want them dried. The drying must be done in a way that the colour is not affected.

(17) observed that packaging materials (leaves) which do not affect the colour, crispiness and marketable quality are *Dorax sp, Alchornea laxiflora* (Esin), are used to preserve bitter kola by placing the leaves inside a nylon which is wrapped within the bitter kola in a container. *Costus lucanusianus* is also used to preserve and package bitter kola and *Spondia Mombin* (Iyeye). *Garcinia kola* seeds do not exhibit orthodox storage behavior and should be treated as carefully as recalcitrant seeds. Because of the moisture content of the fleshly fruits harvested, the seeds should be stored properly and with good storage materials in a short time. If the seeds are to be stored in a moist condition, it is vital that the store areas are ventilated frequently. Most of the raw materials from tree crops get rotten during storage because they were not properly processed. This leads to wastage of the resources and time spent in collection.

Traditionally, bitter kola is stored in layers of red clay soil (22). But in ancient days, harvested bitter kola was stored in a pit dug inside the soil. The seeds are packed into the pit and later covered with soil till when the farmer needs them either for utilization or commercial purpose. For crops that do not produce seeds all the year round, it is important that the products are stored to make them available whenever needed. Many products are wasted during peak harvest and glut due to shelf-life of the seeds. Therefore, effective and adequate storage will ensure that the seeds consumed are of good quality. It reduces the usual wastage and loss of crops that are already produced due to poor processing and harvesting. It also ensures availability of raw materials to agro-based industries.

Pests and diseases attack *Garcinia kola* seed and pod by reducing its quality, market value and nutrient. The pests affect the seeds through piercing and sucking and method. They affect the pod when it is harvested late or when there are wounds or bruises on the pod. They cause microbial deterioration of the fruits when the fruits are not properly harvested: such as when the fallen fruits are left till when it rots or fly larvae affects it; then it will cause soil borne pathogenic fungi.

Some pests that affect bitter kola are:

(i)   Weevils: This is the major pest of kola which affects the fruit and the seed either during harvesting or storage. Examples of weevils include: *Sphororhinus divareti, S. quadricristalus, S. simiarum, Balanogastric kolae,* etc. The weevil penetrates through wound or damaged fruits. They lay eggs in the seeds. The weevil attack cause serious losses up to 50-70% (31). The weevils are controlled by removing the initial attack and thorough inspection of the seeds then removal of all infested seeds before storage.

(ii)   *Caratitis kolae:* This pest attacks the kola in the mature stage. It causes burrowing which create hole easy for penetration of weevils.

(iii)   Aphids (*Pseuducoccus citra*)

(iv)   Scale insects (*Planococcoides njalesis*)

Diseases of *Garcinia kola* are:

a)   Fungal root (*Fusarium* sp): These disease attacks the plants suddenly on the leaves. It leads to plant death. The roots of the plants are covered with brown rhizomorphs and fruiting bodies may be formed at the base of the truck. These diseases can be prevented and controlled by removing the logs, stumps and roots of the infected ones.

b)   Leaf spot disease: They attack only immature leaves, mainly on the later part of the rainy season. The symptoms are brown angular spots, especially on the tips of the leaf, bushy appearance of the plant. The fungus responsible is the *Pestolatia sp* and *Glomerella sp.*

c)   *Penicillin spp, Diplodia macnopyrens, Fusarium moniliform var Subblutinans, Fusarium solani,* etc are fungi disease resulting from fruit and nut.

Fusarium spp and *Penicillin spp* are common infection of nuts which can be prevented when the nuts are allowed to attain their restive stage and transpiration prior to storage.

The major chemical elements found in bitter kola are potassium and phosphorus. The seed has significant higher values for sodium, potassium, copper and cobalt. Due to the activity of flavonoids and other bioactive chemical compounds, they are used in folk medicine, therapeutic benefits (16), (27) and (29).

Garcinia kola is a non-timber forest that is mostly utilized in Africa (2). Virtually all the parts can be used for medicinal purposes. The fruit is used in the treatment of jaundice; the extract can prevent Ebola virus from replicating itself (8). The extract from the bark, stem and seed inhibit the growth of *Plasmodium falciparum* according to (33).

The seed is used in the treatment of bronchitis and throat infection, catarrh, abdominal, colicky pain, improving singing voice etc. When mixed with honey, it reduces and makes a cough syrup and they are highly medicinal (5). The split, stems and twigs are used as chewing stick in many parts of Africa, which offers dental care. The seed is believed to expel snake where they are kept. It posses aphrodisiac and purgative properties and has shown great potential as substitute for hop in tropical beer brewing (7).

Bitter kola holds a high position of cultural importance among all the Nigerian tribes and it has shown anti-inflammatory, anti-microbial and antiviral properties and posses anti-diabetic and anti-hepatotoxic activities and it is an effective plant that derived medicine alternative to synthetic drugs (6).

The objectives of this study are:

To determine different processing method of *Garcinia kola* in Imo State using questionnaire, to isolate and identify pathogen that affect *Garcinia kola* during storage, to identify the phyto-chemicals present in *Garcinia kola* using quantitative analysis and to determine the effect of processing on the storability of *Garcinia kola.*

2. Materials and Methods

The experiments were conducted to determine the best method *Garcinia kola* can be processed and packaged so as to extend its shelf life. The study was carried out in the laboratory of Federal University of Technology, Owerri, Nigeria under room temperature and relative humidity in the green house for a period of one week for processing and five weeks for storage and packaging. Fresh samples of *Garcinia kola* pulps harvested from a local farm at Ngokpola in Imo state were processed in three different ways:

(i)   The pulps were cut immediately as it was fresh. The seeds were extracted from the pod and stored for five weeks in different baskets, using 3 packaging materials and control was also set up.

(ii)   The pods were kept outside under a shaded and cool area. It was allowed to decay and soften for a week, before extracting the seeds and stored for five weeks, using different packaging methods and baskets with control as well.

(iii)   The pods were soaked in water for a week for it to ferment, then the seeds were extracted from the pulps and stored for five weeks using different packaging such as dry plantain leaves, dry cocoyam leaves, polythene bag and control (on package) was also set up.

The following was done during storage:

a)   The initial weight of the pods was taken before processing.

b)   The weight loss of the seeds was taken as they were processed and stored.

c)   The pathogens associated with the seeds were identified.

d)   The pyhto-chemical content of the seeds and pods

were also determined.

e)   The percentage ash content was determined.

All equipments used in this project work were carefully sterilized. The Petri dishes were washed with distilled water and packed into a round metal box and put in the autoclave set at 121°C for 15 minutes. The inoculating needle used was sterilized by dipping it in a lacto phenol. The Petri dishes were taken to the fume chamber (autoclave), and allowed to stay for 15 minutes. They were allowed to cool and the medium was poured into them.

2.1. Preparation of the Medium

The medium used for this experiment was Potato Dextrose Agar (PDA). 125g of Irish potatoes was weighted against 500ml of water. It was washed with distilled water, then peeled and chopped into tiny pieces and then boiled for 30 minutes in a 1000ml beaker. The Irish potatoes were filtered using muslin cloth into a conical flask. 10g of glucose and 10g of agar were added into the supernatant and shake thoroughly for it to completely dissolve to a homogenous mixture. The supernatant in the conical flask was corked with cotton wool and wrapped with aluminum foil to avoid evaporation. The medium was heated for 20 minutes and allowed to cool for 40 minutes before pouring it into the sterilized Petri dishes.

2.2. Isolation and Inoculation of Pathogen

A small portion of sample from each treatment was cut and put into the Petri dishes (different) containing the medium, with an inoculating needle. The Petri dishes were carefully labeled and put into transparent polythene bags to avoid contamination, and put in an incubator (all these processes were carried out in an inoculating chamber) and allowed to stay for two days. Sub-culturing was done till pure culture was obtained. Different micro-organisms were isolated. The various isolates were assigned to their various genera using the illustrated genera of imperfect fungi manual (9). This was done by comparing characteristics of the isolates with those identified species.

2.3. Examination of Cultured Medium

A drop of lactic acid was placed at the center of a clean grease free microscope slide. A bit of the fungal organism was dropped on the slide with the aid of with the inoculating needle on a slide The slide was viewed under a super Tek microscope with 60 magnifications, under low light intensity. The associated micro-organisms were observed, drawn and identified. The entire microscopic colony were noted and differentiated.

2.4. Determination of Phyto-Chemicals

Four Phyto-chemicals were determined using quantitative analysis from different processed sample and pulp. They include:

a) Tannin: 1.0g of sample from each processed treatment and pulp was weighed and put into a conical flask that contains 50ml of water which was used to dissolve the sample. It was filtered through no 44 Whatman filer paper into a 100ml volumetric flask. It was pipette with 50ml of distilled water and 10ml of diluted extract into a conical flask, followed by 5.0ml follin-Denins reagent and 10ml of saturated Na_2CO_3 solution, and then it was diluted with distilled water to a volume.

The tannic acid was calculated, using the formula

$$\text{Tannic acid (mg/100g)} = \frac{C \times \text{extract volume} \times 100}{\text{Aliquot} \times \text{weight of sample}}$$

Where C = concentration of tannic acid

b) Alkaloids: 2.0g of sample from each processed treatment and pulp was weighed into a conical flask that contains 10% acetic acid. The weight of the filter paper was recorded before it was use to filer the supernatant, after ammonium has been add. The weight of the filer paper was recorded, after oven drying.

Percentage alkaloids were determined using the method of (24).

c) Cyanide: 1.0g of sample from each processed and pulp was weighed and grind. The sample was put into conical flask of 100ml distilled water to hydrolyze it for 1 hour. The sample was transferred into 250ml round bottom flask. 10ml of 2.5% NaOH was added and it was put in Soxhlet flask. The round bottom flask with condenser on top was mounting. 2ml of 6 molar ammonium hydroxide, 1ml of 5% potassium iodine and 25ml of the sample was titrate with 0.02 molar silver nitrate.

$$\text{Percentage cyanide} = \frac{T \times \text{Equivalent weight} \times 100 \times N}{\text{Aliquot weight of sample}}$$

Where T = Titre value and N = Normality

d) Saponin: 1.0g of each processed treatment and pulp was weighed and put into a container in an oven. It was brought out from the oven and crushed into a fine particle. The initial weight of filer paper was taken and recorded. The sample was put into a Soxlet to extract the fat, which contain petroleum spirit. It was filtered with a fresh 100ml of 20%ulra volume aqueous solution of ethanol and washed the organic layer twice and poured into a pre-weighed beaker and evaporates to dryness on a boiling water bath. It was allowed to cool and the weight was taken.

$$\text{Percentage saponin} = \frac{\text{Final weight}}{\text{Initial weight}} \times \frac{100}{1}$$

2.5. Determination of Ash Content

1.0g of sample from each processed treatment and control was weighed into a porcelain crucible of known weight. The crucible was place in a muffle furnace at 200^0C and the temperature was gradually increased to 600^0C. It incinerates for 5 hours and light grey ash was observed. It was allowed to cool in desiccators under room temperature and weigh.

Percentage ash content was calculated using the formula

$$\text{Percentage ash} = \frac{\text{Weight of ash}}{\text{Weight of sample}} \times \frac{100}{1}$$

2.6. Determination of Moisture Content

1.0g of sample from each processed treatment and control was weighed into a container of a known weight. It was put in an oven with a constant weight of 105^0C for 4 hours. After drying, the sample was cooled in desiccators and re-weighed.

Percentage moisture content was calculated using % moisture =
$$\frac{Final\ weight}{Weight\ of\ sample\ before\ oven\ drying} \times \frac{100}{1}$$

Questionnaires were administered to 20 people each in the 3 geo-political zones in Imo state, namely Okigwe, Orlu and Owerri. Within each zone, 20 people were split into 10 for packaging and processing in 3 local Government Areas. After this study, inferences were made based on package materials used; processing methods, storage, harvesting, gender, age etc and results obtained were analyzed through the use of percentage frequency. Some seeds were stored under the three different storage materials and control for five weeks. They were replicated, using weekly readings giving a total of 12 treatments in a 2 factor factorial experiment in randomized complete block design in the laboratory. The data was subjected to analysis of variance (ANOVA) at 5% probability level.

3. Results

From Table 1, results showed that harvesting method was statistically significant at 5% probability level. 60% of *Garcinia kola* was harvested by allowing it to fall on the ground and then picking the pulps, 30% by climbing the tree and harvesting, and 10% by plucking the fruits.

Table 1. *Distribution of Respondents According to Harvesting and Processing Methods, Package And Storage Materials*

Treatments	Frequency (%)
Harvesting Methods	
Fallen pulps that are picked	60
By climbing the tree	30
Processing Method	
Kept outside and allowed to decay.	80
Cut the pulp fresh	20
Packaging Materials	
Dry plantain leaves	25
Polythene bag	45
Cocoyam leaves	30
LSD$_{0.05}$	4.725

Investigation revealed that 80% of the pulps are processed by keeping them outside on the ground and allowed to decay before extracting the seeds. They are washed and air dried in a basket, 20% for cases where the pulps are cut fresh, the mesocarp around the seeds scraped off with knife or hand and washed before drying. The result showed that 45% of the respondents use polythene bag as their packaging material, because it retains moisture and crispiness of bitter kola. 30% of the respondents use cocoyam leaves and 25% use dry plantain leaves Table 1.

Result also shows that fungal pathogens isolated from *Garcinia kola* seeds from the harvesting methods and packaging materials used were *Aspergillus sp, Penicillum sp Diplodia sp and Fusarium sp* respectively and they attacked the seeds of stored products. More microorganisms were associated with the control treatments followed by those packaged with polythene bags and *Aspergillus* and *Penicillum* species were of high activity than all the other microorganisms. Also, fallen pulps that are picked and processed had higher organisms in comparison with those obtained from climbing the tree or by plucking. Table 2.

Table 2. *Microorganisms Identified with Gasinia kola with Different Packaging Materials and Harvesting Methods*

Different Package	Microorganisms *Aspergillus sp*	Identified *Penicillum sp*	*Diplodia sp*	*Fusarium sp*
Packaging materials				
Polythene bag	√√	√ √		
Dry plantain leaves	√	√	√	
Cocoyam leaves	√			√
Control				
Harvesting methods	√√	√√	√√	√√
Method 1	√	√	√	√
Method 2	√	√		
Method 3	√	√		

KEY: Method 1 = Fallen pulps that are picked; Method 2 = By climbing the tree; Method 3 = By plucking the fruit; √ = Organism present; √√ = Organism highly present

Table 3. *Effect of Storage Materials on Phyto-Chemical Constituents of Fresh Pods Cut Immediately At Harvest And Seeds Processed*

Different Package	% Ash	% Tannin	%Alkaloids	% Cyanide	% Saponin
Polythene bag	3.375	23.01	70.08	37.53	50.78
Dry plantain leaves	2.614	23.08	70.26	37.74	50.54
Cocoyam leaves	2.961	23.10	70.32	37.61	50.72
Control	1.627	23.02	70.15	37.63	50.6

LSD$_{0.05}$ = 0.0125

Table 4. *Effect of Storage Materials on Phyto-Chemical Constituents of Pods Kept on Ground Under Shade and Allowed to Decay*

Different package	% Ash	%Alkaloid	%Tannin	%Cyanide	%Saponin
Polythene bag	3.264	70.36	23.13	37.33	50.53
Dry plantain leaves	2.844	70.01	23.32	37.64	50.85
Cocoyam leaves	2.585	70.58	23.09	37.55	50.70
Control	1.601	70.23	23.14	37.70	50.73

LSD$_{0.05}$ = 0.150

Table 5. *Effect of Storage Materials on Phyto-Chemical Constituents of Pods Soaked In Water and Allowed to Ferment*

Different package	% Ash	% Alkaloid	% Tannin	% Cyanide	%Saponin
Polythene bag	3.328	70.23	23.18	37.55	50.29
Dry plantain leaves	2.653	70.44	23.22	37.34	50.61
Cocoyam leaves	2.884	70.26	23.05	37.04	50.72
Control	1.732	70.31	23.30	37.80	50.66

$LSD_{0.05} = 0.172$

Results of the phyto-chemical contents on processing methods and packaging materials revealed that the seed from pods processed by keeping outside under a shade and package with cocoyam leaves had higher Alkaloid contents of 70.58, while high ash content was obtained from pods that were processed by cutting immediately it was harvest and package with polythene bag 3.375. Table 3 – 5. Pods that was soaked in water and allowed to ferment recorded statistically significant difference on phyto-chemical constituents at 5% probability level. The ask content was lowest in controlled experiment followed by that packaged with dry plantain leaves when those with polythene bags were high. However, saponin content of control was highest, when that packaged with polythene bags were low Table 5.

The result of processing methods and packaging materials on moisture content at 5 weeks of storage shows that the pods processed by keeping outside on the ground under and allowed to decay as well as the pulp soaked in water and allowed to ferment before extracting the seeds, were statistically significant on moisture content at 5% level of probability. Table 6. The seeds packaged in polythene bag recorded highest moisture content irrespective of the processing methods, than that packaged with either cocoyam leaves or dry plantain leaves, when control was lowest. Result showed that there was a decrease in moisture value of the seeds with weeks of storage, while weeks of storage were statistically significant at 0.05 level of probability. Table 6.

Table 6. *Means Value of Main Effect of Processing and Packaging on Moisture Content at 5 Weeks of Storage*

Different processing	Polythene bag	Dry plantain leaves	Cocoyam leaves	control	Total package	Mean
Process 1	213.4	212.3	211.6	196.0	833.3	208.3
Process 2	226.8	222.7	222.5	210.9	882.9	220.7
Process 3	224.1	213.6	213.6	212.9	869.2	217.3
Total	664.3	648.6	625.7	619.8		
Mean	221.4	216.2	217.6	206.6		

$LSD_{0.05} = 0.7457$

4. Discussion

The results obtained from the 3 different processing methods used in this project work shows that, the best ways

for processing *Gasinia kola* is to keep the pods outside on the ground for 1 week and allow it to decay before extracting the seeds and also soaked pods in water for 1 week and allow to ferment before extracting the seeds. These two processing methods are preferable because pest and diseases cannot penetrate into the seeds easily and they retain moisture.as proposed by (1).They are highly significant at 5% level of probability test. When the seeds are extracted from the pulps, it is easy to wash off the mesocarp sheaths (6). The two processing does not cause any wounds or bruises on the seeds, unlike when the pods are cut fresh, it will not be as smooth as those extract from decayed pod. The pods hat were cut fresh bring molds during storage on the seeds because the sheaths around them are not properly removed. Mean separation for blocks (weeks) using LSD at 5% level is significantly different at probability test for 5 weeks of storage.

The results of the moisture content as assessed for 5 weeks of storage shows that, the 4 different packaging materials used to store seeds extracted from the different processing used shows that those store in polythene bag, cocoyam and dry plantain leaves was highly preferable and were significantly different at 5% level of probability, the moisture content decrease within the weeks of storage. This is true because as the week progresses, the breakdown in physiological activities of the *Gasinia kola* increases and cells of cells break down and produce more accumulated fluids and water. Also microbial activities/ respiration increase and more water accumulate. (25); (10).

They act as a pigment against predator in the seed and pulp. Alkaloid, Tannin and Saponin act as an anti-nutrient which makes bitter kola to be good in treating stomach disorder and to stop vomiting. (10), (21). Cyanide is an anti-oxidant. It is poisonous in the body. It destroys cells when much of it, is in the system. It causes abortion, damage and disorder in the organ. The result does not mean that the phyto-chemical content (Alkaloid) will be high when packaged with cocoyam, because the chemical acts as a pigment against predator in the body. They impact colour to plant which enhance pollination. They act as an anti-nutrient which makes nutrients unavailable to disease infestation. These chemicals fight against any disorder in the stomach, stop vomit and used in food formulation (15) and (28). When bitter kola is eaten much, it is not good in the system due to cyanide content that act as an anti-oxidant. It is poisonous in the body and affects the male organ. Chemicals add value to the pulp, which hitherto is discarded, as a potential source of nutritionally valuable and industrial raw material (6).

The loss of viability of kola nut seeds due to reduction in moisture content is caused by poor storage materials, improper packaging and poor processing. Seeds package with cocoyam and dry plantain leaves stored well but there is a decrease in moisture within the weeks of storage.

Pathogen invades *Garcinia kola* seeds if the processing methods and packaging materials used affect the seeds. The results showed the pathogen that affect stored *G. kola* seed and they are *Penicillum sp.*, *Aspergillus sp.*and *Diplodia sp*, irrespective of the processing method used. These pathogens leads to quick deterioration and decrease in nutritive value of the seeds. Some pathogens invade the seeds mainly through harvesting in line with (14), (11) as well as (3).

5. Conclusion

The investigation of different processing of *Garcinia kola* revealed that there should be awareness and encouragement of the dealer or farmersin gender participation and to embark on simple and easily handled processing through relatively simple and available method, to reduce injuries on the seeds, add value and improve quality of the processed seeds. Farmer/marketers of bitter kola should adopt use of good packaging materials, storage and other management, which will improve the shelf life of *Garcinia kola* seeds and ensure moisture retention and long preservation of seeds. Phytochemical adds nutritional and medicinal value to the seed and pulp and make them a potential source of nutritionally valuable nutrients and industrial raw materials.

Pods processed when they are kept outside on the ground and those soaked in water gave the best quality produce, while seeds packaged with dry plantain and cocoyam leaves produce well and retained moisture seeds of high quality, better shelf life and low associated microorganisms.

References

[1] D.A. Adaramoye, E.O. Farombi, E.O. Adeyemi and G.O. Emerole. Comparative study of antioxidant properties of flavonoids of *Garcinia kola* seed. *Professional Medical publication.* 21(3).2005. pp. 8 – 15.

[2] A.A. Adebeisi. A case study of Garcinia kola nut production-to-Consumption system in J4 area of Omo forest reserve, South west Nigeria. *Forest production, Livelihood and Conservation* 2: 2004. Pp.115-132.

[3] G.O. Adegoke, M.V. Kumar, K. Sambadah and B.P. Lokesh. Inhibitory effect of *G. Kola* on lipid peroxidaticn in rate liver homogenate. *Indian J. Exp Biol.* 36(9): 1998. 907-910.

[4] G.O. Adegoke, F. Olojede, G. Engelhardt P.R. Wallnoefer. Inhibition of growth and aflatoxin production in *Aspergillus parasiticus* NRRL 2999 by *Garcinia kola. Advances in food Sciences* 18, 1996. 84 - 88.

[5] A.A. Adesanya, K.A. Oluyemo, D. A., Olusord, O.V. Ukwenya. Micromorphometric and stereological effects of ethanolic extracts of *Garcinia cambogva* seeds on the testes and epididymides of Adult Waster Rats. *Intl. J. Alt. Med.* 5(1):2007. 1-9.

[6] F.E. Afolabi, C.B.; David. Chemical Composition of bitter kola seed and hulls.Department of Agricultural, Food and Nutritional Sciences, *University of Alberta.* 4: 2006. Pp. 395-400. Canada.

[7] P.E. Ajebesone and J.O. Dina. Potential African substitutes for hops in tropical beer brewing. *The Journal of Food Technology in African.* 9(1):2004. Pp. 13-16.

[8] T. Agyili, M. Sacande and C. Kouame. *Garcinia kola* hecked. Millendum seed bank project KEW. *Forest and landscape*

Denmark. *Horsholm Kongercy 11.* DK-2920 Horsholm. Seed leaflet No. 113. 2006.

[9] H.L. Barnette and B. B. Hunter. Illustrated Genera of Imperfect fungi. *The American Phytopathological Society.* 3rd Edu. St. Paul Minnesota USA, 1998. Pp. 218.

[10] M.I. Dosunmu and E.C. Johnson. Chemical evalustion of the nutritive value and changes in ascorbic acid content during storage of the fruit of bitter kola *(Garcinia kola). Food Chem..,* 1995, 54, 67-71.

[11] L.C. Emebiri and M.I. Nwufo. Effects of fruit type and storage treatments on the biodeterioration of African Pear *(Dacryodes edulis* (G. Don.) H. J. Lam.). *Int. Biodet,* 26, 1990. Pp. 43-50.

[12] E.O. Farombi. African indigenous plants with chemotherapeutic potentials and biotechnological approach to the production of bioactive prophylactic agents. *African J. Biotech.,* 2, 2003. Pp. 662-671.

[13] FDA. FDA/CFSAN/OPA: Agency Response letter: GRAS Notice No. GRN000025 on the use of *Garcinia kola* seed in distillation. Ibanga, I.A.(1993). Use of *Garcinia kola* as bittering agent in brewery industry. *B. Sc. Project, University of Calabar, Nigeria.* 1999.

[14] M.M. Iwu. Handbook of African Medicinal Plants, *Boca Raton CRC Press* 1993. p. 437.

[15] M.M. Iwu, A.R. Duncan and C.O. Okunji. New antimicrobials of plant origin. 1999. P 457-462. In: J. Janick (ed.). Perspectives on new crops and new uses. *ASHS press.*

[16] A. Gyimah (2000). Effect of Pre-treatment Methods on Germination of *Garcinia kola* Heckel seeds. *Ghana. Journal of Forestry:* 9, 2000. Pp.39-44.

[17] J. Kayode and B.M. Ojoo. Assessing botanicals used in the storage of farm produce in Akure Region. *Ethno botanical leaflets.* 13: 2009. Pp.603-10.

[18] R.W. Keay(1989). Trees of Nigeria. Clarendon Press, Oxford. 1989. Pp.476.

[19] D.O. Ladipo. Physiological/Morphological growth rate and fruit/nut yields in *G.kola* tree on acid soil of Onne, *Port-Harcourt. ICRAF* 1995. In-House Report.

[20] O. Leo. Earn decent Income exporting bitter kola *(G.kola). Smartmoney:* 2 (1): 2009. 20 – 26.

[21] I.I. Madubunyi. Antimicrobial activities of the constituents of *Garcinia kola* Seeds. International Journal of Pharmacognosy 33, 1995. Pp.232-237. Contact: Dep. Vet. Physiol. *Pharmacol., Fac. Vet. Med; Univ. Nigeria, Nsukka.*

[22] I. Miranda, I. Dosunmu amd C.J. Ekarika. Chemical evaluation of the nutritive value and changes in ascorbic acid content during storage of the fruit of bitter kola *(Garicinia kola). Food Chemistry.* Vol. 54. 1995. Pp. 67-71.

[23] J.C. Okafor. Mass propagation of species for immediate Utilization paper. Presented at the meeting of under utilized crops of Nigeria. 4-8 May 1998. *NACORAB moor plantation. Ibadan, Nigeria.*

[24] I.O. Okerulu and C.J. Ani. The phytochemical analysis and antimicrobial screening of extracts of *Tetracarpidium conophorum. Journal of Chemical Society of Nigeria.* 26 (1):2001. Pp. 53 – 55.

[25] M.O. Ofor, C.A. Ngoboli and M.I. Nwufo. Ethno botanical uses and trade characteristics of *Garcinia kola* in Imo State, Nigeria. *Int. J. Agric. Rural Dev.* 5: 2004. Pp.140-44.

[26] C.E. Okoro. Development of hop substitutes from tropical plants. *J. Agric. Technol.* 1: 1993. Pp.30-35.

[27] C.O. Okunji, T.A. Ware, R.P. Hicks, M.M. Iwu, D.J. Skanchy. Capillary electrophoreses determination of bioflavarious from *Garcinia kola* in the traditional Africa. *Medical formulation planta Medic.* 68: 2002. Pp.440 - 4.

[28] G.C. Onunkwo, H.C. Egeonu, M.U. Adiukwu J.E. Ojile, A.K. Oluwasulu. Some physical properties of tabulated seed of *Garcinia kola* (HECKEL). *Chemical and pharmaceutical Bulletin* (Tokyo) 52, 2004. Pp.649 – 53.

[29] P.G. Pieita. Flavoronids, as antioxidants *J. Nat. prod.* 63, 2000. Pp.1035-1042.

[30] M. Ruizperez, O. Ndoye, A. Eyebe. and D. Lema. Women and the forest trade: a gender analysis of non-World forest products Markets in the humid forest zone of Cameron. Vol. 50. 1999.

[31] B.W. Smith. Foliar diseases: pp. 55-57. In: Compendium of Peanut Diseases. *American Phyto-pathological Society*, 1994. Pp: 77.

[32] Seed Information Data (SID).2004. Http://www.rbgkew.org.UK/Data/Sid.

[33] L. Tona, N.P. Ngimbi, M. Tsakala, K. Mesia, K.A. Cimanya, S. Spers, T. De Brugne, L Peiters, J., Tatto, A.J. Vlietinck. Antimalaria activity of 20 crude extracts from nine African medicinal plants used in Kinshaba, Congo. *J. Ethnopharmacol,* 68; 1999. Pp.193-203.

13

Spatial Patterns of Nutrient Distribution in Dalingshan Forest Soil of Guangdong Province China

Egbuche C. T.[1,2,*], Su Zhiyoa[2], Anyanwu J. C.[3], Onweremadu E. U.[4], Nwaihu E. C.[1], Umeojiakor A. O.[1], A. E. Ibe[1]

[1]Department of Forestry and Wildlife Technology, Federal University of Technology Owerri, Imo State Nigeria
[2]College of Forest Ecology, South China Agricultural University Guangzhou, China
[3]Department of Environmental Technology, Federal University of Technology Owerri, Nigeria
[4]Department of Soil Science Technology, Federal University of Technology Owerri, Nigeria

Email address:
ctoochi@yahoo.co.uk (Egbuche C. T.)

Abstract: Spatial nutrients that includes OM, Avail.K, Avail.P and TN distribution and the influences on vegetation patterns in Dalingshan was the cardinal focus of this study. Ecological data (moisture content, bulk density and topography) were considered. One way ANOVA was statistically tested of spatial distribution of major nutrients across 4 plots which indicated non significant at $p = 0.05$ level, TN ($p = 0.0216$), OM ($p = 0.00004$), Avail.K ($p = 0.00216$) respectively. Furthermore one way ANOVA was tested on acidity level (pH) measured against the nutrients distribution TN ($p = 0.0031$), OM ($p = 0.0004$), Avail.K ($p = 0.0216$) respectively at non significance level but available phosphorous was significantly different ($p = 0.6412$). The study revealed unique spatial patterns of soil nutrient distribution in Dalingshan and species abundance while vegetation census posed a new direction of study that may be adapted for a broad range of regional vegetation and floristic modeling. This paper suggests that forest soil nutrients and vegetation interaction can be utilized for further studies on multifactor ecosystem responses towards regional ecological restoration.

Keywords: Spatial Patterns, Soil Nutrient, Vegetation Cover, TWINSPAN, Dalingshan Guangdong Province China

1. Introduction

Soil nutrient spatial distribution and soil fertility is of great concern to the management of forest ecosystems. It is an important concept towards biodiversity and multiple values of soil, vegetation cover and agricultural land management. This has resulted to the quest for more knowledge of soil nutrients distribution within spatial pattern in relation to both local and regional vegetation pattern. Plant/vegetation community structure and variability are often attributes of biotic interactions which [1] reported as one primary cause for failure of grazing management in non-equilibrium systems. Generally, spatial patterns of vegetation and abundance of plant species are important aspects which require evaluation of models that can be adopted to describe vegetation dynamics. These literature reports are vital in the assessment of spatial patterns of soil nutrient distribution in relation to abundance

in vegetation cover. Vegetation abundance and canopy are significantly related to spatial nutrient distribution, slope aspects and plant diversity. Studies have been conducted and dedicated to the aspects of temperature, soil moisture availability and tree growth, such studies includes [2], [3] and [4]. However, slope aspects are also considered important factor in distribution of ground-flora species [5], [6]. Generally, more studies are required at this time to understand the spatial patterns of soil nutrients and vegetation variability. Studies designed to investigate soil fertility, nutrient spatial distribution and the effects to vegetation variability are major considerations that are incorporated in the mechanisms of vegetation abundance, species and vegetation canopy. These are important in ecological theories and plant nutrient models. Nutrients in soils can strongly influence the distribution of trees in forest ecosystems and thereby contributes to their health. [7] reported that distributions of soil nutrients in relations to profiles and range of ecological conditions are determinants of plant distribution and domination which is relative to environmental factors such as topography. This study in Dalingshan, Guangdong Province China will help in understanding and estimate soil nutrient parameters in relation to plant species. It is important to understand the connections between plant diversity, vegetation growth and plant community within a spatial environment of a region and can be used in plant models and growth simulation. The main objectives of this study were to evaluate the spatial distribution and patterns of soil nutrient in Dalingshan of Guangdong Province China and to assess the distribution of nutrients in relationship to vegetation cover and abundance thereby infer how soil nutrient will influence the availability/abundance of plant species in the study area.

2. Methodology

Soil Sampling

Physical examination and collection of soil samples from the site grid formation of the soil profile and depth design for soil sample collection was taken from 0-25cm, 25cm-50cm, 50cm – 75cm and 75cm-100cm forming 25 spot points. The random soil sample collection was carefully collected across the site to achieve equitable distributive sample constituent, bulked and treated as a single samples for the purposes of laboratory analyses. This field sample collection was completed in October 2005. Biogeochemical parameters were measured from the sample to determine the spatial pattern trend and evaluate the nutrient level across the site. Soil physical and chemical contents were analyzed using methods of [8] that includes variables of soil texture, gravel, moistures, pH, Available phosphorous, Available Nitrogen, Total Nitrogen, Available Potassium, Organic matter and Exchangeable cation, Electrical conductivity and Hygroscopic water. These analyses were conducted at the Forest Ecology laboratory of the South China Agricultural University Guangzhou.

3. Results

Laboratory results of major nutrients of the study site were further tested to determine the spatial distribution over the plots and which gave insight on how it affected the distribution (fertility) and occurrence of vegetation species. All statistics were considered significant at p=0.05.

3.1. Spatial Variability and Distributions of Nutrients Across TWINSPAN Plots

Greater concentration and evaluation of basic nutrients towards spatial vegetation (plant species) distribution were assessed as strategic results. Having assessed the species abundance and cover using the TWINSPAN, four basic nutrients were measured against acidity concentration (pH)

and spatial variability (TWINSPAN of 18 grids on 4 plots). The major nutrients evaluated were Organic Matter (OM as in Figure 1), Total Nitrogen (TN), Available Phosphorous (Aval.P), and Available Potassium (Aval.K).

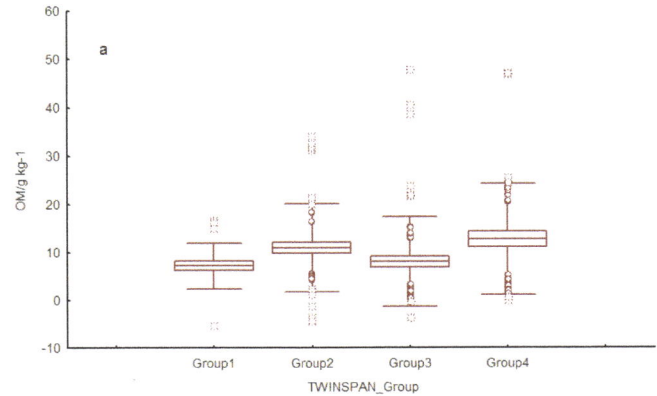

Figure 1. Organic matter variability measured across TWINSPAN 4 groups

The evaluation showed no significant difference of organic matter distribution among the groups (p=0.275 and F= 3,194).

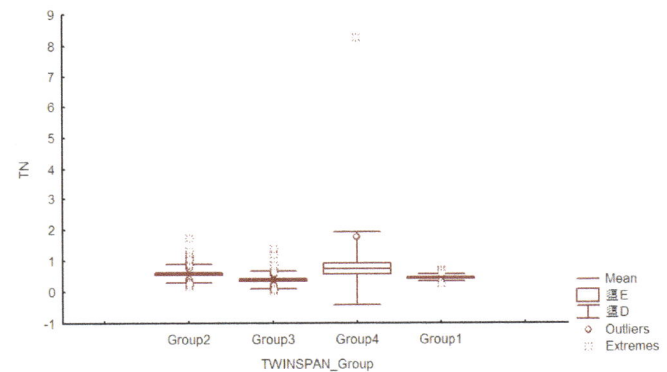

Figure 2. Total Nitrogen variability measured across 4 TWINSPAN groups

The evaluation indicates no significant difference of total nitrogen distribution across groups (p=0.00005, F=3,194).

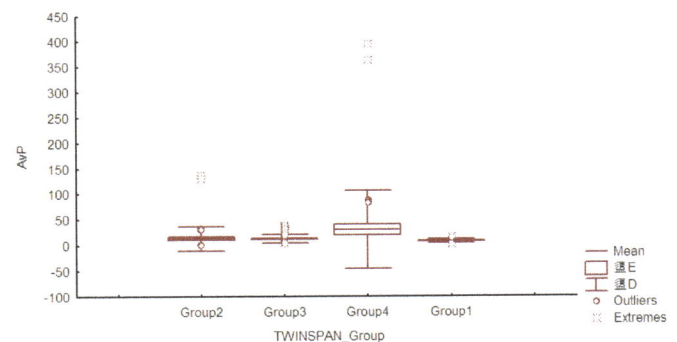

Figure 3. Variability of available phosphorous across TWINSPAN 4 groups

Evaluation shows no significant difference of available phosphorous distribution across groups (p=0.0322, F= 3,194)

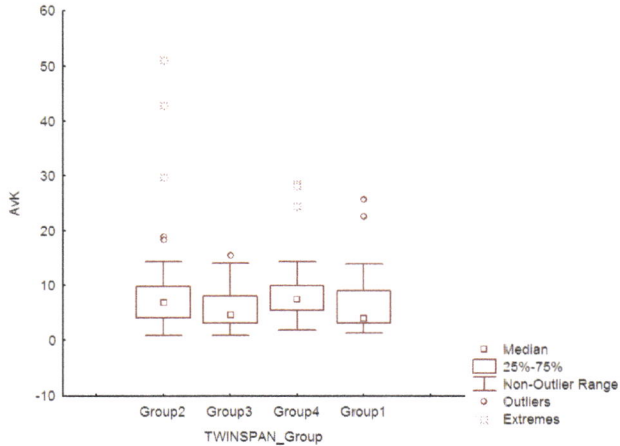

Figure 4. Available Potassium distributions variability across TWINSPAN 4 groups

Evaluation shows no significant difference of available phosphorous distribution across groups (p=0.0047 and F=3,194).

3.2. Nutrient Contents and Distribution in Relation to pH

Furthermore the basic nutrients evaluated across the site distribution were tested in relation to site spatial acidity level. The results were shown as pH variability evaluated across four TWINSPAN group nutrients. They are pH against available potassium, pH against available phosphorous, pH against organic matter and pH against total nitrogen.

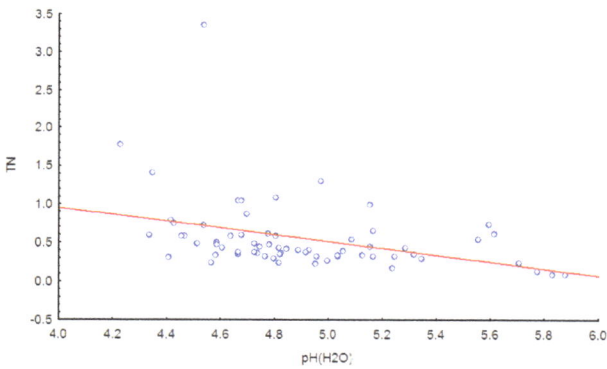

Figure 5. pH distributions in relation to total nitrogen distribution across TWINSPAN 4 groups. The evaluation indicated no statistical differences among the 4 groups where p= 0.0031

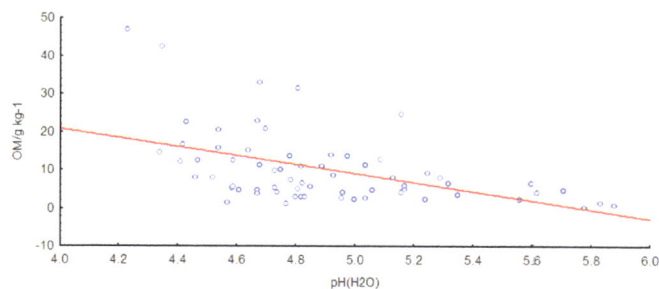

Figure 6. pH distributions in relations to organic matter distribution across TWINSPAN 4 groups. The evaluation indicated no statistical differences among the 4 groups where p= 0.00004

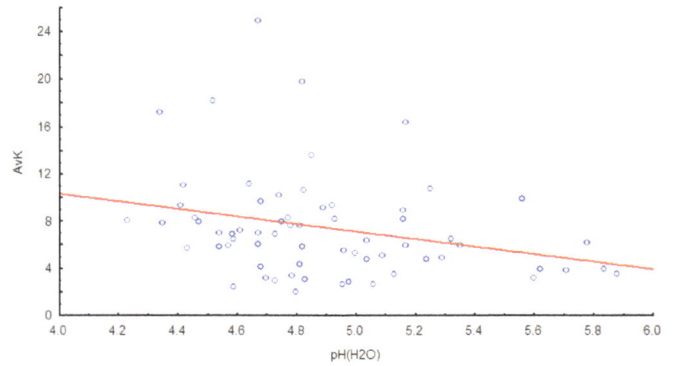

Figure 7. pH distributions in relation to available potassium across TWISPAN 4 groups

The evaluation indicated no statistical differences among the 4 groups where p= 0.0216

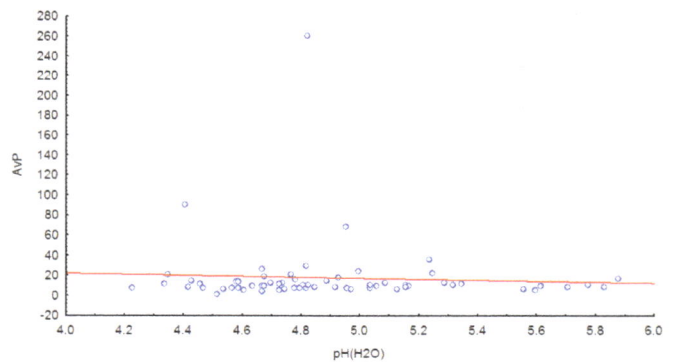

Figure 8. pH distributions in relation to available phosphorous across TWISPAN 4 groups

The evaluation indicated statistical differences among the 4 groups where p= 0.6412

4. Discussion

Spatial Nutrients Characteristics

Field investigation and TWINSPAN analysis over 4 grid dendogram conforms to the spatial variability of nutrient patterns. The basic nutrients evaluated are organic matter (OM), Total Nitrogen (TN), Available Potassium (Avail.K) and Available Phosphorous (Avail.P) in distribution indicates non significant difference across the groups. OM (p=0.275), TN (p= 0.00005), Avail.K (p=0.00047) and Avail.P (p=0.0322). This result accounts for shaping the plant species and diversity patterns. The dominance of species and frequencies were associated by ecological regional factors and as well nutrients which accounts for major factors of soil fertility. The four nutrient value (variables) were used to foresee the variability in the site; though proved non significant respectively. Further statistical test of acidity level (pH) evaluated against major nutrients conforms to and similar to those of other researchers that identified pH vary significantly with ground aspects. But in this study TN (p=0.0216), OM (p=0.00004), Avail.K (p= 0.00216) respectively indicates non significant difference, however pH evaluated against Available Phosphorous (p=0.6412) indicated significant difference. This is in line with [9]. Soil spatial nutrient distribution accounts for growth rate and nutrients efficiency is at recent times attracts greater research attention towards understanding soil fertility mechanisms and supported by [9]. It was documented that Nitrogen is a trait that becomes very useful under the conditions of intense competition for soil resources that characteristize diverse late successional communities. The non-significant of nutrients variables measured across the TWINSPAN plot therefore agree to the theoretical fact that there is strong relationship across all species and statistically affirm the influence of pH level on Phosphorous (p=0.6412) at measured p=0.05 level. This study is supported by the fact that available data on spatial distribution of soil resources is limited though there is variability that ranges from the root zone, as reported by [10] and [11a] [11b]. In overall investigation, it was observed that spatial patterns determine the characteristics of plant species abundance. Generally, this investigation of spatial nutrient indicates and emphasized that nutrient supplies and pH-value are closely correlated and may account for species abundance among the low/even gradient [12] and [13] documented influences of pH-value and species richness of ground of ground vegetation. Certain plants species are found on soils where nutrifiers are active due to weak acidic to neutral pH value of the soil, in another perspective [14] reported that calcium is most important exchangeable action in the soil that influences soil pH-value and controls the availability of other nutrients. Soil moisture and other ecological factors are considered not only pre-requisite for the distribution of species but also responsible for site heterogeneity. These are supported by [15], [16] that the impact may differ at regional scale.

5. Conclusion

Soil nutrients are reflected in the site properties and nutrient uptake which are related to changes in nitrogen availability, since nitrogen is considered to be most important growth-limiting factor in most soils though this nutrient spatial distribution investigation in Dalingshang, confirmed most nutrient tested were equilibrium in distribution but available phosphorous differ in plots evaluated in the content of pH level.

Acknowledgements

We acknowledge and grateful to the College of Forest Ecology, South China Agricultural University Guangzhou, China and the graduate students that participated in the various field/ laboratory soil sample evaluation.

References

[1] Oba G., Stenseth N.C and Lusigi W. J: 2000: New perspectives on sustainable grazing management in arid zones of Sub- Saharan Africa. Bioscience 50: 35 – 51

[2] Tajchman, S. J. & Lacey, C. J. 1986. Bioclimatic factors in forest site potential. Forest Ecol. Manag. 14: 211–218

[3] Hicks, R. R. & Frank, P. S. 1984. Relationship of aspect to soil nu-trients, species importance and biomass in a forested watershed in West Virginia. Forest Ecol. Manag. 8: 281–291.

[4] Munn, L. C. & Vimmerstedt, J. P. 1980. Predicting height growth of yellow-poplar from soils and topography in southeastern Ohio. Soil Sci. Soc. Am. J. 44: 384–387

[5] Huebner, C.D., Randolph, J.C., & Parker, G.R. (1995). Environmental factors affecting understory diversity in second-growth deciduous forests. American Midland Naturalst, 134 (1). Retrieved January 21, 2006 from JSTOR database.

[6] Hutchins, R. L., Hill, J. D. & White, E. H. 1976. The influence of soils and microclimate on vegetation of forested slopes in eastern Kentucky. Soil Sci. 121: 234–241.

[7] Esteban G. Jobbagy and Robert B.Jackson (2000) The vertical distribution of soil organic carbon and its relation to climate and vegetation Ecological Applications, 10(2), 2000, pp. 423– 436 q

[8] Ben-Shahar, R. 1990. Soils, vegetation and herbivores in the Sabi-Sand Wildtuin, Transvaal, S.A. Ph.D. thesis, University of Oxford.

[9] Olivero Adele. M. and Hix David. M.: Influence of aspect and stand age on ground flora of South Eastern Ohio forest ecosystems. Plant ecology 139: 177 – 187. Kluwer Academic Publishers.

[10] Palmer M.W. and Dixon P.M. 1990, Small-scale environmental heterogeneity and the analysis of species distributions along gradients. Journal of Vegetation Science 1:57-65

[11] Jackson R. B. and Caldwell M.M. 1996. Integrating resource heterogeneity and plant plasticity: modeling nitrate and phosphate uptake in a patchy soil environment. Journal of Ecology 84: 891-903. b) Jackson R.B. and Caldwell M.M. 1993. The scale of nutrient heterogeneity around individual plants and its quantification with geostatistics, Ecology 74: 612-624.

[12] Brunet, J., Falkengren-grerup, U., Tyler, G., 1996. Herb layer vegetation of south Swedish beech and oak forests. Effects of management and soil acidity during one decade. Forest Ecology and Management 88, 259–272.

[13] Leuschner C, Hertel D (2003) Fine root biomass of temperate forests in relation to soil acidity and fertility, climate, age and species. Prog Bot 64:405–438.

[14] Pausas J. G., Austin, M. P., 2001; Patterns of plant species richness in relation to different environments: an appraisal. J. Veg. Sci 12, 153-166

[15] Christensen, M., Emborg. J., 1996; Biodiversity in natural versus managed forest in Denmark. For.Ecol.manage. 85; 47-51.

[16] Pausas J. G., Austin, M. P., 2001; Patterns of plant species richness in relation to different environments: an appraisal. J. Veg. Sci 12, 153-166.

14

Socio Economic Conditions of the Hatchery Labors in Chanchra Area of Jessor District in Bangladesh

Tahmina Siddika, Ripon Kumar Adhikary, Md. Hasan-Uj-Jaman[*], Shoumo Khondoker, Nazia Tabassum, Md. Farid Uz Zaman

Department of Fisheries & Marine Bioscience, Jessore University of Science & Technology, Jessore, Bangladesh

Email address:
tahmina.siddika.sathi@gmail.com (T. Siddika), ripon03@yahoo.com (R. K. Adhikary), hasan100401@gmail.com (M. Hasan-Uj-Jaman), shoumo100429@gmail.com (S. Khondoker), naziakeka@gmail.com (N. Tabassum), farid100422@gmail.com (M. F. U. Zaman)
[*]Corresponding author

Abstract: The study was conducted on the hatchery labors in some selected area of Jessore Sadar Upazila in Bangladesh under the district of Jessore by using a logical questionnaire on socio-economic condition of hatchery labor. The main objective of the study is to know the socio-economic condition of the hatchery labors in some selected area of Chanchra region of Jessore district. The specific objectives are to know the different labor categories in hatcheries of Chanchra region and to know the socio-economic condition of hatchery labors. A total of 40 hatchery labors were selected and interviewed. It was found that the average age group of hatchery labor is 26-35 years, predominantly all of them are male and any presence of female labor was not found. It was found that about 95% labor use tube-well water and 5% use deep tube-well water. Regarding mean of transport, 65% labor move by on foot. It was found that about 75% hatchery had no medical treatment facility. However, further study about the socio-economic condition is needed and institutional, organizational, technical and credit supports are needed for their better socio-economic and sustainable livelihood.

Keywords: Fish Hatchery, Hatchery Labour, Socio-economic Status, Sustainable Livelihood and Bangladesh

1. Introduction

Fisheries sector is playing a vital role regarding employment generation, animal protein supply, foreign currency earning and poverty alleviation Hossain *et al.,* [1]. Bangladesh is blessed with huge water bodies in the form of pond, natural depressions (haors and beel, lakes, canals, rivers and estuaries covering an area of 45,75,706 hectares (ha.) DoF [2]. The inland water bodies are rich in freshwater fish species comprising 260 indigenous, 12 exotic and 24 fresh water prawn species DoF [2]. Fisheries sector contributes 4.39% to the national GDP and almost one-fourth (22.76%) to the agricultural GDP Bangladesh Economic Update [3]. In recent years, this sector performs the highest GDP growth rate in comparison to other agricultural sectors (crop, livestock and forestry).The growth rate of this sector over the last 10 years is almost steady and encouraging;

varying from 4.76 to 7.32 percent with an average of 5.61 percent DoF [4]. The sector's contribution to the national economy is higher than its' 4.39% share in GDP, as it provides about 60% of the animal protein intake and more than 11% of the total population of the country is directly or indirectly involved in this sector for their livelihood DoF [4].

Fisher folk are considered as one of the most backward sections in our society. Information on socio-economic framework of the hatchery labors forms a good base for planning and development of the economically backward sector. Lack of adequate and authentic information on socio-economic condition of the target population is one of the serious impediments in the successful implementation of development programme. A socio-economic condition is sustainable when it can cope with and recover from stress and shocks and maintain to enhance its capabilities and assets both now and in the future.

Fish hatcheries and nurseries were not grown in symmetry in all region of Bangladesh. Jessore district is the center point for fry and hatchling production as it is a potential and profitable business avenue in this region. At Jessore the numbers of total hatchery are 35 and 361 nurseries. The total production from these hatcheries and nurseries were respectively 52,639 kg hatchlings and 60, 10,600 kg fry in 2013 (Upazila Fisheries Office, at Jessore Sadar 2013). In this aspect, hatchery labor plays a vital role and they are one of the most valuable communities in Bangladesh. They are poor by any standard. They lead a miserable life. The living condition the standard of living of hatchery labors would mainly depend upon the income earned by them. It will also depend upon the number of employment days available in a month.

The study is needed to portray the socio-economic condition of hatchery labors in Chanchra region of Jessore districts with a view to obtain some knowledge about their social and economic condition and to provide some guidelines which would be economically acceptable. The main objective of the study is to know the socio-economic condition of the hatchery labors in some selected area of Chanchra region of Jessore district. The specific objectives are to know the different labor categories in hatcheries of Chanchra region and to know the socio-economic condition of hatchery labors.

2. Materials and Methods

2.1. Study Area

The study area was Chanchra dalmill, Sabjibag complex and Kotbel Tola in Chanchra region of Jessore district in Bangladesh. Ma Fatema hatchery, Hamja hatchery, Suvro hatchery situated in Chanchra Dalmill, Rupali, Sonali, Anan hatchery situated in Sabjibag complex, Kapotakho, Chaudhuri Mausso, Mudumoti hatchery situated in Kotbeltola. The location of Chanchra region is presented in figure.

Figure 1. The location of Chanchra region.

2.2. Field Survey and Observation

For collecting data on socio-economic condition of hatchery labors, survey was conducted in three places in Chanchra region. 40 labors were randomly selected. Personal interview were conducted with various aspects of effectiveness.

2.3. Questionnaire Survey

To collect data on socio-economic condition of the selected area, the most important matter is to prepare the questionnaire. The questionnaire was developed in logical sequences, so that the labors can answer chronologically.

2.4. Data Collection

2.4.1. Primary Data

Primary data were collected from the fishermen by researcher himself. Several visits were made to the study area to collect appropriate information related to objectives of the study.

2.4.2. Secondary Data

Secondary data were collected from a previous research, Newspaper, Upazila Fisheries Offices and DoF and DFID [5] to clarity.

2.5. Data Analysis

The study used both primary and secondary data in conducting the analysis. Collected information obtained from the survey were accumulated, grouped and interpreted according to the objectives as well as parameters of interested. Some data contained numeric and some contained narrative facts. The data were then presented in graphs and tabular forms. For processing and analysis purpose, MS Excel and MS Word have been used.

3. Results

A total of 40 labors were interviewed from the study areas. Various indicators were collected in different aspects of livelihood characteristics of the labors. A detailed analysis were made on the following way and presented in the section.

3.1. Age Group and Gender Type

From the survey, it was found that 4 labors (10%) belong to under 18 age, 10 labors (25%) belong to between 18-25, 18 labors (45%) belong to group between 26-35, 6 labors (15%) belong to group 36-45 and reaming 2 labors (5%) were belong to above 45 age. Fig. 2 shows that highest number of the labors was between 26-35 age groups.

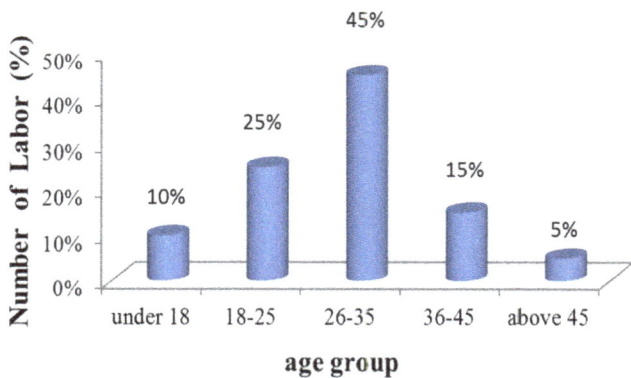

Figure 2. Age group of the labor.

In addition, all labors were male and there is no female contribution. For hatchery working, male gender is very suitable because hatchery operation is very hard. For this reason, there had 100% male contribution and 0% female contribution.

3.2. Type of Labor and Working Hour

From the interview, it was found that many labors work as permanent worker and some labors work as casual labors in peak season. Sometime day basis, contractual labors can be found and this variation created on the basis of working condition. About 90% labor work as permanent and 10% labors work as (contractual, casual, day basis). Labor type of study area in Fig. 2.

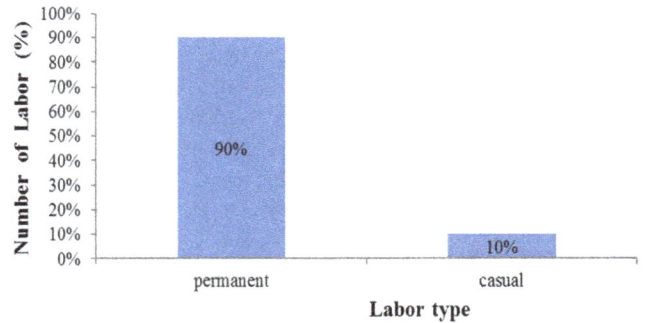

Figure 3. Type of Labor.

It was also found that all permanent labor work in 24 hour but basically they work strength in 8 hour, sometime this working duration should be 12 hours. Few labors work about 2-4 hours for their side income.

3.3. Income Range and Communication Status

Mainly income range of the hatchery labor is based on monthly basis. The highest number of labor were found which had income range in between 5000-8000, about 53% labors and about 29% labor involved in between 8500-12500 and few labor like (18%) belong to group between 12500-16000. Income range of study area is shown in fig. 4.

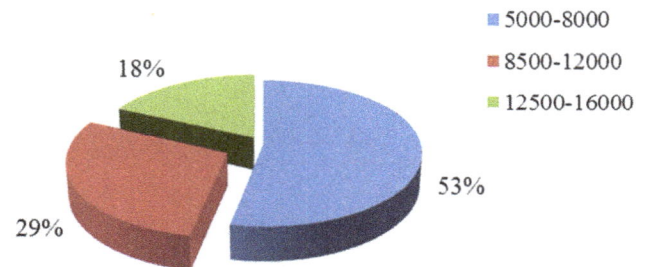

Figure 4. Income range (based on monthly basis).

On the other hand, it was revealed that all labor had mobile phone facility for their communication feature. Because now-a-days mobile phone medium is so much easy and cheap way for communication facilities for all class of labors.

3.4. Unemployment Situation and Side Work

As the survey focused that, hatchery labor had no unemployment period, every time they involved in any type of work which include in hatchery activity. But in the peak season, they did so hard work than the other period. From the survey; it revealed that hatchery is a very busy working sector. Side work means involvement in another work except their main work, but hatchery labor had no easy opportunity to involve in another work except hatchery activity. Their side work mainly included in hatchery related. From the side work, they earned 200-3500 BDT in a month. Some labor made poly bag. Some work as delivery helper, some labor earned from boksis. Few well labors have own fish business out of hatchery (Table 1).

Table 1. *List of side work.*

Side work	Amount of BDT
Making poly bag	750-900
Earning from boksis	250-400
Delivery helper	700-3000
Selfish farming business	Above 2000

3.5. Educational Status and Religious Status

From the survey, we found that most of the labor was literate. About 57% labor had passed primary level, about 13% had secondary level, there had no any S.S.C, HSC educational level labor in hatchery, and about 30% were illiterate. Fig. 5 showed educational status. Religion can play a very important role in the socio-cultural environmental life of labors. In the study area, 95% labors were Muslims and 5% labors were Hindus. Religious status from the questionnaire interview is shown in the Fig. 6.

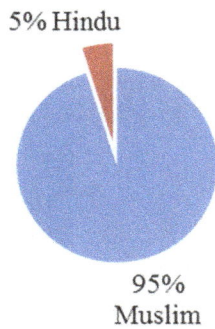

Figure 5. *Educational Status of hatchery labour.*

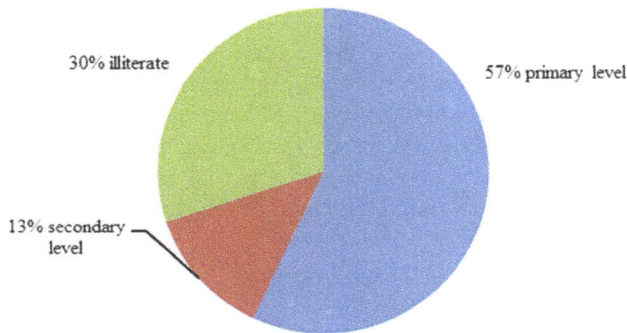

Figure 6. *Religious Status of Hatchery Labors.*

3.6. Family Type, Size, House Ownership and Land Ownership

In the study area, it was found that average labors live in joint family (group of labors related by blood or by law) and some labor live in Nuclear family (married couples with children). About (22%) labor live in nuclear family. The highest percentages obtained in the 4-5 members and this amount was 50%. About (28%) included in 6-8 members (Fig. 7) showed family size.

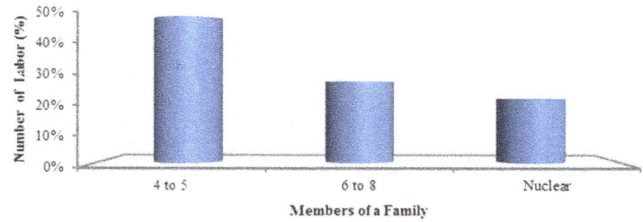

Figure 7. *Family Size of hatchery labors.*

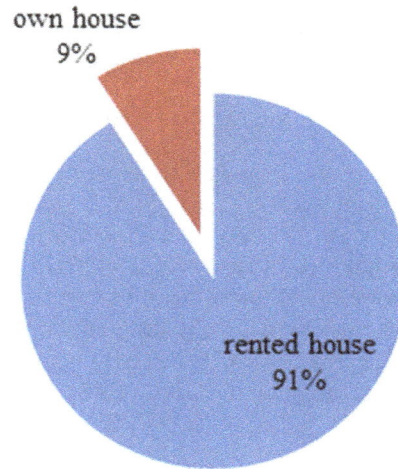

Figure 8. *Ownership of the house.*

In this area most of the families were living in rented or shared house. The survey result also revealed that about (91%) labor live in rented or sheared house and about (9%) labor live in own house (Fig. 8) shows ownership of the house. Highest percent labor had no land ownership, only few labors had little land ownership. About 95% labor had no land ownership and about 5% labor had little land ownership.

3.7. Drinking Water Availability, Electricity Facilities and Supporting Agencies for Hatchery Labor

In the survey, it was found that all labor had pure drinking water facility in their hatchery and their living house. So they were free from water causing diseases.

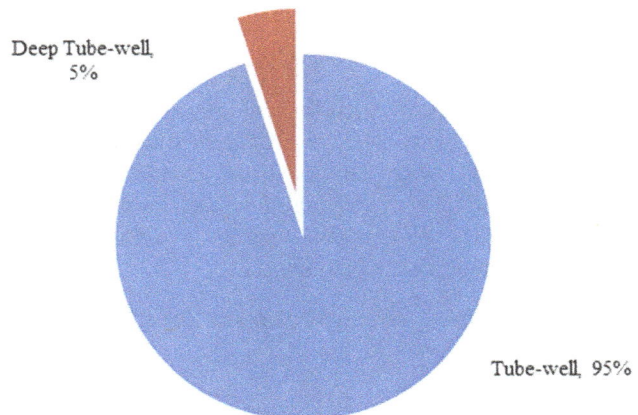

Figure 9. *Drinking water facilities of hatchery labor.*

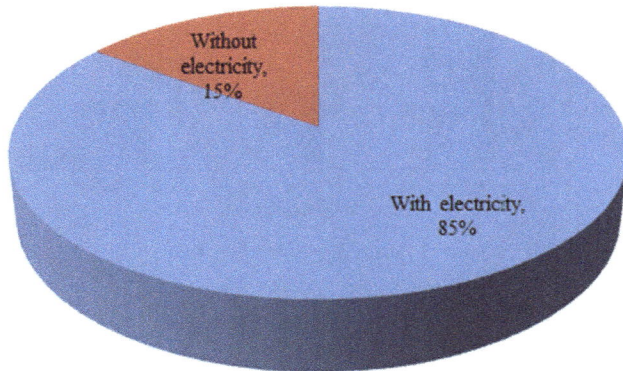

Figure 10. Electricity facilities of hatchery labor.

Maximum labor use tube-well water, some labor use deep tube-well water. About 95% use tube-well water and 5% use deep tube-well water (Fig. 9). In addition, it was found that majority households (85%) had electricity facility, only a few labor (15%) were found to no facilities electricity (Fig. 10). In this area labor were helped by different organization like BFRI, World Fishing Center, Mutsho Odhidoptor gave technical support, training facilities. BFRI gave support 2 times in a year.

3.8. Mean of Transport and Media of Entertainment

In the study area, maximum labor live in hatchery nearest place, for this reason average labor come in the hatchery by on foot, or bicycle. Some labor comes by rickshaw or van, this labor live in far distance from the hatchery. About 65% labor come by on foot, about 25% labor used bicycle for moving and about 10% labor used other way like rickshaw or van.

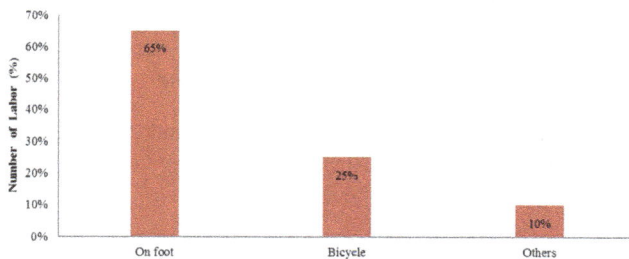

Figure 11. Mean of transport of hatchery labor.

Entertainment must be need for removing boring condition or acquire some new energy for all class of labors. For this reason, hatchery labor enjoys their entertainment and maximum labor use television for their entertainment. Because nowadays the television has become more popular entertainment media than other media.

3.9. Nutrition Status, Medical Treatment Facilities and Health Condition

From the survey, It was attempted to collect information on extent of food and nutrition intake which for socio-economic assessment of the labors, health in particular. The survey revealed that maximum labor took nutrias food.

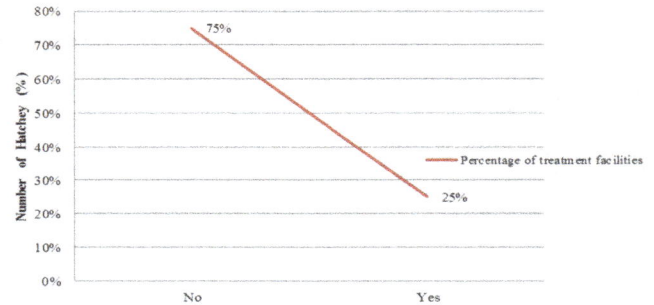

Figure 12. The medical treatment facilities in the hatchery labour.

All labor intake fish meal, vegetable and fish obviously nutrias food. The hatchery labors were suffered from different diseases, like fever, dysentery, gastric, diarrhea etc. To prevent this diseases condition, they require medical treatment facilities. The study shows that maximum hatchery had no medical treatment facilities; about (75%) hatchery had no medical treatment facilities. For this reason, they took quack treatment against diseases and about (25%) hatchery had this facilities Fig. 12. It was found that maximum labor were medium healthy, a few labor become un-healthy. For the hard working, labor must need healthy body, for this reason maximum labor had medium healthy body.

4. Discussion

The total study shows that the highest numbers of hatchery labors (45%) were belonged between 26-35 years age. Kostori [6] reported that majority (36%) fishermen of the Chalan Beel were belonging to age group of 20 to 30 years. Most of the labors of selected area were Muslims (95%); beside this only 5% labors were Hindus. But Raju [7] found 14.33% Hindus and 85.67% Muslims in Sailkupa Upazila. From this survey I found that female worker are not very interested with hatchery work rather they involved in household and agriculture work. Verite [8] reported that approximately 46.6 percent of respondents were male and 53.3 percent of respondents were female. From this study, 57% labor had passed primary level, about 13% had secondary level, there had no any S.S.C, HSC educational level labor in hatchery, and about 30% were illiterate but Bappa, et al., [9] reported that the illiteracy level of majority fishermen (60%) in the study area was found illiterate and only a small portion (10%) passed SSC/HSC examinations. In the study area, it was found that 22% labor live in Nuclear family, 88% live in joint family. The highest percentage of family size was obtained 4-5 members' family (50%). Mahbubullah [10] states 44% of the family's size to be at 6 to 8 in members. The present survey shows that majority percentage of the labors had electricity connection (85%) and minority percentage of the labors had no electricity connection (15%). Kostori [6] reported 48% labors had no electricity facility in a community of chanlan Beel under Tarash Thana in Sirajganj district. In the study area there were 57% primary level educated, 30% were illiterate. DoF [4] in Chanda beel and Ahmed [11] in Tangail found literacy rates were 45% and 69%, respectively. The 95% were found to use tube-wells, others (5%) using deep tube-wells. From the study it can be said that hatchery labor are conscious about their health because safe drinking water is a very important factor for good health. Ali, *et al.,* [12] where 88% fish farmers used their own tube-well in Rajshahi district. But Ahmed [11] mentioned that most farmers used pond water due to lack of tube-well in coastal area. In the survey area it was found that average number of labor like (53%) mean monthly income was 5000-8000. Pravakar, *et al.,* [13] reported that annual income of fish farmers varied from 24,000-1,00,000 BDT. From the study shows that only 5% of the labors had land ownership and the majority 95% of the labors were landless. Islam, *et al.,* [14] found that 2% fishermen were landless. From the survey, it was found that all permanent labor work in 24 hour but basically they work strength in 8 hour. Verite [8] reported that nearly half (47.1 percent) of all respondents reported having worked 11-15 hours, with an absolute average among all workers interviewed of 10.5 hours. The total study shows that 90% labor worked as permanent worker and 10% labor worked as casual worker (on the basis of type of work). Verite [8] reported that 92 percent of contract workers interviewed reported working 7 days per week compared to 76 percent of permanent workers interviewed.

5. Problems and Recommendations

The following problems have identified by asking and relevant questions to the labors of the area:
- This is so many hard working.
- Maximum don't get sufficient amount of BDT.
- Insufficient medical treatment facilities in the hatchery.
- Lack of female worker opportunity.
- Lack of proper sleeping condition.

The recommendation for these problems are-
- Need to appropriate relax time for removing their tiredness mode.
- Give sufficient amount of BDT to fulfill their needs.
- Better job opportunity should be created for female in the hatchery.
- To ensure proper sleeping situation for better work and good health.

6. Conclusion

Chanchra area of Jessore district is one of the hatchery regions in Bangladesh. Many people survey this hatchery as a hatchery labor. This hatchery labor plays a great role and do hard work for hatchery development. As a socio-economic point of view, hatchery labors lid a simple life due to lack of sufficient amount of money. They do so many hard works and get insufficient facilities from the hatchery. Maximum, they do not get opportunity to do any kind of side work out of hatchery. They require adequate training facilities to improve their working skill, better amount of BDT and increase awareness among labors to improve their socio-economic condition.

Acknowledgement

The authors would like to thank the chairman of Fisheries and Marine Bioscience Department, Jessore University of Science and Technology, Jessore, Bangladesh and the hatchery owner of these hatcheries in Jessore Bangladesh for permeating the research team to carry out the quality assessment process in the laboratory.

References

[1] Hossain, M. S., Khan, R. H., Akter, A., (2011). Feasibility of underground water for fish culture in the southern region of Bangladesh: a case study from Laxmipur and Chittagong area. J. Agrofor. Environ. 5 (2): 121-123.

[2] DoF (Department of Fisheries), (2008). Matshaya Sampad Unnayon Ovigun, Ministry of Fisheries and Livestock. Government of the Peoples Republic of Bangladesh. pp. 79- 81.

[3] Bangladesh Economic Update (2012). Poverty Reduction and Economic Management, South Asia Region. Funded by the World Bank.

[4] Department of Fisheries, (DoF) (2013). Fisheries Statistical Yearbook of Bangladesh 2011-2012. Fisheries Resources Survey System (FRSS), Department of Fisheries, Bangladesh. (29): 44.

[5] Department of Fisheries (DoF) and Department for International Development (DFID) (2002). Management options for the shrimp fry fishery, A regional stakeholder workshop in Khulna.

[6] Kostori, M. F. A., (2012). Socio-economic condition of Fishermen of the ChalanBeel under Tarash Thana of Sirajganj in Bangladesh. Bangladesh Research Pulications. Journal 6 (4):393 402.

[7] Raju, A., 2002. Livelihood status of fish farmers in Sailkupa Upazila under Jhenaidah District. M. Sc thesis. Fisheries and Marine Resource T echnology Discipline, Khulna Uni, Bangladesh. pp. 53.

[8] Verite (2012). Research on Indicators of Forced Labor in the Supply Chain of Shrimp in Bangladesh.

[9] Bappa SB, Hossain MMM, Dey BK, Akter S and Hasan-Uj-Jaman M (2014), Socio-economic status of fishermen of the Marjat Baor at Kaligonj in Jhenidah district, Bangladesh. Journal of Fisheries 2(2): 100-105.

[10] Mahbubullah, M., (1986) Case study of polder and estuarine fisheries community in Bangladesh. In Socio-economic Study of Tropical Fishing Community in Bangladesh. A report for Food and Agricultural Organization (FAO), Rome. pp. 12-14.

[11] Ahmed, N. U., (1999). A study on socio-economic aspect of coastal fishermen in Bangladesh, Department of Fisheries. Bangladesh Agricultural University, Mymensingh, Bangladesh.

[12] Ali, M. H., M. D. Hossain, A. N. G. M. Hasan and M. A. Bashar (2008). Assessment of the livelihood status of the fish farmers in some selected areas of Bagmara Upazilla under Rajshahi district, J. Bangladesh Agri. Uni., 6(2): 367-374.

[13] Pravakar, P., Sarker, B. S., Rahman, M. and Hossain, M. B. (2013). Present Status of Fish Farming and Livelihood of Fish Farmers in Shahrasti Upazila of Chandpur District, Bangladesh. American-Eurasian J. Agric. & Environ. Sci. 13 (3): 391-397.

[14] Islam, M. R., Hoque, M. N., Galib, S. M., Rahman, M. A., (2013). Livelihood of the fishermen in Monirampur Upazila of Jessore District Bangladesh. Journal of Fisheries.1 (1): 37-41.

15

Variability in selected Properties of Crude Oil – Polluted Soils of Izombe, Northern Niger Delta, Nigeria

Ihem E. E.[*], **Osuji G. E., Onweremadu E. U., Uzoho B. U., Nkwopara U. N., Ahukemere C. M., Onwudike S. O., Ndukwu B. N., Osisi A. S., Okoli N. H.**

Department of Soil Science and Technology, Federal University of Technology, Owerri, Imo State, Nigeria

Email address:

eeihem2000@yahoo.com (Ihem E. E.)

Abstract: We investigated the variability in some soil properties influenced by crude oil-polluted soils of Izombe in Northern Niger Delta of Nigeria in 2013. A free survey technique was used in the field sampling with nine profile pits dug in the site. Routine soil analysis was conducted on some physico-chemical properties including heavy metals. Soil data were subjected to analysis of variance using proc mix-model of SAS software at $P \leq 0.05$. Results showed that soils were dark grayish brown to red in colour. Soils of the studied area were also deep (>100cm), well drained and having percent sands (>80%). Soils from crude oil-polluted site showed lower pH (<3.92) than the unpolluted soils with pH >4.00. Soil organic matter, C:N ratio, TEA and percent Al. Sat, were appreciably higher in soils affected by crude oil pollution. Unaffected soils by crude oil pollution exhibited higher TN, P, TEB and B.Sat. Heavy metal concentrations in the polluted sites were relatively higher than their unaffected counterparts and were significant ($p \leq 0.05$). Further studies should be conducted on some other properties and in owner-managed farm establishments.

Keywords: Variability, Crude Oil, Soil Quality, Tropical Soils

1. Introduction

Other than agricultural practices, oil exploration is a major economic activity in the Niger Delta areas of Nigeria. This has resulted in the pollution and contamination of agricultural lands for farming (terrestrial and aquatic environment). Soils and water polluted with crude oil poses a serious threat to living organisms within the environment. Crude oil contains heavy metals that may be phytotoxic to plants and injurious to animals [1]. It could also cause acidification of the soils [2]. Apart from factors such as previous farming practices, parent materials, topography, crude oil pollution results to variations in soil properties. However, variations of soil physical and chemical properties, nutrient levels and water content occur at field scale [3]. Crude oil reduces the fertility of the soil by making plant essential nutrients unavailable. [4] reported that sites polluted with crude oil contain metals such as Barium (Ba), Cadmium (Cd), Chromium (Cr), Lead (Pb), Nickel (Ni), Copper (Cu) and Zinc (Zn) and their mobility depends on the concentration and soil properties. Farmers are

now amending and remediating soils affected by crude oil with the hope of optimizing agricultural outputs across fields. These attempts are made to regenerate soil fertility in the study area which has been lowered by the influence of crude oil spillage, using sewage sludge [5], water hyacinth residues [6]. Municipal solid wastes [7]). cassava peel, cattle dung and poultry droppings [8]. These efforts improved quality for increased productivity. Similarly study has been conducted in some other locations outside the study site. However, studies in this region have shown that most of the valuable lands in the studied site are either temporarily or permanently lost to oil exploration activities in form of flow-station, disposal pits for burying oil and land covered with spilled oil [9. Based on this, we investigated the influence of crude oil pollution on the variability of selected soil properties.

2. Materials and Methods

A field experiment was conducted at oil exploration sites during April, 2009 to March, 2010 at Izombe, Owerri –

Nigeria. The experimental site is located within Northern Niger Delta region of Nigeria (Latitudes $5^0 20^1$ and $5^0 41^1$ N and longitudes $6^0 37^1$ and $6^0 49^1$ E). Soils of the area are derived from coastal plain sands and are dominated by ultisols. The site belongs to the lowland area of Nigeria, with a humid tropical climate with mean annual rainfall of 2250mm and mean annual temperature range of 26 – 31^0C. The main socio-economic activity of the study site is arable farming, hunting and oil exploration activities. Land preparation includes slash and burn system with conventional tillage system.

2.1. Field Studies

A free survey technique was used to locate the study site. Nine (9) pedons of about 200cm depths were dug covering polluted sites (7) pedons and unpolluted sites (2) pedons. After horizon delineation soil samples were taken from the component horizons; air-dried and made to pass through a 2mm sieve prior to laboratory analysis.

2.2. Laboratory Analysis

Particle size *distribution* was estimated by hydrometer method [10].

Bulk density was determined by core procedure, [11]. The soil was transferred from the sample holders of core sampler to a container and placed in an oven at 105^0C and dried to a constant weight. The weight of soil was recorded and bulk density calculated by the formula of [12] as follows:

$$\text{Bulk density} = \frac{Ovendryweightsoil}{samplevolume}$$

Soil pH was determined electrometrically in a soil solution ratio of 1:2.5 [13].

Total nitrogen was estimated using the modified micro-kjeldahl digestion method [14] and sodium copper sulphate catalyst mixture [15].

Organic matter was measured as described by [16] Nelson and Somnars, (1982). Organic matter was calculated by multiplying organic carbon by [17], factor" of 1.724.

Available phosphorus was determined by using the molybdemum blue colour Bray II method [18].

Exchangeable Bases were determined from Ammonuim acetate (NH^4OAC) leachates of the soil [19].

Exchangeable Acidity was determined by leaching the soil with 1NKCl and titrates with 0.05 NaOH solutions [20]. Effective cation exchange capacity was estimated by the summation of the total exchangeable bases (TEB) and exchangeable acidity (TEA). It is expressed in Cmol/kg Soil [21].

Heavy metal concentrations were measured individually with atomic absorption spectrometer (AAS) after wet digestion with concentrated H_2SO_4 for Cr, mixture of HNO_3 and HCL for Hg, and Cd and HNO_3 for Ni, V and Pb respectively [22].

Statistical Analysis

Data were subjected to analysis of variance (ANOVA) using the Statistical Analysis System (SAS) version 2008 model.

3. Results and Discussion

The results of the morphological properties of the soil studied are summarized in Table 1.

Table 1. Morphological properties of Studied Soils

Horizon	Depth (cm)	Colour	Structure	Drainage	Boundary
Ap	0 – 23	2.5 YR $^4/_2$	W gr	ewd	Clear
AB	23 – 86	2.5 YR $^5/_2$	M gr	ed	Smooth
Bt$_1$	86 – 130	2.5 YR $^4/_9$	M sb	ed	Gradual
Bt$_2$	130 – 195	2.5 YR $^4/_8$	Si b	wd	Diffuse

W gr = Weak granular, m gr = Medium granular, m sb = Medium sub-angular block, Si b = Sub – irregular blocky, ewd = Excessively well drained, ed = Excessively drained, wd = well drained.

Table 2. Some physical properties of the Studied Soils

Location	Depth (cm)	Sand (%)	Silt (%)	Clay (%)	BD(g/cm^3)	SCR
Polluted soils	0 – 23	80.23	7.80	8.17	1.42	0.95
	23 – 86	84.05	7.65	8.39	1.44	0.91
	86 – 130	81.20	8.90	9.99	1.47	0.89
	130 – 195	80.34	9.40	9.61	1.57	0.98
SED		2.94	1.21	0.85	0.03	0.33
P value		0.5433[NS]	0.4292[NS]	0.0973[NS]	<0.0001	0.4808[NS]
Unpolluted soils	0 – 23	83.17	6.17	10.67	1.37	0.58
	23 – 86	80.16	7.50	12.33	1.45	0.61
	86 – 130	77.67	7.83	14.50	1.49	0.54
	130 – 195	79.33	10.33	10.67	1.55	0.97
SED		5.56	4.11	3.16	0.08	0.74
P value		0.076[NS]	0.8397[NS]	0.6289[NS]	0.2919[NS]	0.3993[NS]

Generally, the soils studied were deep (>100cm), well drained and with cleared horizons differentiation. The deep pedons with distinct horizonation is an indication that the soils have undergone pronounced weathering [23]. Results of soil physical properties are shown in Tab. 2. Results showed that soils in the studied site were sandy (>80%). No significant difference (P<0.05) was shown in the percentage sand between polluted and unpolluted soils. Soil texture is an inherent property and may not have been influenced significantly by crude oil pollution in the study area. The bulk density ranged from $1.37 - 1.57$ g/cm^3 with a mean of 1.48 g/cm^3 for polluted soils and 1.47 g/cm^3 for unpolluted soils. There was no particular trend in bulk density distribution among the pedons but with higher values occurring in polluted soils. Crude oil is known to increase bulk density in soils perhaps due to aggregate disintegration. [24] reported that oil spillage increases bulk density due to aggregate disintegration. The chemical properties of the studied soils are presented in Table 3. The results showed that the soils of the area were acidic, with a mean pH value of 3.92 in polluted soils and 4.00 in unpolluted soils. The stronger soil reaction of the polluted soils could be attributed to the impact of crude oil pollution. Oil in soils tends to

decrease the pH generally, making it unsuitable for crop production. This situation could be compounded by the high rainfall, leaching which results in washing away of basic cations from the soils and the acidic nature of the parent materials in the studied site. The resultant effect of these may be the preferential removal of basic cations through leaching resulting in the accumulation of exchangeable acidic cations (Al^+ and H^+) in the soil absorption complex of polluted soils [25]. Higher values of Organic matter (2.00%) and C/N ration (18.87) in crude oil polluted soils explained presence of carbon in the petroleum hydrocarbon discharged and deposited on the polluted soils during crude oil spillage. This finding is in line with earlier work done by [26] where low organic matter and C/N ratio in unpolluted land units confirmed high mineralization process in the organic matter and also due to high temperature and excessive high rainfall which characterize the study area [27]. Values of total nitrogen, phosphorus, ECEC and percent base saturation were consistently lower in polluted soils compared to the unpolluted counterparts, which revealed that crude oil pollution encouraged nutrient elements imbalance as well as phosphorus fixation among other elements [28]. Results of some heavy metals in the site are shown in Table 4.

Table 3. *Some Chemical Properties of the Studied Soils*

Location	Depth (cm)	pH (1NKCl)	OM (%)	TN (%)	C/N (%)	Av.P (mg/kg)	Ca (Cmol/100g)
Polluted soils	0-23	4.08	3.75	0.1	20.61	11.79	0.9
	23-86	3.95	1.72	0.06	16.94	7.09	0.74
	86-130	3.89	1.51	0.05	18.53	5.02	0.93
	130-195	3.73	1.01	0.05	19.42	5.13	0.88
	SED	0.1	0.94	0.019	3.73	0.9	0.192
	P Value	0.0114	0.0329	<0.001	0.7952	<0.001	0.123
Unpolluted soils	0-23	4.05	2.79	0.17	24.57	1.15	1.18
	23-86	4.25	1.66	0.14	19.28	9.1	1.13
	86-130	4.11	1.39	0.13	16.96	7.87	1.74
	130-195	3.6	1.21	0.12	22.69	7.57	0.83
	SED	0.39	0.59	0.02	0.1446	0.873	1.01
	P value	0.5568	0.1375	0.1666	0.9528	0.9703	0.8599

Table 3. *continued*

Location	Mg (Cmol/100g)	K (Cmol/100g)	Na (Cmol/100g)	H$^+$ (Cmol/100g)	Al^{3+}	ECEC	Al.sat	B.sat
Polluted soils	0.57	0.07	0.64	1.23	0.89	4.11	24.24	50.81
	0.63	0.057	0.46	1.32	0.96	4.07	24.41	46.86
	0.87	0.107	0.46	1.36	0.76	4.13	23.17	45.35
	0.9	0.071	0.51	1.19	0.7	4.13	22.26	50.45
	0.2147	0.036	0.16	0.316	0.07	0.33	0.406	3.87
	0.09	0.5451	0.6235	0.9336	0.0024	0.9971	0.9517	0.4284
Unpolluted soils	1.09	0.18	0.64	0.57	0.62	5.96	11.04	65.54
	1.05	0.15	0.46	0.88	0.69	5.98	12.78	43.99
	0.93	0.18	0.46	0.51	0.55	5.13	15.32	62.4
	0.67	0.23	0.51	0.48	0.45	4.76	11.92	66.03
	0.087	0.16	0.16	0.08	0.09	0.59	3	2.27
	0.0208	0.97	0.674	0.001	0.18	0.245	0.566	0.002

OM = Organic matter, TN = Total nitrogen, C/N = Carbon-nitrogen ratio, Av.P = Available phosphorus, Ca = Calcium, Mg = Magnesium, K = Potassium, Na = Sodium, H = Hydrogen, Al = Aluminum, ECEC = Effective Cation Exchange Capacity, Al.sat= Aluminium saturation, B.sat= Base saturation.

Table 4. Some heavy metals concentration in the Studies Site

Location	Depth (cm)	Hg (mg/kg)	Cd (mg/kg)	V (mg/kg)	Cr (mg/kg)	Pb (mg/kg)
Polluted soils	0 - 23	0.04	7.52	0.09	7.25	5.35
	23 – 86	0.04	6.74	0.94	7.41	4.66
	86 – 130	0.02	5.84	0.61	6.99	4.55
	130 – 195	0.02	5.65	0.59	6.95	4.50
SED		0.003	1.03	0.22	1.12	0.45
P value		<0.003	0.2665	0.0066	0.9741	0.2173
Unpolluted soils	0 - 23	0.02	0.63	0.39	3.48	4.68
	23 – 86	0.01	3.20	0.29	3.81	4.67
	86 – 130	0.02	2.38	0.13	3.10	1.70
	130 – 195	0.01	2.97	0.10	3.48	1.29
SED		0.005	0.06	0.05	0.08	0.29
P value		0.4850	<0.001	0.0027	0.005	0.001

Hg = Mecury, Cd = Cadmium, V = Vanaduim, Cr = Chromium, Pb = Lead

Higher values of heavy metals (Hg, Cd, V, Cr, and Pb) were observed in polluted soils compared to the unpolluted soils. [29] reported that crude oil contains heavy metals and possibly added to the soil during oil spillage but below their critical levels for crop production [30].

4. Conclusion

The study revealed that the study soils were deep and well drained with high proportion of percent sand. Oil exploration activities had a meager effect on particle sizes, influences soil properties by increasing the values of Om, C/N ratio, % H and Al saturation. Soil pH, N, P, % B. sat are decreased as influenced by crude oil pollution. Concentration of heavy metals in the polluted soils were compounded by crude oil pollution and showed a significant difference (P<0.05). The study indicated that crude oil exhibited a negative influence on soil productivity in the Niger Delta region of Nigeria, hence required improved agronomic practices and crude oil remediation for optimum agricultural production.

References

[1] FEPA (1991). "Guidelines and Standards for Environmental pollution control in Nigeria" Federal Environmental protection Agency.

[2] Iwegbue, C.M.A., Isirimah, N.O., Igwe, C. and Williams, E.S. (2006). Characteristic levels of heavy metals in soil profiles of automobile mechanic waste dumps in Nigeria. Environmentalist 26:123-128.

[3] Osuji, I.C and Onajake, C.M (2004).The Ebocha oil spillage II.Fate of associated heavy metals six months after.Agricultural Journal and Environmental Assessment Management (AJEAM – RAGEE) VOL. 9: 78 – 87.

[4] Mbagwu, J.S.C., Oti, N.N., Agbin, N.N. and Udom, B.E (2001).Effects of Sewage Sludge application on selected soil properties and yield of Maize and Bambara Groundnut. Agro-Science 2: 37 – 43.

[5] Oguike, P.C and Mbagwu, J.S.C (2001). Effects of Water hyacinth residues on chemical properties and productivity of degraded tropical soils. Agro-science 2: 44 – 51.

[6] Mbagwu, J.S.C and Piccolo, A. (1990).Carbon, nitrogen and phosphorus concentration on aggregate of organic waste – amended soils.Biol. Wastes.31: 97 – 11.

[7] Onweremadu, E.U. (2007c). Characterization of degraded ultisol amended with cassava peel, cattle dung and poultry droppings in Southeastern Nigeria. J. plant Sci. 2(5): 564 – 569.

[8] Ekoko, G. (1997). Save our land, Delta State University Press, Asaba.

[9] Gee, G.W. and Or, D. (2002). Particle size analysis. In: methods of soil analysis, Dane, J.H and G.C. Topp (Eds) Part 4. physical methods. Soil sci. Amer. book series 5, ASA and SSSA, Madison, Wisconsin 255-293.

[10] Black, G.R. and Hartge, K.H. (1986).Bulk density. In: Klute, A. (ed) methods of soil analysis, Part 1Physical and mineralogical methods 2nd ed. Agronomy No. 9. ASASSSA, Madison. WI, pp 363 – 375.

[11] Black, G.R. and Hartge, K.H. (1986).Bulk density. In: Klute, A. (ed) methods of soil analysis, Part 1Physical and mineralogical methods 2nd ed. Agronomy No. 9. ASASSSA, Madison. WI, pp 363 – 375.

[12] Hendershot, W.H.; Laland, H. and Duquette, M. (1993).Soil reaction and exchangeable acidity. In: soil sampling and methods of analysis.

[13] Bremner, J.M. (1996). Nitrogen – Total. In: Sparks, D.J (ed) Methods of soils Analysis, Part 3, chemical heavy metals in a spent oil polluted typicpaleudult. Southeastern Nigeria. Int. J. Agric. And Rural Development 11(2): 77 – 81).

[14] Bremner, J.M. and Yeomans, J. C (1988). Laboratory techniques in J.R. Wilson (ed) Advances in Nitrogen cycling in Agricultural ecosystems. C.A.B. intWilling for, England.

[15] Nelson, D.W. and Sommers, L.E. (1982).Total carbon, organic carbon and organic matter. In: page A.L., Miller, R.H. and Kenney, D.R. (eds). Methods of soil analysis, part 2. Ameri. Soc. Agron: Madison, Wisconsin 539-579.

[16] Olson, S.R. and Sommers, L.E (1982).Phosphorus. In: Page A.L., Miller, R.H and Keeuey, D.R (eds). Methods of Soil Analysis, part 2 Amer. Soc. Agron, Madison Wisconsin 403-430.

[17] Thomas, G.W. (1982). Exchangeable bases. In: Methods of Soil analysis, part 2. Page A.L., Miller, R.H and Keeney, D.R (eds). Pp. 159 – 165 (American Society of agronomy Madison, W.J

[18] McLean, E.V. (1982). Aluminium. In: page, A. L. soil analysis part 2, American Society of Agronomy, Madison, W.I. pp. 978-998.

[19] Bruce, A.M. and Whiteside, P.J. (1984).Introduction to atomic absorption Spectrophotometer 3rd ed. PyeUnicam Limited England.

[20] Onweremmadu, E.U., Oti, N.N and Uzoho, B.U. (2005). Evaluation of selected soil chemical parameters in two crude oil-spilled sites in Southeastern Nigeria. Journal of Sustainable Tropical Agric. Research, 16: 33 – 42.

[21] Foth, H.D. (1984). Fundamental of Soil Science John Willey and sons, New York. 359pp.

[22] Akamigbo, F.O.R. (1993). Causes, impact and implication of gully erosion in Southeastern Nigeria. Paper presented at the 29th Annual Conference of Agric. Soc. of Nigeria, held at Federal University of Agric. Umudike, Abia State, Oct. 31-Nov.3, 1993

[23] McOliver, K. (1984). Introduction to Petroleum Geology. England 465pp.

[24] Omeke, J.O., Eshette, E.T., Ihem, E.E., Etwuoku, A. and Nnaji, G.U. (2009).Characteristics and fertility status of arable soils proximal to selected gas flaring sites in Delta State Region of Nigeria. Int. Journal of Agric.And Rural Dev. Vol. (12) 34 – 39.

[25] Foth, H.D. (1984). Fundamental of Soil Science John Willey and sons, New York. 359pp.

[26] Akinlabi, O.A. (1981). Effect of oil industrial activities in the North Apoi oil field. Federal Ministry of Housing and Environment., Vol. 5 Nig. 4-10.

[27] FEPA (1991). "Guidelines and Standards for Environmental pollution control in Nigeria" Federal Environmental protection Agency.

16

Ground Water Level Fluctuation and Its Impact on Irrigation Costing at Jessore Sadar of Bangladesh

G. M. Abdur Rahman[1], Champa Bati Dutta[1], Md. Jamal Faruque[2], Mehedi Hashan Sohel[3], Abu Sayed[4, *]

[1]Economics Discipline, Khulna University, Khulna, Bangladesh

[2]Bangladesh Agricultural Development Corporation (BADC), Khulna, Bangladesh

[3]Department of Soil Science, EXIM Bank Agricultural University Bangladesh, Chapainawabgonj, Bangladesh

[4]Department of Agricultural Engineering, EXIM Bank Agricultural University Bangladesh, Chapainawabgonj, Bangladesh

Email address:

rahman82bd@gmail.com (G. M. A. Rahman), champa.dutta@econ.ku.ac.bd (C. B. Dutta), mjfaruque@gmail.com (Md. J. Faruque), mehedibau113@gmail.com (M. H. Sohel), abu_982@yahoo.com (A. Sayed)

[*]Corresponding author

Abstract: The study was conducted to access the ground water level fluctuation and impacts on irrigation cost of Jessore sadar and Jhikargacha upazilla. Questionnaire survey and Key Information Interview (KII) were done to collect primary data from local farmers, pump operator, DAE officials and personnel from BADC. The secondary data were collected from BADC. From study area it was found that the Maximum ground water level varies 5.1 to 9.35m at the month of April to May where minimum ground water level varies 1.15 to 4.88m at October to November from 2004 to 2013. The trends of maximum fluctuation level increased 5.65 to 9.35m and 5.1 to 8.36m at Jessore sadar and Jhikargacha respectively, in April 2004 to April 2013. From the study it was also found that ground water level fluctuation mostly affect the STW irrigation where 100% of STW at the study area have faced pump failure and for the remedial measure 100% taken measure to deep set method. Pumping hour for both STW and DTW has increased in 2013 than 2009. It was found 99% of the respondents were aware of lowering of ground water level in their agriculture fields, 85% of farmers complained that they did not get enough water during the dry season irrigation period. Over 95% of the respondents were well aware about excessive pumping. Most important thing was that 95% of the respondents replied that they didn't get training for irrigation and have not enough knowledge for irrigation efficiency and the crop water use. To obtain desired economic benefits from groundwater resource, the management of ground water is essential.

Keywords: Ground Water Table, Fluctuation, Irrigation, Water Quality, Jessore

1. Introduction

Groundwater is the main source of irrigation which is an important parameter to increase crops production. But the amount of extraction is increased day by day due to increase of population, food insecurity, and poor water management and below average rainfall is putting unprecedented pressure on groundwater [1]. With the continuous abstraction, the water table in many areas started declining and the STWs were no more capable of pumping under suction mode during the peak irrigation period. It has been observed that ground water fluctuate mostly on dry season as because of no rainfall at that time. So the lowering of the water table lead to the drying up of more shallow wells, requiring deeper tube wells, and increased pumping cost. As per report of the International Rice Research Institute (IRRI), irrigation efficiency in Bangladesh is the lowest in the region, where the cost of irrigation is $117.60 per hectare compared to $25.58 in India, $17.94 in Thailand and $17.98 in Vietnam. If water was managed properly, Bangladesh could save additional amount of money equivalent to one-sixth of the total Bangladesh Budget for the Fiscal Year (FY) of 2003-04 (USD 8,962 million) [2] and also similar to the ADP budget of Bangladesh for FY 2009-10 (USD 4072 million).

A recent study shows that groundwater level in some areas of Bangladesh falls between 5-10 m in dry season and most of the tubewells fail to lift sufficient water [3]. Researchers and policymakers are advocating sustainable development as the best approach to today's and future water problems [4, 5]. They shows that frequent shortage of water in the region has had impacts that can be ranged as economical, social and environmental [6, 7 & 8]. Hossain [9] reported that hundreds of shallow and hand tube wells become inoperative due to over exploitation of groundwater, lack of adequate groundwater recharge during wet season and low specific yield of upper aquifers. This situation can be exacerbated by 2030 when water demand will be doubled due to expected increase in dry season agriculture. Now time has come to think over the issue to meet the challenge. Some works already have been done in different parts of Bangladesh. Now it is needed to conduct location specific comprehensive study for realistic assessment of ground reserves and their use at present condition. For this purpose a study was conducted to assess the ground water level fluctuation and impacts on irrigation cost for planning groundwater resource in this area.

2. Methodology

2.1. Study Area

The study area lies in the physiographic unit of high Ganges floodplain. It lies under the AEZ no. 11 (High Ganges river floodplain). The land is very fertile. The area is most suitable for rice, vegetables and upland crops. A total land area of this district is 7287711 acre where cultivable land is 415635 acre. Of the cultivable land there is about 70%

under irrigation and about 80% of total irrigated area is covered by ground water irrigation systems [10]. This is one of the large irrigation districts in Bangladesh. Average maximum temperatures vary from about 25°C to 35°C in April and the minimum temperature in varies from 12°C to 14°C in late December and early January. The maximum rainfall is recorded in August and September.

2.2. Data Collection

Jessore sadar and Jhikargacha upazilla were selected purposively from the study area. Secondary data for groundwater table was collected from BADC. For each upazilla 5 observation wells were selected randomly out of 9 at Jhikargacha and 10 at Jessore sadar. Six (6) questionnaire survey and (3) Key Information Interview (KII) were conducted to collect primary data from local farmers, pump operator, DAE officials and personnel from BADC. Semi structured questionnaire survey used to assess the impacts underline the ground water fluctuation on irrigation system, costing of irrigation, farmers concept on ground water fluctuation and the problems faced in this regards, their thinking to reasons behind the ground water fluctuation, crop diversification, production etc.

3. Results and Discussion

3.1. Ground Water Level Fluctuation

Average yearly ground water level fluctuation of Jessore sadar and Jhikargacha are shown in Table 1. The table shows that the lowest average ground water level fluctuation were 3.48 m in Taposhidanga of Jessore sadar and 3.94 m in Padmapukur of Jhikargacha in 2004. The maximum ground water level depths were 6.41 m in Ichhali of Jessore sadar and 6.56 m in Bish Hari, Jhikargacha during 2010.

Table 1. Average yearly ground water level fluctuation of Jessoresadar and Jhikargacha.

Name of the place		Yearly average ground water level fluctuation (m)									
Upazilla	Union	2004	2005	2006	2007	2008	2009	2010	2011	2012	2013
Jessore Sadar	Tapaswidanga	4.50	6.10	5.92	6.02	4.48	6.36	5.72	6.46	6.13	6.87
	Ichhali	3.96	6.00	5.50	6.53	4.55	6.31	5.81	6.63	6.11	7.07
	Bijoynagar	3.90	5.98	5.29	5.99	4.46	6.17	5.70	6.51	6.17	7.02
	BirNarayanpur	3.30	5.95	5.53	6.46	4.30	6.29	5.78	6.69	6.13	7.11
	Shekhati	4.15	6.07	4.67	5.26	4.46	6.26	5.63	6.49	6.09	6.97
Jhikargacha	Bankara	3.97	4.14	4.38	4.17	4.11	4.93	6.13	5.51	5.99	5.96
	BishHari	4.15	4.99	4.92	4.83	4.74	5.38	6.56	5.92	6.07	5.97
	Padma Pukuria	3.94	4.91	4.92	4.92	4.76	5.28	6.30	5.68	6.16	6.13
	Mohammandpur	4.41	5.37	5.34	5.17	4.83	5.41	6.44	5.82	6.06	5.93
	Shankarpur	4.33	5.16	5.71	4.87	4.62	5.36	6.50	5.55	6.12	6.04

Figure 1 shows the average ground water level fluctuation in the observation well at Jessore sadar and Jhikargacha over the time period. The trends of average ground water level increased gradually from 2004 to 2006 where except in Sheikhati mouza observation well where it is shown abrupt

increased from 2004 to 2005 and then gradually decreased to 2008. After 2008 the average ground water level increased rapidly and going to the peak at 2010 in both upazilla and then small decreased to 2013.

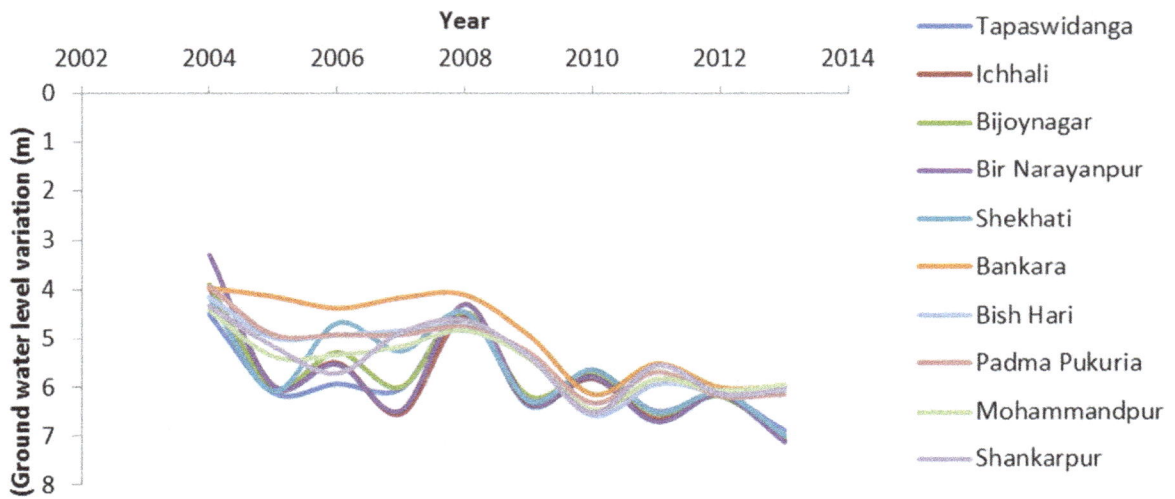

Figure 1. *Average ground water level fluctuation trends over the time period in Jessoresadar and Jhikargacha.*

Figure 1 also shows that groundwater level variation (maximum water level-minimum water level) in the observation well of Jessore sadar and Jhikargacha respectively. It was seen that groundwater level variation ups and down in consecutive year except at Jhikargacha where shown rapid increase from 2004 to 2006 and rapid decrease to 2007 and then follows the same. But from the analysis it was found that groundwater level variation more increase at Jessore sadar where minimum variation in 2004 and

maximum in 2013 for all of the observation well. However in Jhikargacha, groundwater level variation shows somewhat different and it was minimum in 2004 and then variation abruptly increase to 2006, and that was the peak variation of all observation wells. In Jessore sadar, minimum water level variation 3.3 m at Bir Narayanpur in 2004 and maximum 7.11 m in 2013 at the same well. As well as in Jhikargacha, 2.5 m at Padma Pukur in 2004 and 6.51 m at Bankara in 2006.

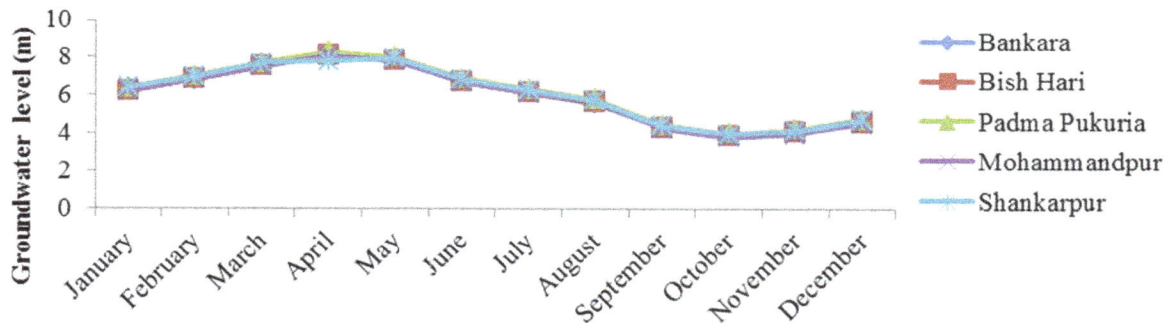

Figure 2. *Seasonal variation of ground water level at Jhikargacha in 2013.*

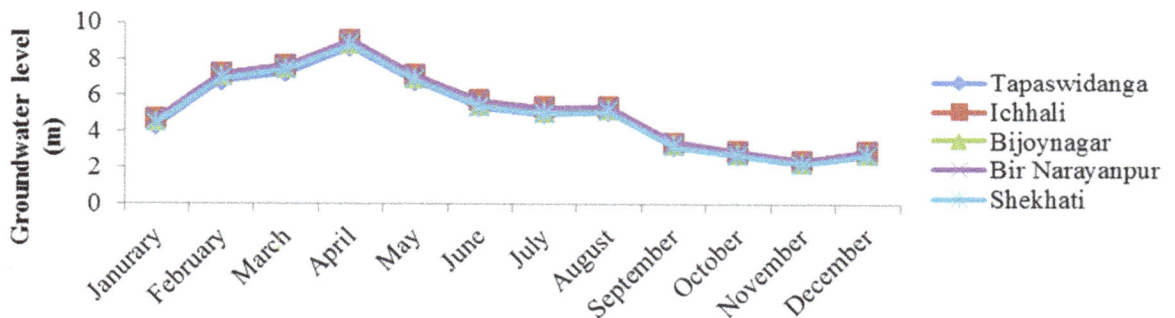

Figure 3. *Seasonal variation of ground water level at Jessore sadar in 2013.*

Figure 2 and 3 shows the seasonal variation of groundwater level in the observation well at Jhikargacha and Jessore sadar in 2013 respectively. From seasonal variation analysis in the experimental well at both areas were shown 2

peak levels where as maximum ground water level lies in the month of March to April and minimum ground water level in the month of October to November. In the study area most of ground water withdrawal was occurred in dry months starting from January and continues up to May and in June some cases in dry season. During the period the recharge was almost nil, the rate of evaporation and evapotranspiration was high. As a result the water level was declined sharply and reaches to maximum depth in April. In rainy season ground water level decrease in correspondence to the ground surface as because during this time begins to recharge to the underground storage. At the same time it was occurred minimum withdrawal of ground water for irrigation, and high relative humidity in the atmospheric reduces the rate of evaporation and evapotranspiration. All these cause a gradual increase in ground water reservoir which was reflected by the change of water table and it was going to minimum in the end of rainy period month of October to November.

3.2. Factors Affect to Ground Water Level Fluctuation

3.2.1. Rainfall

The table 2 shows that the average rainfall trends over the time period of 2004 to 2013 at Jessore sadar and Jhikargacha. The average rainfalls in 2004 to 2005 were decreased suddenly and then gradually increased to 2007 and after then decreased to 2012. From analysis of average ground water level and average rainfall it was found that ground water level fluctuation is related to rainfall as because rainfall is most important factor for ground water recharge. When the rainfall was higher the ground water level was lower. Figure 4 shows the seasonal rainfall variations in Jessore during 2004, 2009 and 2013.

Table 2. *Average ground water table, rain fall and river flow of Jessore sadar and Jhikargacha.*

Para meter	Years									
	2004	2005	2006	2007	2008	2009	2010	2011	2012	2013
GWT	4.340	5.398	5.545	4.741	4.541	5.687	6.036	5.812	5.736	5.154
Rainfall	207.50	125.1	147.2	177.6	161.0	133.7	103.3	94.4	89.6	144.5

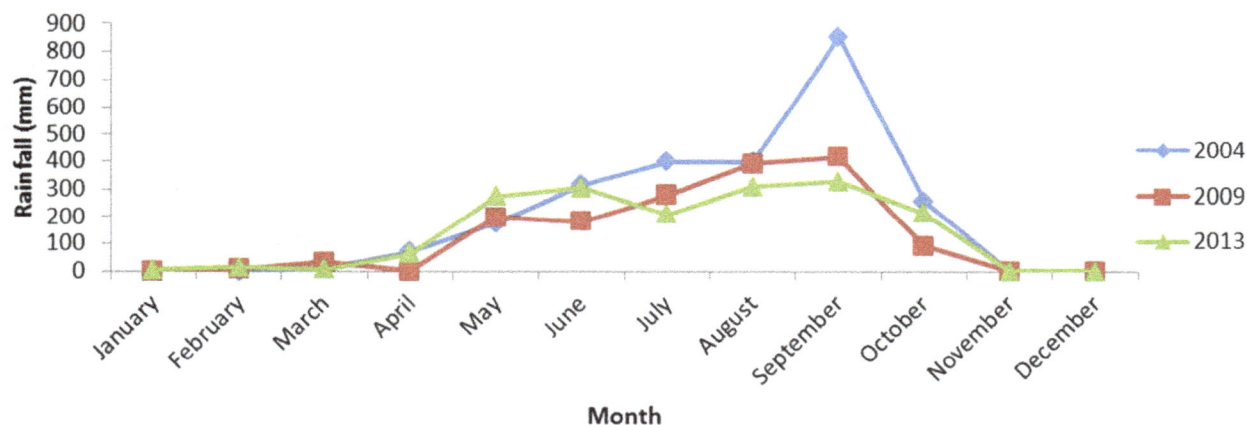

Figure 4. *Seasonal rainfall variations in Jessore.*

3.2.2. River Flow

Jhikargacha upazilla lies in catchment area of river Kapatakha. Table 2 shows that the average river flow discharge varies during time period 2006 to 2013 and it is highest in 2008 and the rapidly decrease to 2010 after then increase gradually with ups and down. Table 2 also shows that ground water level fluctuation trends have shown significant in aspect of average river flow discharge in 2010 highest in ground water level and then its gradually decrease according to the river flow with ups and down.

3.2.3. Welling Intensity

According to table 3 it found that DTW were increased steadily in both areas and it were 277 to 289 in 2005 and 297 to 337 in 2013 respectively at Jessore sadar and Jhikargacha. In case of STW, it was increased 8220 to 8917 from 2005 to 2013 in Jessore sadar. But in Jhikargacha the number was increased 8383 to 8448 from 2005 to 2009 after then decreased to 8175 in 2013. From the study it was seen that

the average irrigable land was about 5 acre for STW and 25 acre for DTW. In consideration of soil characteristics the command area would be 8-10 acre for STW and 45-50 acre for DTW. In time correspondence the number of irrigation equipment was increased but the irrigated land was remaining same. So the water withdrawn from ground water was increased that ultimately affects to ground water level and in dry season during April-May it reaches peak. Irrigation practice in the study area is mostly from ground water source, a greater portion of which is being abstracted through STWs Shahid [11] and Shahid and Behrawan [12] stated that the contribution of ground water has increased from 41% in 1982-83 to 77% in 2006-07and surface water has declined accordingly.

Table 3. *Numbers of DTW and STW over the time period.*

Study Area	DTW			STW		
	2004	2009	2013	2004	2009	2013
Jessore sadar	277	292	297	8220	8331	8917
Jhikargacha	289	322	337	8383	8448	8175

3.3. Correlation GWTD, Rainfall and River Flow

Table 4 shows correlation between GWTD and Rainfall in Jessore sadar. From the statistical analysis it was found negative correlation between rainfall and GWTD where the statistical correlation value is 0.900 that is strongly correlated and shows highly significant (Annex A3). It means if the rainfall increases then the GWTD decreases because rain water infiltrates to the earth and recharge to ground water table that leads to decrease depth of GWT. Table also shows correlation among GWTD, Rainfall and River flow in Jhikargacha. From the statistical analysis it was found negative correlation between GWTD and rainfall where statistical value is 0.822 that is also strongly correlated and shows highly significant. In case of GWTD and river flow, it was also seen negative correlation and the value is 0.535 that means medium correlation and it also shows significant but lesser significant than rainfall. In that cases if the river flow increases then there has occurred recharge to GWT which leads to decrease depth of GWT also. So rainfall is the important factor in both areas for GWT recharging.

Table 4. Correlation between GWTD and Rainfall in Jessore sadar.

	Jessore sadar		Jhikargacha		
	GWT	Rainfall	GWT (Avg)	Rainfall	River Flow
GWT	1		1		
Rainfall	-0.900082467	1	-0.822335596	1	
River Flow	-	-	-0.535602603	0.456343195	1

3.4. Ground Water Level Fluctuation and Impact on Irrigation

3.4.1. Impact on Pumping Systems

In dry season particularly during April to May ground water level fluctuation was high which gone 9.3 m maximum at Jessore sadar in April 2013 and 8.3m at Jhikargacha at April 2013 (Fig 2 and 3). The fluctuation level was highly significant for the centrifugal pump because of in practical ground water can be withdrawn within 7.5 m. From the observation it was found that 100% of STW at that area have faced pump failure and for the remedial measure 100% taken measure to deep set method. In deep set method, pump head was lowered 2.5 to 3 m under the ground surface by digging out within the bore. About 60% DTW were faced unable to withdraw of water. The reason here mentioned for instance of DTW that those tube well were installed during 1980 to 1990, at that time pumping head was designed for 12 to 16 m. For these circumstances well pumping has created drawdown that goes 1to 2m and causes failure to discharge. For that situations 100% cases adjusted pump dynamic head with lowering of the pump into the casing. It was seen that the trends of maximum fluctuation level increasing 4.5 m in 2004 and increased to 9.3 m in 2013 at Jessore sadar where as for Jhikargacha it was 4.5 m in 2004 and 9.6 m in 2013. So both the cases it was alarming for the irrigation in relating to irrigation pump because for STW method used 85% for irrigation sector. Moreover, it associated to increase cost in terms of pumping system management and also reduced the irrigation coverage. It was seen that for Deep tube well that are installed during 1980 to 1985 by BADC also facing pump failure because at that time turbine was set in the depth of 10 m to 15 m. In that cases withdrawing water from the well creating drawdown which going beyond that level. For instance BADC, made the correction by lowering the pump head.

3.4.2. Impact on Irrigation

From the study it was found that irrigation pump mostly used in during Boro-season and the peak use in the month of March-April. Table 5 shows the average pumping hour for DTW and STW respectively during 2013 and 2009. It was found pumping hour for DTW 1086 to 1256 and 655 to 922.5 for STW during period 2009 to 2013. During T-aman season the utilization of pump depends on rain. So over the time period pumping hour for both STW and DTW were increased. In addition irrigation costing was increased due to pump failure including deep set transplant for STW and pumping head adjustment for DTW. According to BADC and DAE, there are no rules and regulation for irrigation in relating to irrigation system and for the water users. For this reason who are able to install pump getting pump without concern of regarding institutions.

Table 5. Average DTW and STW pumping hour.

Crops season	Month	Average DTW pumping hour for 33 acre irrigated land		Average STW pumping hour for 8.5 acre irrigated land	
		2013	2009	2013	2009
Boro season	January	212	180	95.5	72
	February	217	195	130	90
	March	401	360	240.5	180
	April	327	295	196	150
	May	21	18	12.5	8
T-Aman Season	August	37	15	48	25
	September	14	10	70	55
	October	17	8	80	45
	November	10	5	50	30
	Total hour	1256	1086	922.5	655

4. Conclusion

In conclusion the groundwater level fluctuation at the study area has shown significant to irrigation. Average groundwater level fluctuation and the maximum water level variation increases day by day putting pressure on irrigation system as well as irrigation costing in regards on pumping hour, command area and the irrigation management. Decreasing average precipitation, unusual nature of river water flow and increasing welling intensity greatly affect to groundwater level fluctuation at the study area. Moreover the lack of knowledge on efficient irrigation management and the crop water use also affect the groundwater level fluctuation. To obtain desired economic benefits from groundwater resource, the management of ground water is essential. The proper management of ground water resources requires adequate knowledge of the extent of the storage, the rate of discharge, the rate of recharge to ground water body and the use of economical withdrawal.

References

[1] N. C. Dey, S. K. Bala, A. K. M. Islam, M. A. Rashid, and M. Hossain, "A study on Sustainability of Groundwater Use for Irrigation in Northwest Bangladesh" The study was carried out with the support of the National Food Policy Capacity Strengthening Programme, Government of the People's Republic of Bangladesh, Dhaka, 2013.

[2] N. C. Dey, S. K. Bala, and H. Seiji, "Assessing the Economic Benefits of Improved Irrigation Management: A Case Study of Bangladesh" *Journal of International Water Association*, pp, 8, 2006.

[3] N. C. Dey, and A. R. M. Ali, "Changes in the Use of Safe Water and Water Safety Measures in WASH Intervention Areas of Bangladesh, a Midline Assessment" Working Paper, No. 27, BRAC-RED, Dhaka.573-84, 2010. Available at: http://www.iwaponline.com/wp/00806.

[4] D. P. Loucks, "Sustainable Water Resources Management" *Journal of Water International*, Vol. 25(1), pp. 3-11, 2000.

[5] X. Cai, C. Daene, M. Kinney, M. W. Rosegrant, "Sustainability Analysis for Irrigation Water Management: Concepts, Methodology, and Application to the Aral Sea region" Discussion Paper, No. 86, International Food Policy Research Institute, USA,2001.

[6] K. Takara, and S. Ikebuchi, 'Japan's 1994 Drought in Terms of Drought Duration Curve', Proceedings of the fifth symposium of Water Resources, pp. 467-477, 1997.

[7] A. K. Sajjan, A. B. Muhammed, and C. D. Nepal, 'Impact of 1994-95 Drought in the Northwest of Bangladesh through Questionnaire Survey', Proceedings of the 2nd annual paper meet of Agricultural Engineering Division, Institution of Engineers, Bangladesh, 31 May 2002.

[8] N. C. Dey, M. S. Alam, A. K. Sajjan, M. A. Bhuiyan, L. Ghose, Y. Ibaraki, and F. Karim, "Assessing Environmental and Health Impact of Drought in the Northwest Bangladesh", *Journal of Environmental Science & Natural Resources*, 4(2), pp. 89-97, 2011.

[9] A. Hossain, "Groundwater Monitoring and Management" Pakistan Water Partnership (PWP), Islamabad, Pakistan, 2000.

[10] BADC, *'Statistical Report on Irrigation in Jessore'*, Bangladesh Agricultural Development Corporation (BADC), Ministry of Agriculture, Government of the People's Republic of Bangladesh, Dhaka, 2012.

[11] S. Shahid, "Spatial and Temporal Characteristics of Droughts in the Western part of Bangladesh", *Hydrology Process*, 22, pp. 2235–2247, 2008.

[12] S. Shahid, and H. Behrawan, "Drought Risk Assessment in the Western part of Bangladesh", *Journal of Natural disaster*, 46, pp. 391-413, 2008.

17

Investigation into the Effectiveness of Selected Bio-Based Preservatives on Control of Termite and Fungi of Wood in Service

Faruwa Francis Akinyele, Egbuche C.T., Umeojiakor A. O., Ulocha O. B.

Department of Forestry and Wildlife Technology, Federal University of Technology, Owerri, Imo State, Nigeria

Email address:

faruwa@gmail.com (F. F. Akinyele)

Abstract: The study focused on the effectiveness of using bio based preservatives as a controlling measure to biodetoriation from fungi and termite for wood in service. Wood samples of *Triplochiton scleroxylon, Gmelina arborea, Ceiba pentandra* used for the study were obtained from the sawmill of Ondo State Afforestation Project (OSAP) Oluwa and processed at the wood workshop of the Department of Forestry and Wood Technology, Federal University of Technology Akure. The sample were converted into 60 mm × 20 mm × 20 mm and seasoned to 12 % moisture content. The samples were treated with bio-preservatives from *Parkia biglobosa, Tridax procumbens* and tar oil obtained via pyrolysis. This study showed that even though *Ceiba pentandra* has the highest retention for the preservatives, it is the most susceptible to fungal and termite attack. *Gmelina arborea* with the lowest retention have resistance to termite and fungal attack, thus, the effectiveness of the preservatives on the wood samples is not only determined by the retention level of each preservative but also the chemical constituent (pH) of the preservative. This study recommends the use of tar oil for preservation of the wood samples against termite where colour is not important while, ethanolic extract of *Tridax procumbens* can be used to preserve wood against fungal attack.

Keywords: Ethanolic Extract, *Tridax procumbens,* Tar Oil, Preservation, Termite

1. Introduction

Wood has been a pre-dominantly constructional material for both domestic and industrial use globally. According to [5], more than 80% of timber product in Nigeria is utilized for construction purposes such as doors, windows, rafters, flooring, panels, purlines, furniture and cabinets making, tool handles, packaging, crates, and bridge carving among others. Despite the usefulness of wood, its service life can be degraded by various biodeteriorating agents; these include fungal infection, termites, insects, marine borers. fire attack and mechanical failure [1]. For effective utilization, wood needed to be protected from the attack on it by decay fungi, harmful insects, or marine borers by applying preservatives.

Wood preservation is the chemical way of conditioning wood to increase its resistance to invading destructive organisms (fungi, insects, marine borers and animals) and deterioration caused by unfavorable environmental conditions to protect the life span of wood in service [4].

The degree of protection achieved depends on the preservative used and the proper penetration and retention of the chemicals. Some preservatives are more effective than others, and some are more adaptable to certain use requirements.

With the current problem of global warming and environmental degradation, Some of the substances used as chemical preservatives have been prohibited due to environmental restrictions and thus, there is a search for alternative techniques which can extend wood service life, and which also at the same time is less harmful to the environment and man [2]. This necessitated the use of natural plant extract (Bio-preservatives) which are less harmful to the environment and more economical to protect wood against degrading agents such as fungi and insect. Biopreservatives constitute a wide range of natural products from both plants and animals which can be useful in the extending shelf life of wood reducing or eliminating survival of pathogenic bacteria and increasing overall quality of the wood products. The objective of this study is therefore, to evaluate the effects of plant extracts as preservative on wood and to evaluate their anti-fungal and anti- termite efficacy.

2. Materials and Methods

Wood samples of *Triplochiton scleroxylon*, *Gmelina arborea* and *Ceiba pentandra* used for the study were obtained from the wood sawmill of Ondo State Afforestation Project (OSAP) Oluwa and processed at the wood workshop of the Department of Forestry and Wood Technology, Federal University of Technology Akure. The wood samples were dimensioned into 60 mm × 20 mm × 20 mm and seasoned to 12 % moisture content. The wood samples were treated with bio-preservatives from extracted from *Parkia biglobosa*, *Tridax procumbens* and tar oil obtain via pyrolysis. The wood samples were divided into three sets based on the type of Field test to be conducted (Termite and Fungi attack test). A total number of 72 wood samples were used for the test (3 from each species treated with bio-preservatives and 3 each as replicate for each species) for each field test (36 each for both Fungi and termite attack). Preparation and impregnation of wood samples were carried out according to ASTMD1413-76 standard. The test samples were marked using a water proof permanent marker for ease identification. Samples were oven dried at $103\pm2°C$ for 24 hours and the oven dry weight recorded as T_1. Treatment of samples was done using bio preservatives. The method of preservative adopted was cold soaking for 72 hours. The samples were drained of excess preservatives, re-weighed and the treated weight was recorded as T_2. Percentage absorption and retention of treated samples was calculated as follow

Absorption = $100.\left(\frac{T_2-T_1}{T_1}\right)$

Retention (Kg/m^3) = $\frac{GC}{V} \times 10$

Where:

$G = (T_2 - T_1)$ = initial weight of the wood before treatment subtracted from

Initial weight of wood plus treatment,

C = grams of preservative in 100g of treating solution,

V = volume of sample in cm^3

2.1. Method of Preservative Extraction Tridax procumbens

Tridax procumbens was collected from the field and air dried at room temperature for 10 days, and pulverized. The ethanolic extract of the sample was prepared by soaking 100g of the powdered sample in 1litre of 95% ethanol for 24hrs.

2.2. Parkia Biglobosa

The seeds of *Parkia biglobosa* were collected from "Oja-oba" market in Ibadan, and soaked in water for about 15 minutes and washed in water to remove dirt. The seeds were dried and then cooked for 24 hours; the extract was separated from the seeds. Tar oil was obtained from wood pyrolysis at 650°C from *afzelia africana* (apa).

3. Result and Discussion

3.1. Retention

Table 1 shows the percentage retention of the Bio-preservatives on the selected wood species. The result shows that for tar oil treatment, *Ceiba pentandra* has the highest retention value (385.41±35.23), followed by *Triplochiton scleroxylon* (228.47±0.03) and *Gmelina arborea* (156.25±63.10). For *P. biglobosa* and *T. procumbens* extracts *Triplochiton scleroxylon* has the highest retention values. The experiment showed that among all the preservatives used, tar oil is the most retained bio-preservatives.

Table 1. The (Percentage) retention of tar oil, Parkia biglobosa, and Tridax procumbens ethanolic extracts on selected wood species.

Treatment	Gmelina arborea	Ceiba pentandra	Triplochiton scleroxylon
Tar oil	156.25±63.10ᵃ	385.41±35.23ᵃ	228.47±0.03ᵃ
Parkia biglobosa extract	0.19±0.12ᵇ	0.26±0.02ᵇ	0.37±0.19ᶜ
Tridax procumbens extract	0.13±0.03ᶜ	0.34±0.06ᵇ	7.32±8.25ᵇ

Value are mean ± SE. Value with same superscript are not significantly different (*), but different superscript means they are significantly different (ns) at p ≤ 0.05.

3.2. Absorption

Table 2 shows the percentage absorption of the preservatives on wood. *Ceiba pentandra* has the highest absorption (113.61±8.02), followed by *Triplochiton scleroxylon* (55.25±10.14), and *Gmelina* arborea (32.80±14.80) respectively for tar oil treatment. In woods treated with *Parkia biglobosa* extract, *Gmelina arborea* and *Triplochiton scleroxylon* had the lowest (54.33) and highest(128.98) absorption respectively, while for wood treated with *Tridax procumbens* extract. *Ceiba pentandra* had the highest absorption (77.23±44.10), while *Gmelina arborea* and *Triplochiton scleroxylon* had lower absorption, with the value of the two wood species having no significant differences (P>0.05).

Table 2. (%) absorption of tar oil, Parkia biglobosa, and Tridax extracts on selected wood species

Treatment/ species	Gmelina arborea	Ceiba pentandra	Triplochiton scleroxylon
Tar oil	32.80±14.80ᶜ	113.61±8.02ᵃ	55.25±10.14ᵇ
Parkia biglobosa extract	54.33±34.19ᶜ	110.42±7.53ᵇ	128.98±24.26ᵃ
Tridax procumbens extract	22.56±8.06ᵇ	77.23±44.10ᵃ	12.55±16.69ᶜ

Value are mean ± SE. Value with same superscript are not significantly different (*), but different superscript means they are significantly different (ns) at p ≤ 0.05.

3.3. Weight Loss

3.3.1. Percentage Weight Loss as a Result of Termite Infestation

Table 3 shows the result for the percentage weight loss that occurred in the wood species after 12 weeks of exposure to termite attack. The Table showed that *Gmelina arborea* has the lowest weight loss, followed by *Triplochiton scleroxylon*, with *Ceiba pentandra* having the highest weight loss for the treated samples, while the untreated wood species has the highest overall weight loss. There is no significant difference within wood treated with tar oil, but for the other treatment, there is significant difference among the wood species. This means that *Gmelina arborea* treated with tar oil had the highest resistance to termite attack (18.40±4.57).

Table 3. % weight loss of wood after exposure to termites attack

Preservatives	Weight loss of selected wood species		
	Gmelina arborea	*Ceiba pentandra*	*Triplochiton scleroxylon*
Taroil	18.40±4.57	46.37±26.01	26.22±8.67
Parkia Biglobosa extract	80.1±87.46	100.00±0.00	100.00±0.00
Tridax procumbens extract	10.50±6.63	100.00±0.00	47.01±46.01
Control	20.13±14.92	100.00±0.00	100.00±0.00

Value are mean ± SE. Value with same superscript are not significantly different (*), but different superscript means they are significantly different (ns) at $p \leq 0.05$.

3.3.2. Percentage Weight Loss as a Result of Fungi Infestation

Table 4 shows that *Gmelina arborea* treated with *Tridax procumbens extract* has the lowest weight loss (8.26±8.86), but there is no significant difference in treatment effect of wood species treated *Tridax procumbens extract*. *Ceiba pentandra* treated with *Parkia Biglobosa extract* have the highest weight loss (61.29±6.66). Generally, there is significant difference among treatment.

Table 4. % weight loss of wood after inoculation of fungi

Preservatives	Weight loss of selected wood species		
	Gmelina arborea	*Ceiba pentandra*	*Triplochiton scleroxylon*
Taroil	14.73±0.56	39.28±9.22	27.83±14.00
Parkia Biglobosa extract	18.94±3.81	61.29±6.66	44.74±3.15
Tridax procumbens extract	8.26±8.86	9.63±3.59	11.16±7.72
Control	26.89±0.68	29.95±2.07	29.70±1.23

3.4. Result for Visual Observation

Visual Observation After Exposure to Termite

This study revealed that wood species treated with tar oil had the highest rating, meaning that it is resistance to termite attack. Figure 1 showed that *Ceiba pentandra* treated with Tridax procumbens extracts was attacked by termite within two weeks of field test. After four weeks of exposure to termite *Ceiba pentandra* treated with *Tridax procumbens* extract have been completely deteriorated and at ten weeks *Ceiba pentandra* and *Triplochiton scleroxylon* treated with *Tridax procumbens* and *Parkia biglobosa* had failed. At the end of twelve weeks of observation untreated *Ceiba pentandra* and Triplochiton *scleroxylon*, also *Ceiba pentandra* treated with *Tridax procumbens* and *Parkia biglobosa* have been completely deteriorated by termite. This shows that *Ceiba pentandra* is the most susceptible to termite attack.

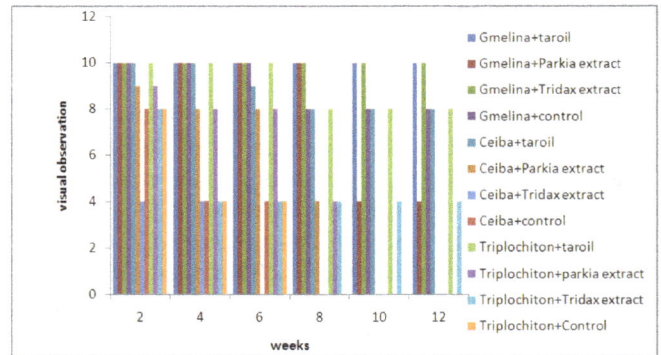

Figure 1. visual observation of wood species exposed to termite deterioration

Visual observation after inoculation of fungi Figure 2 showed that there is no growth of mycelium on wood species treated with tar oil, wood treated with *Parkia biglobosa* reveal growth of mycelium.

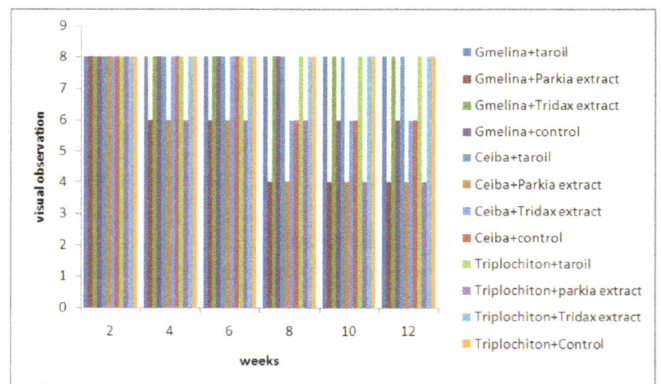

Figure 2. Visual observation after inoculation of fungi on wood species

4. Discussion

4.1. Retention

The result in Table 1 showed that the retention of preservatives on the wood species is significantly different. This agrees with the result obtained by [7] that wood exhibit variation in absorption and retention of chemical preservatives when treated with wood species. This variation in biocide retention may occur due to variation in the anatomical structure of wood species. Also, this view is supported by [5] where he reported that the amount of preservative retained in the cellulose material possibly varies due to differences in anatomical structures of the various wood species. However, for the preservative treatment, tar oil is highly retained by the three wood species selected when compared with retention of other preservatives. This may be due to its oil based attribute and its viscosity.

4.2. Absorption

As shown in Table 3, *Triplochiton scleroxylon* has the highest absorption. This correlates with the work of [3] where he reported that anatomical structure of wood exhibit a lot of influence on the absorption and retention of wood preservatives. The differences in absorption could be as a result of extractives deposited in the wood samples.

4.3. Weight Loss After Exposure to Termite Attack

The result revealed that wood treated with tar oil had the lowest weight loss. This shows that tar oil is effective for preservation of wood against termite attack but *Tridax procumbens* and *Parkia biglobosa* had the highest weight loss, this indicate that these two treatment are not effective against termite attack. The effectiveness of tar oil may be because is highly toxic as it is found to contain carboxylic acid in greater percentage as reported by [8]

4.4. Weight Loss After Inoculation of Fungi

This study showed that untreated wood samples and wood samples treated with *Parkia biglobosa* extracts have means with the lowest weight loss, while wood species treated with *Tridax procumbens extract* have the highest mean for weight loss, followed by wood species treated with tar oil. From this study *Ceiba pentandra* have the highest weight loss, followed by *Triplochiton scleroxylon*, and *Gmelina arborea* have the lowest weight loss, this may be due to variation in the anatomical structure and the natural durability of wood.

Visual observation for termite deterioration. This indicates that *Parkia biglobosa* and *Tridax procumbens* extracts are not effective for the preservation of wood against termite attack. Tar oil is the most effective preservative of the selected preservatives on the wood species but the limitation is that it impacts colour on the wood, this is in conformity with the work of [1] that pyrolytic oil and Jatropha oil are effective for the preservation of wood. Tar oil can be recommended for the protection of the selected wood species against termite attack.

4.5. Visual Observation for Fungi Deterioration

This study shows that wood samples treated with *Tridax procumbens* extract does not show the growth of fungi mycelium, this indicate that the extract is capable of protecting wood samples against fungal attack, this is in agreement with the findings of [6], while working on the antifungal property of *Tridax procumbens* L. against three phytopathogenic fungi. This is supported by similar work carried out by [9], he reported the antifungal activities of crude medicinal plant species including *Ocimum sanctum, Ricinus communis and Jatropha curcas*. Wood samples

treated with parkia biglobosa revealed growth of fungal mycelium. It was observed that *Ceiba pentandra* is more susceptible to fungi deterioration and this might be due to the differences in the physical properties of the two wood samples.

5. Conclusion

This study showed that even though Ceiba pentandra has the highest retention for the preservatives, it is the most susceptible to fungal and termite attack. But Gmelina arborea with the lowest retention have resistance to termite and fungal attack, so the effectiveness of the preservatives on the wood samples is not only determined by the retention level of each preservative but also the chemical constituent of the preservatives.

In conclusion, after twelve weeks of field test, for the termite deterioration tar oil is found to be the most effective preservative, and for the fungal deterioration *Tridax procumbens* extract was found to be the most effective.

This study recommends the use of tar oil for preservation of the wood samples against termite where colour is not important, and *Tridax procumbens* from ethanolic extract can be use to preserve wood against fungal attack.

References

[1] Areo O. O (2002): Preservation of small round logs of Tectona grandis exposed to termite attacks using copper chromium arsenate (CCA). An HND submitted to the Federal College of Forestry, Ibadan.

[2] Arldo C.Bernardis and Orlando popoff (2009): Durability of *pinus elliottii* wood impregnated with quebracho Colorado (schinopsis balansae) bio-protectives extracts and CCA. Maderas Ciencia technologia 11(2):106-115.

[3] Desch, J.M. (1981): Timber Its nature and behaviour, van Nostrand Reinhold Co. Ltd, Bershire, England. 170pg.

[4] Elegbede M.O (2002): A profile of wood preservation industry in Ibadan and Lagos Metropolis. A project submitted to Department of Wood and Paper Technology in the Federal College of Forestry, Ibadan Oyo state pg 2-4

[5] Ogbogu G.U (1996): The State of Wood Treatment Technology in Nigeria Institute of Nigeria

[6] Sandeep Acharya and Srivastava R.C.(2010): Antifungal property of *Tridax procumbens L.* against three pythopathogenic fungi Arch. Arch. Pharm. Sci. & Res Vol. 2, No.1, pg 258-263.

[7] Schutlz, T.P, Nicholas, D.D, Henry, W.P, Pittman, C.U, Wipf, D.O and Goodell, B. (2005): Review of laboratory and out door exposure efficacy results of organic biocide; antioxidant combinations, an initial economic analysis and discussion of a proposed mechanism. Wood and Fiber Science. 37(1); (175-184).

[8] Sipilae` K, Kuoppala E, Fagernae`s L (1998). Characterization of biomass–based flash pyrolysis oils. Biomass Bioenergy ; 14(2): 103–13

[9] Siva, N., Ganesan, S., Banumathy, N. and Muthuchelian
     (2008): Antifungal Effect of Leaf of Extract of Some
     Medicinal Plants Against *Fusarium oxysporum* Causing Wilt
     Disease of *Solanum melongena L.* Ethnobotincal Leaflets 12:
     156-163.

18

Evaluating the Spatial and Environmental Benefits of Green Space: An International and Local Comparison on Rural Areas

Luan Cilliers

Unit for Environmental Sciences and Management, North-West University, Potchefstroom Campus, South Africa

Email address:

luan.cilliers@gmail.com

Abstract: Many issues exist from isolated planning of urbanized areas and environmental areas. Current approaches focussing on the integration of Urban Planning and Urban Ecology seek to address such issues. Urban Ecology practice aims to describe the study of the joined relationships between humans and nature. Urban Ecology thus forms a major part of Urban and Spatial Planning, with regard to the objectives of sustainable planning and development, green infrastructure planning, and resilience. Green spaces support sustainable human settlements by means of the different benefits which nature provides, referred to in this research as ecosystem services. Green spaces, in this sense, are fundamental areas in human settlements, in need of intentional and structured planning approaches to enhance sustainability and said environmental benefits. Rural settlements in South Africa experience various problems and challenges in terms of planning for the environment through green spaces (as well as sustainability), mainly as a result of the fragmentation of these rural areas, the existence of lost spaces, urbanisation, urban sprawl and poverty. This research attempted to address the challenges of integrated planning and green space provision in a local rural context; evaluated the spatial and environmental benefits of green space; and enhanced the importance of planning for such benefits in rural South African areas. A local and international comparative study was conducted in order to evaluate the green space planning of South African rural areas in terms of international approaches. The comparative study also served as guidance for new green space planning approaches and recommendations in South African rural context.

Keywords: Ecosystem Services, Rural Areas, Green Spaces, Green Infrastructure Planning, Resilience

1. Introduction

Urbanized environments are often being studied individually and separate from its surrounding natural environment [1]. Current approaches focussing on the integration of Urban Planning and Urban Ecology seek to address these issues of integrated planning. Urban Ecology is an emerging interdisciplinary field that aims to understand how human and ecological processes can coexist in systems dominated by humans and help societies with their efforts to become more sustainable [2].

The term 'urban ecology', because of its unique focus on both humans and natural systems, has been used in various ways to describe the study of (a) humans in human settlements, of (b) nature in human settlements, and of (c) the joined relationships between humans and nature [2].

Studies that aim to enhance environmental planning in urbanized areas have indeed increased over the years internationally, especially by the means of the new effective method, green infrastructure planning [3]. In South Africa however, similar studies were conducted on macro scale and not on specific micro areas that endure specific problems caused by the lack of integration [3]. A lack of such a study on micro scale that addresses specific environmental problems unique to the specific area, thus exist in South Africa. There is also a lack of such a study within rural areas that desperately need change in the way the environmental planning (if any) is conducted.

The environment in human settlements is dependent on the people (in terms of planning and conservation) but the people (society) are also dependent on the environment (in terms of certain benefits which are provided by the said green spaces and environment). Human settlements depend on a healthy environment that continuously provides these benefits known as ecosystem services [4]. When the benefits that nature provides in the specific area are identified and the value of these benefits is understood, the movement towards creating a sustainable human settlement can progress even more.

In an attempt to bridge the objectives of Urban Ecology and Spatial Planning, this research aims to evaluate the spatial and environmental benefits of green spaces; and how such spaces can be planned in especially rural areas to enhance the said benefits.

2. Delineation of Case Studies

South Africa does not appear to be as evolved in integrating urban and rural areas with the natural environment as is the case internationally [3]. It is thus important for South African Planners to consider the countries which are indeed more evolved and successful in this regard. Following a best practices approach enables local planners to consider strengths and opportunities of an integrated planning approach, while seeking ways to adopt international approaches to fit the local context and challenges.

2.1. International Case Studies

Planning in Sweden is conducted from a framework that respects the need to preserve the natural environment [5]. The city of Stockholm is creating policies and using planning in such a way that develops a more sustainable society, a better quality of life for the people, and a quality natural environment [5].

The areas in and around Stockholm were thus selected to serve as international case studies to review, evaluate and compare; thus contributing to the identification of applicable approaches for South Africa. The following two micro case studies in Sweden were studied.

Table 1. International case studies.

Case study	Type of area	Significance of case study
Hågaby	Eco-village (rural area)	A resistant, flexible human settlement integrated with green structures.
Hammarby Sjöstad	Eco-city (urban area)	Integration of environmental goals in order to provide a healthier and conserved environment.

The reason for the evaluation of a rural and urban area respectively, is to obtain diverse results from diverse area typologies, thus providing diverse planning approaches.

2.2. Local Case Study

The Vaalharts area, located in the Northern Cape Province of SA, consists of a beautiful rural landscape with scattered rural settlements and is best known for its water irrigation scheme [6]. The Vaalharts area was thus selected as the local case study as its characteristics represent those of an average rural South African area. The residents in the rural settlements in the Vaalharts area are mainly dependent on the environment

(mostly agriculture and water). This contributes to the suitability of this area as the local case study in this research.

Each of the above mentioned international and local areas was firstly studied by means of physical surveys and site analysis, and secondly through structured questionnaires provided to key informants and professionals.

3. Ecosystem Services

Parks and other green spaces play multiple roles in making human settlements more sustainable [7]. This is mainly due to the different benefits which nature provide, known as ecosystem services. Parks and green spaces are thus fundamental areas in human settlements that need intentional planning as it provides the opportunity to enhance sustainability and the appearance of environmental benefits [8].

In order to define the concept of ecosystem services, it is important to understand that healthy ecosystems are the foundation of sustainable human settlements and in order for a settlement to be 'healthy' it depends on the natural environment that continuously provides a range of benefits [4]. Ecosystem services are thus the benefits that humans derive directly or indirectly from ecosystem functions.

3.1. Ecosystem Categories

Ecosystem services can be divided into four categories, namely Provisioning services, Regulating services, Habitat or supporting services and Cultural services.

3.1.1. Provisioning Services
Ecosystem services which are mainly material or energy outputs from ecosystems, such as food, water and medicinal resources [4].

3.1.2. Regulating Services
Services provided by regulating the quality of air and soil, or providing flood and disease control [4]. Examples are natural waste-water treatment, erosion prevention and local climate regulation. Fig. 1 illustrates examples of (a) food as provisioning service and (b) wetlands as natural waste-water treatment.

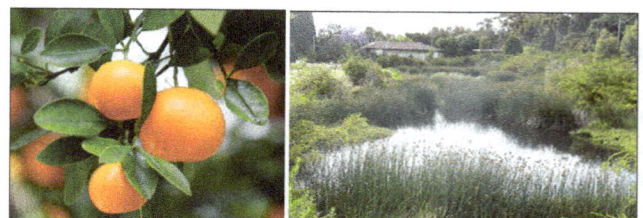

(a)   (b)

Figure 1. Provisioning and regulating services.

3.1.3. Habitat or Supporting Services
Ecosystems provide living spaces for plants or animals and can also maintain a diversity of plants and animals [4].

3.1.4. Cultural Services

(a)   (b)

Figure 2. Habitat/supporting and cultural services.

This is the non-material benefits that people obtain from contact with ecosystems. This includes aesthetic, spiritual and psychological benefits such as physical activities, recreation and tourism [4]. Fig. 2 illustrates examples of (a) a lake as natural habitat and (b) nature as inspiration for aesthetic activities.

3.2. Providing Ecosystem Services

An understanding and consideration of ecosystem services is necessary in order to create and maintain a well-planned and managed environment. An integrated planning approach is needed, incorporating green spaces as part of the spatial planning process, referred to as green infrastructure planning [9]; and creating an environment which can respond to change or disturbances and ensure sustainability for future generations, referred to as resilience [10].

Table 2 indicates theoretical principles regarding spatial planning and ecosystem services, which are important to include when planning for green spaces in order to enhance environmental benefits.

Table 2. Theoretical principles in green space planning.

Principles	Significance
Determining the status quo of the environment	Ensures an overview of the area's environmental strengths, weaknesses and opportunities.
Focus of the key stakeholders	The planning focus should include a focus on the provisioning of ecosystem services
Interconnection of green spaces	Ensures no isolation between the green spaces.
Consider green spaces as green infrastructure	Green spaces receive an equal amount of attention just as other infrastructure during planning processes.
Providing multifunctional green spaces	Green spaces that provide a variety of functions (heterogenic spaces).
Providing different ecosystem categories	Ensures a variety of ecosystem services and thus a quality environment.

Evaluating the planning of green spaces in the international and local case studies in terms of the above mentioned theoretical approaches, indicates a clear link in terms of the provisioning of environmental benefits internationally, while the local South African approach needs to be revised in order to be aligned with theoretical objectives.

4. Current Challenges in SA

The current reality of South African rural areas faces a number of environmental challenges. These challenges originate mostly from historic spatial patterns and occurrences in South Africa.

4.1. Existence of Lost Space

Lost spaces lead to the existence of deteriorated parks which serve no purpose and are in need of redesign because they make no positive contribution to the surroundings and users [11].

In the local case study of Vaalharts, green spaces are mostly left unplanned and isolated with no vision, usage or maintenance. This contributes to the existence of lost space which provides no environmental benefits.

4.2. Fragmentation of Human Settlements

Fragmented settlements encourage people to live isolated lives thus discouraging the use of public green spaces [12]. The Vaalharts area is located far from urban areas which creates a lack of economic activities in order to support planned green spaces. The fragmented environment in terms of human settlements, thus have a negative effect on the planning and maintenance of green spaces.

4.3. Urbanisation

Urbanisation pressurises the planning and maintenance of human settlements and especially the green spaces in these settlements [13]. The Vaalharts area is showing an increase in population as more people migrate to the informal human settlements. This could mainly be due to the low living costs in this area as well as the abundance of water and other natural resources. The unplanned green spaces are in effect developed into housing and other infrastructure to fill the growing housing need.

4.4. Urban Sprawl

The growing population causes urban sprawl to take place which in turn leads to the fragmentation of natural green spaces as urban sprawl follows no specific growth pattern [14]. The problem in the Vaalharts area concerning urban sprawl, is that the informal settlements are growing unplanned and unmaintained, filling the green spaces that had potential for the provisioning of ecosystem services.

4.5. Poverty

The need of the poor communities for natural resources causes them to use the natural environment and green spaces in unsustainable ways, thus causing degraded areas [13].

In the Vaalharts area a lack of finances for the planning of green spaces exists. This alludes to a lack of knowledge on ecosystem services. Unsustainable, unplanned and unmaintained green spaces with no specific functions are the result.

5. Policies and Legislation

Policies and legislation are important for the inclusion and guidance of planning for the environment. Fourteen policies and legislation were evaluated, whereof only four indicated a weak level of support for the planning of green spaces. A total of ten policies and legislation indicated a medium to high level of support.

The question however is whether these policies and legislation are comprehensive and sustainable to guide the planning and provision of green spaces in South African rural areas. The White Paper on environmental management policy for SA (1998) was concluded to be one of the policies with the strongest level of support for green spaces that provide environmental benefits specifically in rural areas. Legislation in broad thus supports the integration of the environment through the use of green spaces with urban and rural areas.

It is important to take note that even though the policies and legislation consist of the right focus and aims, it is of great importance that the local communities are informed, educated and involved in order to be able to implement and maintain such approaches. It is thus evident that rural settlements in the Vaalharts area are in need of education and more public participation in terms of environmental planning policy and legislation.

6. Green Space Planning: International vs. Local Approaches

Certain gaps exist in the local South African approach to planning and provision of green spaces. Accordingly the local South African approach to planning of green spaces was evaluated by means of a gap-analysis against the international best practices. Table 3 captures the key findings from the site analysis and questionnaires of the international case studies with the local case study.

Table 3. International best practices and gap-analysis in SA.

International best practices	Gap-analysis in the Vaalharts area
Economic activity of urban centres supports the green spaces.	Green spaces are not situated near economic activities as few economic activities exist in the rural settlements.
Direct environments consisting of natural green areas integrate easily with a settlement's green spaces thus creating more green spaces.	The environments directly bordering the rural settlements are mostly natural green areas with indigenous vegetation. Potential thus exists for the integration of green spaces.
The interconnection of green spaces in settlements.	Most green spaces are isolated from each other, but the rivers flowing through a few of the settlements cause some interconnection of green spaces to occur.
The planning and design of other aspects such as housing and transport has an influence on the green spaces.	Green spaces are not integrated with the surroundings and are planned isolated from the surroundings.
The conservation of the natural indigenous vegetation and animals when planning for green spaces is crucial.	The green spaces in the Vaalharts area consist mostly of natural indigenous vegetation as well as animals (in the game reserves).
Cooperation between various professionals is important when planning for green spaces.	A strength exists in the cooperation of a variety of professionals in order to plan for a more sustainable area.
Variations in the design of elements in the areas contribute to creating interesting and attractive areas.	Green spaces appear to be mostly homogeneous.

It is evident that a number of gaps exist in the green space planning approaches in the Vaalharts area, and ultimately in South Africa. However, a number of strengths can be noted in the Vaalharts area which contributes to its potential of enhancing its provision of ecosystem services. Table 4 lists these strengths.

Table 4. Strengths in the local case study.

Vaalharts is situated in an area rich in natural green areas and indigenous vegetation as well as animals.
The area is highly accessible.
A variety of land use types exist in the area.
Residents of the settlements in the area are dependent on the environment.
Cooperation of different professionals within the planning process exists.
The abundance of water ensures the provision of sustainable green spaces
The Vaalharts area has potential to support ecosystem services.

7. Planning Recommendations

Recommendations are made to provide new and innovative approaches to planning for the environment in rural areas in South Africa. These recommendations thus focus on proposing approaches which (a) comply with suitable theoretical approaches, as well as best practice approaches found in international case studies; and (b) enhance and provide different ecosystem service types from the different ecosystem categories within local rural areas. The recommendations are discussed subsequently.

7.1. Shifting the Focus to Ecosystem Services

The education of municipalities, key stakeholders and residents on the importance of ecosystem services, is the starting point to a shift in the focus when planning for settlements.

A method that was used in the international case study of Hammarby Sjöstad, is the education of residents by the use of placing educational information signs along the green spaces. These educational signs inform the residents on unique environmental qualities in the specific area. This can be incorporated with hiking trails and other recreational activities in green spaces. This approach can be implemented within the local rural context of South Africa as well. Fig. 3 illustrates such information boards in the international case studies.

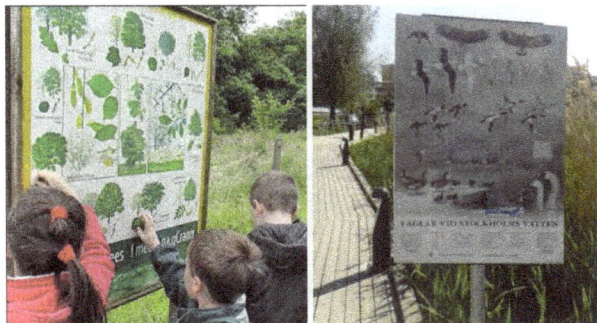

Figure 3. Educational information signs in parks.

7.2. Interconnecting Green Spaces

The rivers and streams running through the Vaalharts area provide opportunity for interconnected green spaces as the rivers physically connect the green spaces. The interconnected green spaces along the rivers and streams should be planned in an integrated manner in order to enhance activities and services such as recreation, conservation of the natural environment and storm water management. A green-avenue that was planned along the canals in the international case study is a best practice example of transforming canals (which created lost spaces along the sides of the canals) to interconnected green spaces that provide benefits and activities to the surrounding areas. Fig 4 illustrates green spaces that formed along canals or natural rivers.

Figure 4. Green-avenues planned along canals.

7.3. Integrating Green Spaces with Surroundings

The green spaces in the Vaalharts area need to be planned according to specific functions which it can provide to the specific surrounding communities within the area. Green spaces located nearby schools and residential areas (accommodating large numbers of children), can be functionally integrated with the surrounding area by planning the green space with a focus which will attract children. The international case studies were found to be located mainly in areas with high numbers of young families and children. Green spaces were thus planned mostly as playgrounds and sport facilities in order to functionally integrate the green spaces to the surrounding communities.

7.4. Providing a Variety of Ecosystem Services

As the quality of the natural environment in the Vaalharts area has great potential to provide environmental benefits, it is recommended that provisioning is made for ecosystem services from all four the ecosystem categories. This ensures

that a variety of ecosystem services are provided, addressing a great variety of the community's needs.

7.4.1. Provisioning Services

It is recommended that parts of the green spaces situated within the residential areas of the rural settlements are planned for the provisioning of food to the local communities. As the poor communities live in settlements of high density, every housing unit does not have its own backyard and thus need space for their own food production. It is thus recommended that community gardens be implemented in parts of the green spaces. Fig. 5 illustrates an example of a community garden in the international case study.

Figure 5. A community garden for residents with no private garden.

7.4.2. Regulating Services

The regulation of storm water in the Vaalharts area can be enhanced through the planning and provision of ecosystem services. The natural vegetation, including the soil, act as filters of water, breaking down waste in water through biological activities as the water flows through vegetated areas. It is thus recommended that built water drainage pipes along the roads be incorporated with the natural environment by removing parts of the built concrete pipes in order to ensure that the storm water flows through vegetated areas as well to filter the water. The water thus has to be directed towards the green spaces where it can be absorbed by trees and plants, or directed to water sources such as dams or streams in the green spaces, as in the cases of the international case studies. Fig. 6 illustrates how storm water is directed to green spaces and canals in the international case study.

Figure 6. A Storm water drainage system directed to a green space and canal.

7.4.3. Habitat or Supporting Services

This area's natural environment is considered a great habitat for African wildlife as a number of nature reserves exist within the area. It is recommended that current protected nature reserves which are located close or adjacent to human settlements be integrated with these settlements' green spaces. This creates a continuous habitat spreading throughout the entire Vaalharts area (This recommendation only considers small mammals, insects and birds that cannot be kept in by fences). Integrating the nature reserves with the settlement's green spaces was proven to be successful in the international case studies. This was done through the development of informative hiking trails stretching from the settlements' green spaces toward the nature reserves.

7.4.4. Cultural Services

Tourist activities should be integrated with green spaces in settlements. This can be done by using the green spaces as special markets or festivals on occasional days in order to sell agricultural products unique to the Vaalharts area. This enhances the usage of green spaces and ensures opportunities to inform tourists on the natural environment. This also provides economic benefits which brings a vital source of income to the Vaalharts area through the usage of the natural environment. The international case study of Hågaby used this method, thus creating an awareness among the tourists about the importance of the environment as well as sustainable living.

Recreational activities such as hiking trails can be implemented to contribute to the residents' mental as well as physical health, and to create an aesthetic appreciation in the area. Fig. 7 illustrates a unique recreational hiking trail that was implemented in the international case study.

Figure 7. A hiking trail developed along a water source in Hammarby Sjöstad.

The sustainability of these recommended approaches is greatly dependent on the cooperation of the local communities, local authorities and professionals involved in the planning process.

8. Conclusion

International approaches of planning for green spaces in order to enhance the provisioning of environmental benefits can be adopted locally in SA. This paper highlights that the following two issues are of great importance in the planning of green spaces to provide environmental benefits in the local South African context: (1) a unique green planning approach can be created for the specific area's characteristics and (2) a team of key role players understanding the importance and role of green spaces within the spatial planning reality should be involved. An integrated area (urbanized or rural) which provides environmental benefits to the community, while simultaneously conserving the natural environment can thus be created.

Table 5 captures the summative conclusions based on the evaluation of the spatial and environmental benefits of green space provision.

Table 5. Spatial and environmental benefits of green space.

Best practices identified in the study	Spatial benefits	Environmental benefits
Educating people on the importance of environmental benefits.	Expands the knowledge and insight regarding sustainability included in the spatial planning process.	Enhancement of residents' quality of life
Interconnection of green spaces.	Creates lively, open and accessible areas for the community.	Provides recreational activities, conservation of the natural environment and storm water management
Green spaces regarded as part of a settlement's infrastructure.	Spatially integrated, quality areas which are accessible	Regulation of storm water from urban infrastructure to green spaces which provides filtered water.
Multifunctional spaces.	Encourages mixed land uses as well as accessibility amongst different uses	Different activities such as gardens for food provisioning, recreational activities (hiking), educational activities (school tours), economic activities (shops and restaurants) and social activities are enhanced through the green spaces.
Integration of green spaces with surroundings.	Spatially integrated areas that are open and accessible	Enhances the usage and functions of green spaces as well as the quality of life, in especially poor communities.
Conservation of indigenous plants and animals in settlements.	Creates spatially attractive areas. Conserving nature influences the areas' functions and thus the physical structure of settlements	Provides a source of income for the area; purifies flowing water, healthier air quality, and habitats for animals

An integrated area (urbanized or rural) which provides environmental benefits to the community, while simultaneously conserving the natural environment can thus be created when a focus is placed on the achievement of spatial and environmental benefits.

Acknowledgements

This research (or parts thereof) was made possible by the financial contribution of the NRF (National Research Foundation) South Africa. Any opinion, findings and conclusions or recommendations expressed in this material are those of the author and therefore the NRF does not accept any liability in regard thereto.

References

[1] M.M. McConnachie and C.M. Shackleton, "Public green space inequality in small towns in South Africa," Grahamstown: Department of Environmental Science, Rhodes University, 2012.

[2] J.M. Marzluff, E, Shulenberger and E. Endlicher., "Urban ecology: an international perspective on the interaction between humans and nature," New York: Springer, 2008.

[3] A. Schäffler and M. Swilling, "Valuing green infrastructure in an urban environment under pressure: the Johannesburg case," Ecological Economics, vol. 86, pp. 246 – 257, 2013.

[4] TEEB, "The economics of ecosystems & biodiversity: ecosystem services in urban management," USA, 2011, http://www.teebweb.org/Portals/25/Documents/TEEB_Manual _for_Cities_Ecosystem_Services_for _Urban_Management_FINAL_2011.pdf. (Accessed 1 February 2014).

[5] A. Nelson, "Stockholm, Sweden: city of water," Sweden, 2006, http://depts.washington.edu/open2100/Resources/1_OpenSpac eSystems/Open_Space_Systems/Stockholm_Case_Study.pdf, (Accessed 12 June 2014).

[6] S.P.R. Labuschagne, J. Van Loggerenberg and J.H. Lombard, "The Role of GIS to Identify Nodes and Guide Sustainable Development in Rural Areas," Potchefstroom: NWU. (Mini-dissertation – Hons), 2013.

[7] J. Byrne and N. Sipe, "Green and open space planning for urban consolidation – A review of the literature and best practice," Brisbane: Griffith University, 2010.

[8] Commission for Architecture & the Built Environment, "Start with the park: creating sustainable urban green spaces in areas of housing growth and renewal," London: Cabe Space, 2011.

[9] U.G. Sandström, on green infrastructure planning in urban Sweden, Planning Practice & Research, vol. 17(4), pp. 373-385, 2002.

[10] J. Ahern, "From fail-safe to safe-to-fail: Sustainability and resilience in the new urban world," Landscape and Urban Planning, pp. 341-343, 2011.

[11] R. Trancik, "Finding lost space: Theories of urban design," Boston: John Wiley & sons, 1986.

[12] J. Barnett, "The fractured metropolis: Improving the new city, restoring the old city, reshaping the region," New York: Harper Collins, 1995.

[13] Department of Environmental Affairs and Tourism, "South Africa environment outlook: A report on the state of the environment," Pretoria: Department of Environmental Affairs and Tourism, 2006.

[14] J.B. McMahan, K.T. Weber and J.D. Sauder, "Using remotely sensed data in urban sprawl and green space analysis," Intermountain Journal of Science, vol. 1(8), pp. 30 - 37.

19

Effect of Tillage Methods on the Growth and Yield of Egg Plant (*Solanum macrocarpon*)

Ibeawuchi I. I.[1], Ihejirika G. O.[1], Egbuche C. T.[2], Jaja E. T.[3]

[1]Department of Crop Science and Technology, School of Agriculture and Agricultural Technology, Federal University of Technology, Owerri, Imo State, Nigeria

[2]Department of Forestry and Wildlife, School of Agriculture and Agricultural Technology, Federal University of Technology, Owerri, Imo State, Nigeria

[3]Department of Applied and Environmental Biology, Rivers State University of Science and Technology, Nkpolu, Port Harcourt, Rivers State, Nigeria

Email address:

ii_ibeawuchi@yahoo.co.uk (Ibeawuchi I. I.)

Abstract: The experiments on the effects of different tillage method (Flat, Bed and Trench) on the yield of egg plant (*Solanum macrocarpon*) were conducted at School of Agriculture and Agricultural Technology (SAAT) Training and Research farm, Federal University of Technology Owerri, (FUTO), Imo State Nigeria. The result showed that plant heights of *Solanum macrocarpon* increased with age of the plant. The apices cutting technique helped to increase the number of branches per plant and the bed tillage method performed significantly better than the flat and trench methods in flower set , fruit set and development. However, tillage methods are location specific and vary with climate, soil type, and crop and management level.

Keywords: Tillage methods, effect of tillage, growth and yield, Solanum macrocarpon

1. Introduction

Solanum macrocarpon otherwise known as the African Eggplant or Gboma is of the *Solanaceae* family. It is a tropical biennial plant that is closely related to the eggplant; it grows to a height of 1-1.5 m and has an alternate leaf pattern with the blade width of 4—15 cm and a height of 10—30 cm. The shapes of the leaves are oval and lobed with a wavy margin. Both sides of the leaves are hairy with simple hairs [1]. Prickles may or may not be present on the leaves depending on the cultivar. When prickles are present they are found more along the midrib and lateral veins. The prickles are straight and can grow up to a length of 13 mm. The flowers have a diameter of 3—8 cm and are located on short stalked inflorescence that can contain 2 to 7 flowers [2]. The lower portion of the plant carries bisexual flowers while the upper portion contains male flowers. The flowers are 2-3.5 cm in length and usually have a purple or pale purple colour, on rare occasions there are white flowers. The fruits are round, the top and the bottom are flattened out and have grooved portions with a length of 5—7 cm and a width of 7—8 cm. The stalk of the fruit is 1—4 cm long and is either curved or erect. At a young stage the color of the fruit is green, ivory, or a purple and white color with dark stripes. When ripe, the fruit turns yellow or a yellow-brown. The fruit contains many seeds and it is partly covered by the calyx lobes. The seeds have a length of 3-4.5 mm, a width of 2-3.5 mm, and the shape is obivoid. *S. macrocarpon* has a large cultivar and varieties which grows in areas of high rainfall found in the tropical and humid regions of West and Central Africa, South East Asia, South America and the Caribbeans. Some cultivars can be found in the savanna and semi-arid region of Northern Ghana, Burkina Faso and their neighboring countries. [1] The cultivars grown there consist of plants with small leaves and fruit and the fruit cultivars are only able to grow in humid coastal areas [3]. *S. macrocarpon* can occasionally be found at higher altitudes but have a slower growth rate and are more robust as it reproduces mostly by self-pollination although out crossing occurs by bees and other insects but this occurs at low frequencies early in the morning [1]. The eggplant is an important vegetable/fruit in our economy, therefore, the experiment on effects of different tillage methods for the production of eggplant was carried out with a view of understanding the best tillage method for the production of eggplant in Owerri Southeastern Nigeria..This will help increase its yield and attract more resource poor farmers to the production of this all important crop.

2. Materials and Methods

2.1. Location

The experiments was conducted at School of Agriculture and Agricultural Technology (SAAT) Training and Research farm, Federal University of Technology Owerri, (FUTO),Imo State Nigeria, located between latitude $5^0 23^{'}$ N and longitude $6^0 59^{'}$ E, at an altitude of 55m. The area has minimum and maximum temperatures of 27^0C -32^0C respectively and characterized by more than 2,500mm annual rainfall and 89%-93% relative humidity [4]. The soils of Owerri belong to the soil mapping unit number 431 that is Amakama-Orji-Oguta soil association [5] and derived from classification the coastal plain sands [6]. The ripped eggplant fruits used was bought from Imo ADP, Owerri Imo State and processed. The seeds were sowed in the nursery first, from where the resulting seedlings were transplanted to the experimental field. The area was cleared of secondary forest and the debris packed. Marking out was done and the different tillage methods prepared. A randomized complete block design (RCBD) was used with three treatments. The treatments were beds, small trenches and flat, there where 6 replications and the treatment were randomly allocated to the different plots using the piece of paper method. The total area used for the experiment was 10x17m = 170 m^2 , each plot measured 2 x3m with a 0.5m gap between plots and 1.0m gap between blocks and guard area of 1.0m round the experimental area. After the tillage methods were prepared the seedlings were transplanted in the evening at a spacing of 0.5 x 0.5m. The seedlings were planted at a depth of 2cm. Dry grasses were used for mulching to avoid excessive evaporation of water from the soil due to intensity of the sun. The crops were watered regularly in the morning and evening after they were transplanted to the field. The apices or apexes of the stems and branches of all the seedlings were cut off two weeks after planting in the field. This technique was applied in order to stimulate the young plants to branch profusely and bear more fruits on additional branches. Weeding was done with hoe at 2, 4, and 6 weeks after transplanting and manure was applied at the rate of 10,000kg[-1.] Data were collected on the following parameters: Plant heights (cm), Number of leaves, Number of branches and number of mature fruits.

2.2. Plant Heights

In each treatment, four plants were randomly selected and measurements of the heights were taken from the basal end of the plant to the apex of the last leaf. The plant heights were obtained using a measuring tape. The mean heights were recorded per replicate and treatment and were also subjected to statistical analysis

2.3. Number of Leaves

The leaves of four randomly selected plants in each plot were physically counted and the mean numbers recorded and subjected to statistical analysis

2.4. Number of Branches

After the apices of the stems of the plants were cut off, the number of the branches that grew out was physically counted from four randomly chosen plants on each plot and the average number recorded and subjected to statistical analysis.

2.5. Number of Flower

The numbers of flowers at four randomly selected plants in each plot were physically counted and the average numbers recorded and then subjected to statistical analysis.

2.6. Number of Fruits

At maturity and fruits were ready for harvest, four randomly selected plants with matured fruits were collected and the mean numbers recorded for each plot and subjected to statistical analysis.

2.7. Fresh Weight of Fruits

Crops were harvested eight weeks (8 weeks) after planting in the field. This was because the species is an early maturing variety. The mature fruits were handpicked from the branches in order to make very good assessment of the difference in field of the plants owing to the treatments they received, the weight of the fruits after harvesting were recorded with the aid of a saltare balance and mean weight recorded for each plot and subjected to statistical analysis. First weight of mature fruits reading were taken randomly from four samples of the test crop (S. macrocapan) in each replicate at 2 weeks, 4 weeks, 6 weeks and 8 weeks after planting. All data collected were statistically analyzed for each crop using the procedure analyzed [7] and presented by [8] for RCBD and significant mean differences detected using Fishers least significance test [F — LSD] at 5% probability.

3. Results and Discussion

Table 1. Effects of tillage method on the average number of leaves of Solanum Macrocarpon at 2, 4, 6 and 8 weeks after planting.

Treatments (Tillage method)	Number of weeks			
	2	4	6	8
Flat	4.33	10.67	16.67	18.17
Bed	4.50	11.50	18.50	22.87
Trench	4.33	9.57	17.17	18.67
LSD$_{0.05}$	NS	NS	NS	2.36

The results in table 1 showed that there were not statistically significantly different (p = 0.05) in the mean number of leaves Solanum macrocarpon at 2, 4 and 6 weeks after planting. However, at 8 weeks after planting, there were significant differences among the three different tillage methods in mean number of leaves of *Solanum macrocarpon*, Bed has the highest mean number of leaves (22.87) and least was flat method (18.4).

Table 2. Effect of tillage method on the average number of branches of Solanum macrocarpon at 2, 4, 6 and 8 weeks after planting.

Treatments (Tillage methods)	Number of weeks			
	2	4	6	8
Flat	1.33	2.33	4.83	6.50
Bed	1.17	2.67	5.83	8.50
Trench	1.17	2.33	5.33	6.83
LSD$_{0.05}$	NS	NS	0.71	1.20

The results Table 2 showed that at 2 and 4 weeks after planting, showed no significant differences. At 6 weeks after planting, there were significant differences in the mean number of branches. The highest mean numbers of branches were recorded for bed followed by trench while flat recorded the least. The trend at 8 weeks after planting showed that bed (8.50) gave the highest mean number of branches for the *Solanum macrocarpon* followed by trench (.6.83) and flat (6.50) respectively. This could be attributed to the well pulverized seed beds that helped better aeration, water infiltration, water retention and nutrient uptake.

Table 3. Effect of tillage method on plant height of Solanum macrocarpon at 2, 4, 6 and 8 weeks.

Treatment(Tillage method)	Number of weeks			
	2	4	6	8
Flat	16.17	23.17	31.63	40.33
Bed	19.68	29.27	41.72	53.68
Trench	16.67	27.37	35.47	43.33
LSD$_{0.05}$	NS	3.26	4.13	4.23

The result showed that plant heights of *Solanum macrocarpon* increased with age of the plant. The result also showed that at 2 weeks after planting, there were no significant differences in the heights of *Solanum macrocarpon* among the different tillage methods used. At 4 weeks after planting, the crops planted on bed were not significantly different (29.27) from those planted in trench (27.37) but was statistically significantly different from those planted on flat (23.17).However, at 6 and 8WAP, the bed method had significantly taller plants than those on flat and trench. This could be as a result of nutrient release which is fast in pulverized soils than in unpulverized ones

Table 4. Effects of tillage method on the number of flowers of Solanum macrocarpon at 6 and 8 weeks after planting.

Tillage methods	Number of weeks	
	6	8
Flat	10.67	5.33
Bed	12.83	5.67
Trench	10.50	5.45
LSD$_{0.05}$	2.01	NS

Field observations showed that flowering started from the sixth week after planting. The result showed that at 6 weeks after planting, there was significant difference (12.83) in the

number of flowers of *Solanum macrocarpon* for the crops on bed. This was followed by flat method (10.67) and trench method (10.50). It was further observed that as the weeks go by, the number of flowers decreased. Flowers increased at 6 weeks after planting and decreasing at 8 weeks after planting and there were no significant difference among the different tillage methods.

Table 5. Effects of tillage method on the mean number and fresh weight of matured fruits of Solanum macrocarpon at 6 and 8 weeks after planting.

WAP Tillage methods	6	8	Mean weight of matured fruits
Flat	1.25	6.33	46.22
Bed	1.77	8.85	89.40
Trench	1.47	6.50	61.50
LSD$_{0.05}$	NS	0.94	6.87

WAP - weeks after planting

The results Table 5 showed the effect of tillage method on the mean number of matured fruits at 6 and 8 weeks after planting and also the weight of matured fruits harvested at maturity .The results showed that the mean number of matured fruits of *Solanum macrocarpon* were not statistically different. However, at 8 weeks after planting, we observed that the number of matured fruits increased and the plants on the bed had higher mean number of fruits (8.85) than trench and flat (6.50 and 6.33) respectively. The fruits on bed had higher mean weight of fruits, were bigger in size and were statistically significantly different from those of flat and trench respectively.

4. Discussion

Generally, the plants on bed did better than those on flat and trench, in plant height, number of leaves, fruit size and weight. The apices cutting technique helped to increase the number of branches per plant. As the branches grew out of the plant, the number of leaves per plant also increased. The experiment showed that branching is an index of leaf proliferation in the production of *Solanum macrocarpon* and also, helps in production of more flowers, fruit set and development. The bed method did better than the flat and trench methods within the period of Solanum production because a well prepared seedbed encourages moisture reserve [9], quick nutrient release and also improves soil drainage which helped the *Solanum macrocarpon* to grow well and taller than the other methods [10].

5. Conclusion

Tillage can either enhance or destroy good soil tilth. It is true that bed preparation breaks up clods and loosens the topsoil, but the stirring action helps to stimulate the microbial breakdown of beneficial soil organic matter. However, seedbed preparation is very location specific and varies with climate, soil type, crop and management level .One has to be careful in over tilling the soil to avoid the negative consequences. The trench and flat however, has their own advantages and disadvantages and may overtime be more productive than an over tilled soil and bed tillage method had the advantage of releasing nutrients faster than the flat and trench tillage methods.

References

[1] Schippers R.R (2000) African indigenous vegetables. An overview of the cultivated species. Chatham,UK, Natural Resources Institute /ACP-EU. TechnicalCentre forAgricultural and Rural Cooperation.2000, 77

[2] Bansu ,K.O., Schippers, R.R., Nkansah G.O., Awusu, E.O., and orchard. J.E.,(2000). Gboma eggplant, a potential new export crop for Ghana. Book of abstract fifth International Salonaceae Conference, Botanical Garden of Nijmegen, July 23-29, pp. 14.

[3] Bunkenya, Z.R., 1994. *Solanum macrocarpon* an under utilized but potential vegetable in Uganda. In Seyani J.H. & Ckiuni A.C. (E Ibeawuchi I.I Onweremmadu E.U, Oti N.N, (2006): Effect of Poultry Manure on green (*amaranthus cruentus*) and water leaf (talinum triangular) an degraded ultisol of Owerri, South Eastern Nigeria java; $5_{(1)}$ 53-6.ditors)

[4] Ibeawuchi I.I Onweremmadu E.U, Oti N.N, (2006): Effect of Poultry Manure on green (*amaranthus cruentus*) and water leaf (talinum triangular) an degraded ultisol of Owerri, South Eastern Nigeria java; $5_{(1)}$ 53-6.

[5] Federal Department of Agriculture and Land Resources(FDALR)1985: The reconnaissance soil survey of Imo State, Nigeria(1:250,000)Soil report 1985, 133.

[6] Lekwa ,G and E P Whiteside, 1986Coastal Plain sands of Southeastern Nigeria .Morphology classification and genetic relationship.Soil Science Soc J 50:154-156.

[7] Wahua T.A.T (1999) Applied statistics for scientific studies. Africa Link Books Pp 129-189

[8] Rilay Janet (2001) Presentation of statistical analysis. Expl. Agric volume 37, Pp115-123

[9] Jitendra, Singh (2012) Basic Horticulture.. Kalyani publishers. New Delhi. Pp 65-75

[10] David, Leonard (1986) Soils,Crops and fertilizer use. United States Peace Corps. Information collection and Exchange. Reprint . R0008. Pp 63-75.

proceeded 13th plenary meeting AETFAT, Malawi. Volume 1. Pp. 17-24

20

Social Upliftment as a Result of Green Space Provision in Rural Communities

Nicolene de Jong

Unit for Environmental Sciences and Management, North-West University, Potchefstroom, South Africa

Email address:

nicolenedejong@gmail.com

Abstract: Rural communities, especially those within the South African context, are faced with daily challenges – most of which are associated with a lack of basic social provisions, needs and spaces. This paper introduces the concept of green space planning as an alternative planning approach to address these social challenges within rural South African communities. It provides a comparative study between social challenges experienced in South Africa and the benefits provided by green spaces; introducing green space provision as an alternative planning method to address social challenges, especially those prevalent in rural South African communities. By providing successful and sufficient green spaces, numerous rural challenges are addressed and an overall upliftment of the communities' mental and physical well-being, social inclusion and quality of life is established.

Keywords: Social, Green Space, Rural, South Africa, Social Challenges

1. Introduction

Rural communities, especially within the South African reality, are widely defined as, and associated with concepts such as 'social backlog'; 'previously disadvantaged'; 'unsafe'; 'socially challenging'; etc.

Rural communities and the development thereof continue to be one of the main priorities within frameworks and constitutions guiding the economic and social development of (especially developing) countries. Reference [1] substantiates this statement by referring to the pervasiveness of poverty and poor delivery of basic services in rural areas as a primary constraint regarding a country's development efforts.

Therefore planning for rural communities and identifying or creating new approaches to rural development should receive a great deal of attention, as this is the core of addressing government's commitment to eradicate poverty [2] in a pro-active attempt at economic restitution and social upliftment.

Government is struggling to deal with rural challenges and related social issues (especially that of safety and security) and numerous strategies were formerly discussed and drawn up (for example the Rural Safety Summit which took place on 10 October 1998 aimed at achieving consensus regarding issues of rural insecurity; as well as crime prevention strategies as defined by the SAPS White Paper on Safety and Security (1998).) However, very little (if any) in-depth research on the possibility of upgrading public spaces into lively green places as solution has been done.

The failure of previous rural development projects implemented during the last three decades [1] leads to the consideration of different approaches that will focus and increase the relatively low development levels in the rural areas of South Africa [3] through interventions that support and enhance livelihood such as sound rural-development planning policies and programmes [3] that are oriented towards the provision of basic needs, the development of human resources and a growing economy in order to generate sustainable livelihoods [2].

In response to abovementioned necessities and social backlogs, numerous (primarily international) approaches have been implemented towards the creation and promotion of safe green spaces in an attempt to provide social benefits by promoting community integration [4]. These 'green spaces' provide more depth and meaning to a community than merely being a natural surface [4] between buildings [5], but fulfil a whole range of functions such as health functions, amenity and social functions as well as ecological functions [6]. It provides a unique platform for community integration [4] that can potentially influence the physical and mental well-being of community members [4] and contribute to their overall quality of life [6]. These green space benefits can be directly associated with and implemented as planning approaches to enhance the rural reality in South Africa and subsequently provide quality and sustainable rural communities.

This paper establishes a link between literature and practical rural issues by providing insight into the current reality of rural communities in South Africa as well as providing a defining background into green spaces and green space benefits. In creating public green spaces for rural communities, issues of safety, inequality, sociability and community coherence are addressed. Through the correct planning initiatives consequently drawn up, overall quality of life of those living in rural communities can be improved, decreasing the social challenges experienced and promoting sustainable rural communities.

2. Rural Social Challenges: South African Communities as Case Study

As mentioned in the introductory paragraph, South Africa as a developing country faces numerous challenges at a continuously increasing rate, especially regarding economic degrading and social backlog [7]. These economic challenges have a direct impact on the social development and overall quality of life of individuals living and working in South Africa, with an *enhanced social degrading* experienced in rural South Africa – hence the continuous existence of 'rural social challenges' [1].

2.1. Background to the Correlation between 'Rural' and 'Social Challenges'

When discussing a complex concept such as "rural", location and the delineation of such an area is of utmost importance. For the purposes of this paper it is necessary to define and comprehend "rural" specifically within a South African context.

The complexity of "rural" is substantiated by the broad spectrum of definitions and descriptions found globally on what "rural" means and what it encompasses. In South Africa, a 'rural community' is commonly perceived differently compared to other countries where the term usually demonstrates or refers to the density of a population and the dependence on manual labour [2]. Another perception of 'rural' is summarized in the LDCE [8] as a "happening in or relating to the countryside, not the city". The European definition of 'rural' refers to any agricultural land and/or areas [2].

When abiding to the first perception of 'rural' as defined by the DRDLR [2] the term "rural" would have defined the whole of South Africa (all-inclusive) up until 1995 (referring to the outcomes of censuses and official surveys) [2] and the other perceptions would not encapsulate the unique and detailed composition of 'rural' areas as found in South Africa.

In reaction to these deficit definitions regarding 'rural' in the South African context, the Rural Development Framework

of South Africa that was originated in 1997, defined 'rural' as "... *the sparsely populated areas in which people farm or depend on natural resources, including villages and small towns that are dispersed through these areas*" [2].

The relevance and importance of the social dimension of planning as a key component of sustainable development was ignored for many years where focus was mainly directed at environmental issues and the integration of the environmental and economic dimension. Due to various developments and advancements the social dimension was recently included as an equally relevant and influential dimension as the aforementioned [9]. The ISRDS [1] further emphasize the increasing accountability of the 'social' component of planning especially regarding its contribution to development and development practices. Scoones [10] provides a sound basis when highlighting the importance of society and people as the anchor of social development – especially their local needs and cultural contexts.

Philips [11] expands 'social' and the irreplaceable importance of people and the society within the 'social'-persona by including various aspects such as safety, culture, housing, labour and community relations. The DRDLR [1] expands this defining list by including social services like food and water, shelter, energy, health and education, and transport and communication services.

In order to present a platform when addressing something as broad and unlimited as 'social', the mixed definition will be accepted conditionally, as defined by Henslin [12]. He defines a social problem (or for the purpose of this paper a social challenge) as "an aspect of society that people are concerned about and would like changed." This refers to a social challenge as something (an aspect, happening or circumstance for example) within the community that one can measure or experience. Simultaneous to this, another key element regarding social problems is that of a subjective concern, which refers to a concern that a number of people have regarding the previously-mentioned condition. Overall, anything people-oriented or contributing to human experience (positively or negatively) can be included as part of 'social' and any happening or circumstance disturbing or disadvantaging 'social' and all its aforementioned aspects can be regarded as 'social challenges'.

The importance of people and the human aspect of social can indeed be seen as another possibility for the eruption of more social challenges. According to Veenhoven and Ehrhardt's [13] theory of liveability, the challenge of social inequality because of differences in dispersion across nations might occur, providing sufficient substance for this paper's focus on enhanced social challenges (as defined and discussed above) as experienced in rural communities - even more that of South Africa.

2.2. Social Challenges Experienced in South African Rural Communities

In the light of the background given in the previous section regarding the defining of 'rural' and 'social challenges', it is evident that the challenges as experienced and identified in the

South African context are primarily rooted in the lack of social development present – a lack much heftier experienced in rural communities. The ISRDS [1] substantiates this statement by referring to the pervasiveness of poverty and poor delivery of basic services in rural areas as a primary constraint regarding a country's development efforts.

In South Africa, rural development is an even more predominant challenge as it is estimated that half of South Africa's population lives in rural areas [14] and that an astonishing three quarters of this country's people living below the poverty line or MLL (minimum living level) live in these rural areas [2].

This serves as statistical proof and support for the interceding statement of this article that associates characteristics such as 'backlog', 'social challenges' and 'disadvantaged' with rural communities and (based on above) especially with regards to South African rural communities.

In support to above, the following table provides a concise and conducive summary of three researched rural communities and the primary challenges identified within these communities.

With reference to former sections of this article in which the concepts of 'social' and 'social challenges' were defined, it can justifiably be derived that the majority of the challenges as identified, coincide with concepts included in the broad spectrum of 'social'. As 'social challenges' were formerly described as any happening or circumstances disturbing the social nature (i.e. people-oriented aspects) of communities, it proves evident that above can be regarded as the blueprint for rural social challenges within a South African context.

3. Green Space Planning

3.1. Defining and Understanding the Context of Green Spaces

A green space is diversely defined. Barbosa et al. [4] defines a public green space as every parcel of land classified as a natural surface, judged to be publicly accessible. 'Natural surface' means the green space should be predominantly natural (i.e. earth, water and living things) with a sense of quality and the presence of several maintained facilities [16], therefore introducing green spaces as the 'glue' between buildings [5].

These green spaces fulfil a whole range of functions such as health functions, amenity and social functions as well as ecological functions [6]. The execution of these functions lies within the inclusion of certain aspects or areas.

Public green spaces can provide social benefits by promoting community integration in a way that private gardens cannot, since social interactions in gardens are focused around a private social network [4]. They also provide the primary contact with biodiversity and the 'natural' environment (refer to previous definition of natural surfaces as defined by Shackleton and Blair [16]) for many people and may therefore influence the physical and mental well-being of those people [4] as well as contribute to their overall

quality of life [6]. The direct managerial role of government is established in Young [17]'s definition of green spaces as "...publicly managed natural resource assets... including street trees, parks, 'natural areas'...".

For the purpose of this paper, green spaces will mean (as compiled and summarized from above) safe and maintained parcels of land that remains predominantly natural, dispersed throughout a community with sustainable sense of quality and the presence of a variety of maintained functions and social facilities for residents.

The following citations summarize the potential benefits and/or influences that may emerge with the successful provision of the former combined definition of 'green space' (i.e. successful provision implies all aspects included in the definition should be adhered to such as safety, natural environment, maintained, quality, variety, etc.):

Shifting the paradigm of spatial segregation into multifunctional landscapes [16] wherein 'quality' regarding these spaces are linked to the 'value' associated with spaces by recognizing the need of these spaces to reflect the changing social, economic and environmental conditions;

Economic benefits like the enhancement and provision of economic prosperity that becomes evident when preserving and enhancing natural ecosystems (i.e. green spaces) [18];

It has a very particular value and plays a unique role in the sustainability and liveability of towns and cities, the provision of which requires appropriate planning approaches, implementation strategies and financial resources [16];

Green spaces address and therefore contribute (in many ways) to sustainable development beyond merely providing recreational value [5] but rather making a valuable contribution across the spectrum of social, environmental and economic benefits [19];

The importance and value of green spaces to the people is further emphasized in the study of Shackleton and Blair [16], wherein a survey resulted in the average of 93% of respondents stating that PGS (public green spaces) are important (substantiated by Ward et al. [20] with 99% of surveyors regarding urban green spaces as important). The reason for their decision and therefore the summary of values associated with green spaces in communities include amongst others [16, 20]:

Recreation and relaxation;

Provision of jobs (creation of such spaces need to be maintained);

Environmental benefits;

Attraction of tourism (also contributes to the potential provision of jobs);

Promotion of human well-being;

Appreciation and exercise (health)

Enhancing quality of life;

Escapism and breathing space;

Preservation for future generations (i.e. sustainability);

Events, concerts and alternative displays;

Education and research; and

Spiritual and moral functions.

Despite these valuable contributions of green spaces, the

Department of Transport, Local Government and the Regions (DTLR) [21] stated that these invaluable resources are often neglected, resulting in a loss of potential and potential benefits. Therefore, in order to enhance the potential and benefits that the green spaces can provide, it is crucial to focus on the quality of the green spaces, and the management of these spaces within any given community.

3.2. Green Spaces and Associated Challenges and/or Deficiencies: South African Context

The challenges in planning and maintaining urban public green spaces in towns in the developing world such as South Africa differ remarkably from those of the developed world as discussed in the previous section on international approaches [16]. Developing countries experience high levels of urbanisation and population growth prohibiting urban planning agencies to keep up and contributing to green space planning in areas which are rather being targeted for land invasion [22]. South Africa, being a developing country, therefore provides an interesting opportunity to examine the distribution of public green space in a developing country in relation to wealth attributes – wealth being correlated to public green space attributes [22].

Table 1. Challenges experienced in various South African rural communities.

Rural Community	Primary Challenges Experienced
Umgababa – KwaZulu-Natal	Poverty; Need for development; Low quality of life; Need to create awareness; Inequality – women subject to poverty, hardship, hostility, abuse and neglect; Lack of ability to use skills – women; Low self-esteem; and Social exclusions as "a major challenge"[15]
Nigel and Zonkizizwe - Ekurhuleni Metropolitan	Uncommitted Community Police Forums; Not representative of all the community organisations; Lack of safety and security; SAPS and EMPD understaffed; Lack of infrastructure; High crime rates; Alcohol abuse; Mob justice/vigilantism – community members taking disciplinary actions into their own hands; Lack of access to transport facilities; Housebreaking, burglary and theft; and Complete lack of public participation and inclusion.
Vaalharts (Northern Cape and North-West Provinces)	Inadequate basic government services; Unsatisfactory municipal services; Inadequate educational and training facilities; Lack of financial resources and incentives; High unemployment; Lack of agricultural needs, knowledge and skills; Infrastructure need – housing, recreational, religious, health and schools, shops, transport and roads; Lack of social services; Safety and security; and Absence of emergency services.

The majority of residents (as responded in a survey conducted in two small towns in the Eastern Cape Province of South Africa) agreed that there was insufficient public green space (PGS) in their respective towns and suburbs and that the local municipality did not do enough in providing efficient and sufficient PGS or maintaining existing spaces [16]. Ward *et al.* [20] agree with 55% of respondents stating that there is insufficient green space planning within urban areas.

Even though residents experience a high level of dissatisfaction with the amount and condition of current PGS and insufficient municipal commitment and funds to adequately maintain these PGS, there still exists a high level of willingness amongst residents to get involved (either through a commitment of time or funds) [16]. The importance of understanding residents' needs and attitudes is therefore important in ensuring that planning and management objectives are grounded in local needs and desires regarding PGS [16] in order to create efficient green spaces they want and would visit.

Improving attributes such as accessibility [4], proximity [16] and safety [4] contribute to the attraction of green spaces, increasing the frequency and duration of visits. By creating public green spaces which are fully maintained and cared for, the feeling of being unsafe is eliminated, contributing to the overall value and success of these public green spaces [16] which enhances the values not only of the green spaces themselves, but the whole of the community in which they are successfully provided.

The case study areas were selected at random based on the relevant data available for the purposes of this paper. The selected case studies are also representative of a broad spectrum of the South African regions including a case study area located in KwaZulu-Natal, Ekurhuleni/Gauteng, and the North West/Northern Cape Provinces, therefore successfully representing a variety of culture groups, differently performing economic areas and covering a large area of the country's extent.

The table merely concludes a small sample of challenges faced in South African rural communities, but the continuous similarities proves that it can serve as a sufficient representative of other rural communities in South Africa. A collective summary of the primary challenges experienced i.e. challenges regarded as "social challenges" based on former definitions as listed above can be regarded as basic human needs; sufficient social services and facilities; need for infrastructure (housing, shops, roads, transport); lack of safe and secure spaces and viable security services; inequality; social exclusion and associated decrease in self-esteem and self-worth; poverty; unemployment; overall low quality of life; and unmaintained (unsafe) spaces, areas and infrastructure.

It deems inevitable therefore that planning for rural communities and identifying or creating new approaches to rural development should receive a great deal of attention, as this is the core of addressing government's commitment to eradicate poverty [2] and addressing various other social challenges reported above. A sustainable and viable option for such "...new approaches to rural development" proves to be

that of green space planning and provision and will be substantiated in sections to follow.

4. Rural Communities and the Social Benefit of Green Space Planning

Role of Green Spaces in Enhancing Social Benefits

Swanwick et al. [19] refer to the social benefits of urban green space planning as having both "...an existence value, because people know it is there, and a use value for a wide range of different activities." This is further supported by Barbosa et al. [4] referring to green space planning as providing purported social benefits by bringing diverse communities together and promoting interactions between people from different socio-economic and ethnic groups.

In collaborating the challenges of South African rural communities as listed and summarized in Table 1 and reviewing them in the light of the benefits associated with providing sufficient and successful green spaces (refer to section 3.1. of this article as to what is regarded as "successful provision of green spaces"), the indubitable positive outcome of providing green spaces in rural communities as a planning approach to potentially address rural social challenges, is evident:

Table 2. Correlation between rural social challenges and green space benefits.

Primary Rural Social Challenges Experienced	Benefit of Green Spaces as Potential Addressor of Rural Social Challenges
Poverty	Economic benefits – enhancement and provision of economic prosperity. Contribute to overall quality of life;
The need for development and a better way of life	Preservation for future generations (i.e. sustainability); Environmental benefits.
A need to create awareness	Plays a unique role in the sustainability and liveability of towns and cities and therefore contribute (in many ways) to sustainable development; Escapism and breathing space.
Inequality	Shifting the paradigm of spatial segregation into multifunctional landscapes. Influence the physical and mental well-being of those people;
Lack to achieve a better way of life	Promotion of human well-being; Enhancing quality of life.
Low self-esteem and social exclusions	Making a valuable contribution across the spectrum of social, environmental and economic benefits.
Safety and security issues (high crime rates, unsafe open spaces and fields, violence, mobs)	Maintained and surveillance areas; On-site officials and maintenance; Unsafe open fields converted into useful and busy social areas.
Lack of facilities (transport, social, personal, hygiene, health, recreational, religious)	Recreation and relaxation; Spiritual and moral functions; Appreciation and exercise (health).
Lack of quality infrastructure (buildings, maintained social spaces, roads, schools housing)	Events, concerts and alternative displays.
Complete lack of public participation and inclusion	Promoting community integration.
High unemployment	Provision of jobs (creation of such spaces need to be maintained).
Inadequate educational and training facilities	Education and research.
Inadequate and unsatisfactory basic government and municipal services	Attraction of tourism (also contributes to the potential provision of jobs.

The table above represents various rural challenges and social benefits of green spaces as found in the relative former sections. This table therefore provides a strategic representation to proof the undeniable similarities between the primary challenges and struggles experienced in rural communities and the social benefits provided by creating

successful green spaces within any given community.

5. Conclusion

When providing sufficient areas of green space that are also accessible, these PGS (public green spaces) will promote connection with the places in which people live and work – therefore eliminating long travel distances and improving accessibility, providing relative easy means for recreation, exercise, relaxation and other human health and well-being related challenges. This is reflected in the perceptions they form about green spaces and will also influence their maintaining and respecting the use of these public green spaces [16], therefore motivating them to personally get involved and contribute in such a manner that they feel a sense of ownership and inclusion within their community. This will subsequently fuel their self-worth and indirectly contribute to the state of their mental and physical well-being, thus increasing the level of the quality of life they currently experience – i.e. providing an overall sense of social upliftment due to the simultaneous addressing of various previously experienced social challenges.

In subsection 2.1 the concepts of both 'rural' and 'social challenges' were elaborated through the inclusion of various definitions, contexts and citations. In all of the above expressions the direct relation between rural communities (especially in South Africa) and the presence of persistent social challenges is established.

This relevance and applicability in South African context is substantiated in subsection 2.2 and listed in Table 1 where various South African communities and their predetermined social challenges are included.

As introduction to the probability and applicability of green space planning as addressor of rural social challenges, the context of "successful PGS (public green spaces)" is defined in subsection 3.1.

The inverse correlation between the rural social challenges of the South African case study communities summarized in Table 1 and the benefits of providing green spaces in communities are drawn in section 3.2.

Section 4 emphasizes the probability of addressing aforementioned rural social challenges by providing these successful green spaces. Table 2 serves as a supportive summary, representing a concise listing of the primary social challenges formerly identified in the rural case study areas (left column). The column to the right represents some of the benefits of providing green spaces as derived from the in-depth discussion in Section 3. The green space benefits (right column) are arranged next to the corresponding social challenge (left column) that it can potentially address and/or uplift.

Therefore Table 2 provides visually presented proof that green spaces can potentially provide upliftment for each identified rural social challenge. This leads to an inevitable decrease in social challenges, culminating in an overall social upliftment for any given rural South African community.

References

[1] Department of Rural Development and Land Reform *see* South Africa (1). Department of Rural Development and Land Reform.

[2] South Africa (1). Department of Rural Development and Land Reform, Integrated Sustainable Rural Development Strategy (ISRDS), 2000, Pretoria.

[3] Department of Rural Development and Land Reform *see* South Africa (2). Department of Rural Development and Land Reform.

[4] South Africa (2). Department of Rural Development and Land Reform, Rural Development Framework (RDF). 1997, Pretoria.

[5] NSDP *see* South Africa (3).

[6] South Africa (3), National Spatial Development Perspective (NSDP), 2006, The Presidency, RSA.

[7] O. Barbosa, J.A. Tratalos, P.R. Armsworth, R.G. Davies, R.A. Fuller, P. Johnson & K.J. Gaston, Who benefits from access to green space? A case study from Sheffield, UK, 2007, Landscape and Urban Planning, 83 (2007): 187-195.

[8] O. Wilson & O. Hughes, Urban Green Space Policy and Discourse in England under New Labour from 1997 to 2010, 2011, Planning Practice & Research, 26(2):207–228.

[9] E. Lange, S. Hehl-Lange & M.J. Brewer, Scenario-visualization for the assessment of perceived green space qualities at the urban–rural fringe, 2007, Journal of Environmental Management, 89(2008):245-256.

[10] P. McGroarty, South Africa's Economic Growth Continues to Decline., 2014, http://www.wsj.com/articles/south-africas-economic-growth-co ntinues-to-decline-1416913563 Date of access: 30 March 2015.

[11] LDCE (Longman Dictionary of Contemporary English), Harlow, England: Pearson Education Limited, 2003, 2272p.

[12] I. Mulalic, Embedding social dimensions into economic and environmental accounting and indicator systems: Some aspects to consider. Copenhagen, Denmark: Statistics Denmark, 2004.

[13] I. Scoones, Livelihoods perspectives and rural development. Journal of Peasant Studies, 2009, 36(1):1-26.

[14] Philips, Liveable Cities: Challenges and opportunities for policymakers, A report from the Economist Intelligence Unit. London: The Economist Intelligence Unit, 2010.

[15] J.M. Henslin, Social Problems, 2003, http://wps.prenhall.com/hss_henslin_socprob_6/0,6624,49456 3-,00.html Date of access: 3 March 2013.

[16] R. Veenhoven & J. Ehrhardt, The cross-national pattern of happiness: Test of predictions implied in three theories of happiness. Social Indicators Research, 1995, 34:33-68.

[17] C. Campbell, Y. Nair, S. Maimane & Z. Sibiya, Supporting people with AIDS and their carers in rural South Africa: Possibilities and challenges, 2008, http://eprints.lse.ac.uk/5471/ Date of access: 28 Feb. 2013.

[18] M. Gopaul, The significance of rural areas in South Africa for tourism development through community participation with special reference to Umgababa, a rural area located in the province of KwaZulu-Natal. Pretoria: University of South Africa, 2006. (Dissertation – Master of Arts).

[19] C.M. Shackleton & A. Blair, Perceptions and use of public green space is influenced by its relative abundance in two small towns in South Africa. Landscape and urban Planning, 2013, 113(2013):104-112.

[20] R.F. Young, Managing municipal green space for ecosystem services, Urban Forestry & Urban Greening, 2010, 9(2010):313–321.

[21] J.O. Odindi & P. Mhangara, Green Spaces Trends in the City of Port Elizabeth from 1990 to 2000 using Remote Sensing. International Journal of Environmental Research, 2012, 6(3):653-662.

[22] C. Swanwick, N. Dunnet & H. Woolley, Nature, Role and Value of Green Space in Towns and Cities: An Overview. Built Environment, 2003, 29(2):94-106.

[23] C.D. Ward, C.M. Parker & C.M. Shackleton, The use and appreciation of botanical gardens as urban green spaces in South Africa. Urban Forestry and Urban Greening, 2009, 9(2010): 49-55.

[24] Department for Transport, Local Government and the Regions (DTLR), Green Spaces, Better Places: Final Report of the Urban Green Spaces Taskforce. London: TSO, 2002.

[25] M.M. McConnachie & C.M. Shackleton, Public green space inequality in small towns in South Africa. Habitat International, 2009, 34(2010):244-248.

The Planning and Development of Child-friendly Green Spaces in Urban South Africa

Zhan Goosen

Unit for Environmental Sciences and Management, North West University, Potchefstroom, South Africa

Email address:

goosenzhangoosen@gmail.com

Abstract: The impact that urban green spaces in urban environments have on the sustainability and quality of life of the residents is phenomenal [3,19,40]. The local reality in South Africa confirms that green spaces, specifically child-friendly green spaces, in urban environments are decreasing because of growing populations and increasing urbanization [34]. Preference is given to provide housing for a growing population, due to the impacts of urbanization, and the development and enhancement of green spaces are often neglected in this regard [30]. Although literature proofs the benefits and need for green spaces in urban areas (also in South Africa), the planning and development of these spaces do not realize in many instances, due to a lack of municipal priorities and funding, driven by the urgent need to provide housing, but also coupled with a lack of understanding of the benefits and importance of planning for green spaces or child-friendly green spaces in urban development. The planning and development of green spaces in the urban environment are investigated as part of this research, along with the benefits that such spaces can provide to communities by focusing on the aspect of child-friendliness. This study evaluated the planning and development of child-friendly spaces in the urban environment of the city of Durban, Republic of South Africa, confirming how ineffective the current child-friendly spaces are. Two international case studies are identified as best-practice cases, namely Mullerpier child-friendly public playground in Rotterdam, the Netherlands, and Kadidjiny Park in Melville, Australia. The aim was to determine how these international child-friendly spaces were planned and developed and to identify tools and planning approaches of the two international case studies that were used to accomplish the goal of providing successful child-friendly green spaces and how it can be implemented in South Africa. The policies and frameworks which influence the study area in Essenwood, Durban, were identified and analysed in order to establish whether or not the planning and development of child-friendly spaces is supported within the chosen area. This research concluded that child-friendly green spaces do however have a positive impact on the urban environment, caters for children's needs and assist in their development and interaction with the natural environment, only if these spaces are maintained.

Keywords: Child-Friendly Spaces, Green Spaces, Urban Area, Built Environment, Open Spaces

1. Introduction

The planning of child-friendly spaces is no new phenomena in the international context, but however, limited in local context. Child-friendly green spaces have a positive impact on the urban environment, caters for children's needs and assist in their development and interaction with the natural environment. The need for these spaces is of utmost importance and the proposed implementation and improvement of child-friendly green spaces should be supported by way of specific policies and legislation, in order to ensure the success and sustainability of these spaces.

A child's interaction with the world is directly affected by their natural environment [22] and therefore the importance to plan and provide sufficient open spaces for children should be emphasized as part of current spatial planning approaches. The core problem which emanated from this research is the lack of qualitative urban green spaces, and more specifically child-friendly spaces in the local South African urban environment. Qualitative urban green spaces in the context of this research implying usable, functional spaces located within the urban environment. Child-friendly spaces in context of this research imply public spaces that are planned and developed specifically for children and their needs. The local reality currently suggests of "green" and "public" spaces that are mostly uninviting and unsafe in general, resulting that children feel uncomfortable interacting with their natural environment or being outdoors [28,37].

The need for these qualitative urban green spaces are also emphasised by statistics that children are not active enough [29,43] finding themselves to be more indoors than outdoors. Children appear to be spending their time watching endless hours of television or playing computer games [6]. One can reason that a less safe neighborhood, with no child-friendly qualitative urban green spaces, contributes to the inactiveness of children. In South Africa overweight and obesity in children, living in urban areas is an increasing problem [43], which has a major effect on their mental and physical health, leading to the risk of developing heart disease and diabetes. A further problem arise as there is no, or limited, policy and legislative frameworks initiating the planning and improvement of child-friendly spaces within local context.

2. Green Space Planning

2.1. Defining Green Spaces

Reference [36] defines green spaces as areas that have continuous vegetated areas and space. Artificially created city parks, botanical gardens, street trees that are isolated and even private gardens are all examples that can be included in defining green spaces [40]. These urban green spaces all affect urban development in a positive way by contributing to sustainable development and improving the quality of life in general.

2.2. Defining Qualitative Urban Green Spaces

Helping to define and support the identity of towns and cities, qualitative green spaces exceed to enhance a city's attractiveness for living in the urban areas, enhance social aspects and improve better neighborhood relations [3,29]. Qualitative green spaces reveres to all open, green areas in an urban or rural environment, which has a function and contributes to the quality of the surrounding area.

2.3. Benefits and Importance of Planning for Urban Green Spaces

More and more people are migrating from rural areas to urban areas resulting in an increase in urbanization. Growing urbanization can place major pressure on the urban environment. The obvious problem is that there is more people and less space. As a result of rapid urbanization, the land set aside for green spaces are rather being used to build and provide houses for the growing population, resulting in the quality of life in the urban environment not being taken into consideration. But an even bigger challenge that the urban areas are facing, is the lack of quality child-friendly spaces, spaces provided specifically for children to interact with their natural environment in an urban setting. The main problem with most of the current green spaces is the lack of facilities and maintenance by municipalities, which results in the space having an uninviting feeling [29].

The provision of green spaces for the inhabitants of urban areas has a positive impact on their health and mental well-being [3,40]. The palpable main benefit is the lowering in temperatures by reducing heat build-ups in urban areas. It has been proven that green spaces are more beneficial than paved open surfaces, for the obvious reason of heat being trapped in surfaces such as for example concrete. These spaces also contribute to improving air quality, by removing pollutants from the air [30].

While there is no doubt that a need exists for the creation of green spaces in urban areas within South African environment, the small pockets of green spaces that do occur in South African towns are unfortunately usually distributed unevenly [23] and are insufficient.

The biggest stumbling block in the provision of green space planning is the financial aspect of developing and maintaining these spaces. According to reference [29] there is an obvious need for these spaces, however the financial provision for such are usually not prioritized in budgets. This is more often than not by reason of the need for more housing to provide for a growing population. Therefore financial limitations prevent the development and maintenance of green space planning in urban environments, where it is most needed [8]. When green spaces are indeed provided for and developed in South African cities and towns, the maintenance thereof is an on-going expense for the local government and typically receives only a small vote in the budget, often insufficient.

3. Child-Friendly Green Spaces

According to reference [22] a child's living environment has an extreme influence on almost all aspects of their lives. Therefore children need open space, in order to have a relationship with their natural environment to develop their skills and natural abilities to their full potential. "The concept of child-friendly environments has been inspired by the concept of child-friendly cities" [28]. The concept refers to developing better conditions for children in the urban environment by focusing on child-friendly green spaces in the urban area. A primary concern in improving the urban environment should thus be children's health and their direct relationship with the natural environment.

3.1. Defining Child-Friendly Spaces

Child-friendly spaces can be defined as public green spaces, which are specifically designed in an urban area for children to enjoy the natural environment, and at the same time have a positive impact on their skill development [18,24,44]. Child-friendly green spaces per definition are mainly green areas, where children feel safe interacting with nature. Access to these parks should be easy, ideally within walking distance from the children's homes [10,28].

3.2. Benefits and Importance of Planning for Child-Friendly Spaces

In South Africa, overweight and obesity in children living

in urban areas is an increasing problem [44], which has a major adverse effect on their health, leading to an increasing risk of heart disease and diabetes. The problem exists when there is a lack of physical activity. This is often because they find themselves in a less safe neighborhood, with a lack of green spaces which then result to children being segregated from their society.

Child-friendly green space development should be incorporated in all influencing policy and legislation documents, supporting the development and improvement of child-friendly green spaces. The problem however in South African Policies and Legislation applicable to green space provision is keeping it sustainable.

By giving children space in their natural environment, their abilities to think and react can be practically observed [4]. It is essential for child kinetics and planning to collaborate when child-friendly green spaces for children in urban areas are planned and developed. These spaces do not only focus on providing a "play" space for children, but influences their developing stages when they interact with the specific facilities and objects provided in the space [9]. Focus should be shifted towards improving different planning approaches to provide child-friendly green spaces.

Qualitative green spaces that are found in and around cities can be utilized to improve these spaces for a specific goal. A majority of the spaces are usually to cater for everyone's needs and in general it can be identified as parks. These spaces can be developed in a more specific way to only provide to the needs of the cities children's, by developing it to be more child-friendly and incorporating child-friendly urban design tools in these spaces [29,30].

A good international example of qualitative green spaces, being developed as places with a purpose for enjoyment, is water parks [5]. Water parks are areas that provide activities and enjoyment to different age groups, attracting the inhabitants of a city (adults and children) to enjoy the outdoor space and place provided for them to their full potential [5]. In and around New York City a variety of water parks can be identified, some indoors and some outdoors, all within driving distance for a fun day in the natural environment of the urban areas.

4. Planning Approaches Focused on Child-Friendly Green Spaces

4.1. Child Orientated Planning Approach

The child orientated planning approach lays great responsibility on development and planning, to improve spaces, by becoming more child-friendly and incorporating a child orientated planning approach [29]. By implementing a child orientated planning approach, benefits can range from promoting healthier lifestyles, improving the social interaction and at the same time improve sustainability of natural spaces in the long term [22].

The key to planning for child-friendly green spaces, is incorporating child-friendly design procedures and participatory decision making [22], where the children of the specific city, town or region combine their inputs with the development and planning of theses spaces [29]. By means of incorporating the children, their wants and needs will directly be satisfied.

Safety, access, integration and green spaces are all important concepts in the child planning orientated approach. These concepts should be integrated in the planning and development of child-friendly green spaces [7,12,22].

4.2. Urban Design Approach

The development or improvement of child-friendly spaces in the urban environment contributes to improving the quality of life for the children in the cities, by providing green spaces, which positively affects sustainable development. Urban Design is a planning approach that plays a major roll when providing spaces and places for the public to enjoy [7]. Reference [20] defines Urban Design as "the art of creating possibilities for the use, management, and form of settlements of their significant parts". Urban design is for the people, adding quality to their life.

Contributing to the development of green spaces, specifically child-friendly spaces in the urban environment, urban design tools such as traffic calming elements, separation, different surfaces and sufficient benches are more examples of tools that can be included in designing and developing the ideal child-friendly spaces [7].

Items, element and physical structures that are provided in green spaces and the layout of it in the specific space, determines the attractiveness of the space for the children, the purpose of why it is there and how it would contribute to a child's life [30]. The urban design approach is thus a process concerned with the use of land and design of the urban environment, with its origins reclined in the movement for urban reform [7]. The urban planning approach can include urban renewal, by adapting urban planning methods and at the same time focus on the massive challenges associated with urban growth.

4.3. Place Making Approach

To fully understand space, its physical, social and symbolic dimensions should be taken into consideration [21]. The biggest challenges in today's cities is to provide quality green spaces for the public, where the spaces have meaning and development has taken place to provide outdoor environments for the inhabitants of the cities. But too often green spaces exist in new developments and cities where ill-planning were the cause and according to reference [38] it becomes "after-the-fact cosmetic treatment".

The place making approach is focused on the importance of lively neighborhoods and inviting public spaces. The approach is both an overarching idea and a hands-on tool for improving a neighborhood, city or region where public spaces should form the heart of every community, in every city. The approach thus inspires people to create and improve their public places, strengthening the connection between them and public space itself [30].

Thus, the importance of planning for green spaces with a certain meaning and goal cannot be clearer, for it positively affects the city and its inhabitants on different levels to achieve sustainability and improve the quality of life for each person in the city [3].

The process of place making has a certain criteria that should be followed to ensure a well-developed and designed space providing places and spaces with meaning. The idea of place making is to develop sustainable and livable areas where people want to go, visit and enjoy.

Place making is an idea, tool and concept for improving a neighborhood, city or region. By incorporating the place making concept and using it as a tool in the developing process of cities or town, the attractiveness is enhanced and sustainability is improved. According to reference [25] place making is how public spaces are collectively shaped, to maximize shared value. Place making involves the planning, management, design and programming of public spaces. Therefore place making is how humanity's future is shaped. Project for Public Spaces (PPS) explains place making as both a process and a philosophy. Place making has grown into an international movement, where green spaces must serve the people of the community as a vital place where function is put ahead of form [25].

For the consideration of any public place or space, PPS has developed The Place Diagram (Fig. 1) as a tool to decide whether it is a successful public place [25].

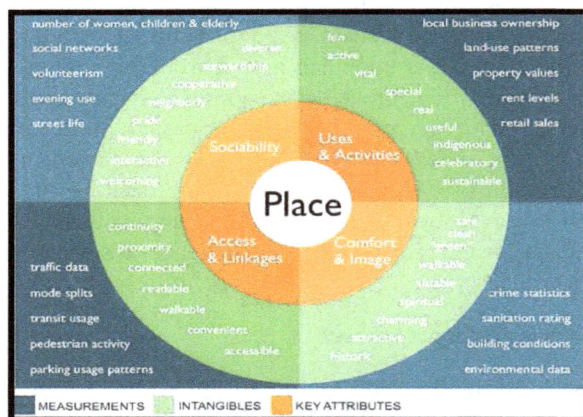

Figure 1. PPS Judgement of any place.

Figure 1 captures the four different attributes of place making as identified by PPS, namely sociability, uses & activities, access & linkages and comfort & image.

There are limited approaches to green space planning and almost none approaches that focus specifically on child-friendly spaces. There is a definite need for more green spaces which are specifically designed and developed for children in these urban areas and the different planning approaches, as discussed, should be followed to ensure that these green spaces will be useable, inviting, safe and developed to its maximum potential to ensure sustainability to the environment and at the same time cater for the needs of the children.

5. Policies and Legislation Guiding the Planning of Child-Friendly Green Spaces in South Africa

5.1. Constitution of the Republic of South Africa 108 of 1996

The rights of children are entrenched in the Bill of Rights in the Constitution of the Republic of South Africa. In the Constitution of the Republic of South Africa 108 of 1996 "child" can be defined as a person under the age of 18. The constitution lays emphasis on the importance of the rights of children and that a child's best interests are number one priority in every matter concerning the child [31].

The democratic values of human dignity, equality and freedom is established in the Constitution of the Republic of South Africa 108 of 1996, where the state must respect, protect, promote and fulfil the rights in the Bill of Rights [31].

Everyone has the right to an environment that is not harmful to their health or well-being, through reasonable legislative and other measures that prevent pollution and ecological degradation, promote conservation, secure ecologically sustainable development and use of natural resources while promoting justifiable economic and social development [31].

5.2. National Spatial Development Perspective: NSDP (2006)

The NSDP provides a framework for future development always focusing on sustainable development. The framework is however a far more focused intervention by the state in equitable and economic growth is a main concern. According to the NSDP (2006) a sustainable development paradigm requires that economic growth and social development are in balance with environmental priorities [35].

Furthermore the NSDP has strong views on protected areas in South Africa. There is however no provision made for the improvement or development of public green spaces, nonetheless child-friendly green spaces in the urban areas of South Africa [35].

5.3. National Urban Development Framework: NUDF (2009)

The NUDF addresses the different challenges and opportunities that the South African towns, cities and city-regions are facing. The framework is focused on environmental sustainability, where the development of greener buildings and renewable energy sources are encouraged.Secondly the framework also focuses on social equity, ensuring that urban and rural areas are not divided, but rather form part of a continues region [11].

The framework also illustrates the problem that South African cities face in terms of urbanization. According to reference [11] cities are not up to the task of managing urbanization, and are poorly equipped to deal with such urban growth. The NUDF states that a shocking 5.8% of children living in cities in the developing world die before reaching the age of 5 [11].

There is no focus on the importance of green spaces for these urban cities. Urbanization is a reality and cannot be stopped, but the environment can be improved by providing better, greener and more sustainable environments for the children and residents of the over populated cities. The main focus should be on providing such spaces.

5.4. National Sport and Recreation Act 110 of 1998

The National Sport and Recreation Act 110 of 1998 provide opportunity to promote and develop sport and recreation and the co-ordination of the relationship between them. The Act requires that funds should be provided annually for the creation and upgrading of basic multipurpose sport and recreational facilities [33].

The National Federations must take full responsibility for the safety issues within their sport and recreational disciplines. Accessibility of such facilities should be taken into consideration and maintenance should be ensured by the beneficiary of the provision of such facilities. Sport and Recreation South Africa must organise and promote programmes aimed to mobilizing the nation to play, but all sport and recreational activities must be conducted in such a way that the environment is not harmfully affected [33].

5.5. Spatial Planning and Land Use Managing Act: SPLUMA (2012)

The Spatial Planning and Land Use Managing Act are set to provide a framework for spatial planning and land use management in the Republic of South Africa. The framework does include the provision of green or public spaces within the municipal boundaries [27].

The objects of this Act are to [27]:

(a) Provide for a uniform, effective and comprehensive system of spatial planning and land use management for the Republic;

(b) Ensure that the system of spatial planning and land use management promotes social and economic inclusion;

(c) Provide for development principles and norms and standards;

(d) Provide for the sustainable and efficient use of land;

(e) Provide for cooperative government and intergovernmental relations amongst the national, provincial and local spheres of government; and

(f) Redress the imbalances of the past and to ensure that there is equity in the application of spatial development planning and land use management systems.

5.6. UNICEF (Unite for children): South Africa Annual Report (2012)

According to reference [39] only 29% of children have access to safe play areas. Children have basic rights, which include basic education, protection, health services and safe outdoor space. Laws alone are not enough to ensure a child's rights, therefore UNICEF supports the children by fighting

for their rights [39].

Layer upon layer the right of children should be established and the rights of children should be protected, but the lack of access to basic services and the rising unemployment statistics are affecting young lives. For UNICEF prioritizing the rights of children in high level discussions on policy, law and budgetary allocations is of utmost importance [39].

5.7. Children's Act 38 of 2005

The primary purpose of the Children's Act 38 of 2005 is to define and protect the rights of children. Such rights should be respected, protected and promoted by the state [32].

Amongst the rights of children is the right to basic education. Provision for early childhood development programs should be made and implemented. Therefore the child's need for development and to engage in play appropriate to such child's age should be recognized. This includes appropriate recreational activities, since such activities play an important role in the child's developing stages [32].

The Children's Act 38 of 2005 also deals with the child's right to health and safety. Children have the right to be protected and such can be achieved by providing a safe and healthy environment for the child, which is conductive to the child's growth and development [32].

5.8. White Paper on Spatial Planning and Land Use Management (2001)

The White Paper on Spatial Planning and Land Use Management states the importance of the usage of land. Land is an asset, scarce and is fragile. When the development of land takes place, high level planning processes that is inherently integrative and strategic, is needed.

The White Paper is set out with an ultimate goal to formulate policies, plans and strategies for land-use and land development that address, confront and resolve the spatial, economic, social and environmental problems of the country.

The spatial planning, land use management and land development norms are:

• Land may only be used or developed in accordance with law;

• The primary interest in making decisions affecting land development and land use is that of national, provincial or local interest as recorded in approved policy;

• Land development and planning processes must integrate disaster prevention, management or mitigation measures;

• Land use planning and development should protect existing natural, environmental and cultural resources;

• Land which is currently in agricultural use shall only be reallocated to other uses where real need exists and prime agricultural land should remain in production.

5.9. Durban Local Agenda 21 (2002)

Durban Local Agenda 21 is aimed on improving sustainability in the urban area. To improve sustainability, the "green" element of the city should be improved. Durban Metropolitan Open Space System (D'MOSS) is focused on the design of an open space plan, where the management of the city's natural resource base can *be protected and guided [13].*

Although the Local Agenda does not make provision for child-friendly green spaces, the program does include the provision of open spaces for the community. By providing these open space in high-density residential areas, quality of live would be improved and jobs could be created. Unfortunately key problems facing the provision of open spaces to the communities are the lack of maintenance funds, poor project management and vandalism [13].

5.10. Spatial Development Framework (SDF) Ethekwini Municipality (2013)

The SDF of eThekwini Municipality is mainly focused on developing the city to become more sustainable. The city is focused on introducing the "green" aspect to the city and improving "green" development on different levels. The Greening Durban Program 2010, is a initiative aiming to ensure that a positive environment is achieved. By focusing on "green" development, sustainability will be improved [15].

The SDF makes provision for providing open/ public spaces, focusing on developing these spaces to be more "green". These spaces provide visual attractiveness, improve quality of live and provide space for recreational purposes. Certain areas are set aside according to the SDF, to be protected. These areas include urban open spaces. With the current open spaces or usually open space that was not planned for, fragmentation is a concern [15]. Therefore open spaces should be conserved and linked.

The SDF seeks to guide a more efficient use of the limited infrastructure, urban space and natural resources. The main goal is to create a city that is more efficient, sustainable and safe. The city should provide high quality public space for the residents, recreational opportunities should be provided and equal access to goods is eThekwini Municipality's SDF main goal and focus. More focus should be set aside for developing the public spaces to become more child-friendly [15].

5.11. Integrated Development Plan (IDP) Ethekwini Municipality (2013/2014)

The main goal set by the IDP of eThekwini Municipality, for the residents, is that all those who live, work, play and invest in eThekwini feel and are safe in private and public spaces. It is important for the eThekwini Municipality to provide a cleanandgreenenvironment,capableofdeliv*ering a range of ecosystem* goodsand services, leading to homely neighborhoods [14].

Contributing to these safe and homely neighborhoods are economically and environmentally sustainable public spaces, which is an essential components of a green and prosperous City. An increase in the use and appropriate design and maintenance of public open spaces also contributes to enhancing neighborhoods and reducing risks [14]. This exists because of people who want to take positive action to make public spaces safer.

The IDP of eThekwini Municipality supports the development of public green spaces, contributing to the environment and the residents of Durban. It addressesspatialsegregationthroughactionssuchasacomprehen siverehabilitationof neighborhoods,creatinghighqualitypublicspacesandfacilitiesi nareasthatwerepreviously underserviced [14].

6. International Case Studies of Child-Friendly Urban Green Spaces

A case study analysis was conducted to evaluate international examples of child-friendly spaces, where the best practices were identified and illustrated in terms of applicability in South Africa. These international case studies include Kadidjiny Park in Melville, Australia and Mullerpier public playground incorporated in Rotterdam, the Netherlands.

A local case study in South Africa was conducted to evaluate the approach and opportunities of planning for child-friendly spaces in local context.

6.1. Kadidjiny Park, Melville, Australia

Melville situated in Western Australia believes that an age-friendly environment benefits the entire community for current and future generations [17]. Melville strives to become an attractive city, with safety being their prime vision. Sustainability plays an important role for the city of Melville, where the consequences of their actions for future generations are taken into account [17].

The city of Melville strives to become more and more child-friendly, believing that when children have sufficient outdoor space for free play in a natural environment, there are overwhelming benefits, including a child's overall physical health and emotional well-being [17].

Melville attempts to provide more and more outdoor green child-friendly spaces for their young citizens. When the children "play" and explore in these spaces provided for them, it is only then when they learn how to cope with life, experience risk taking, face challenges, begin to understand the world around them and develop a sense of belonging [17]. Therefore the City of Melville is considering making positive changes towards providing more natural environments for the children.

Melville incorporated different guidelines such as talking to the children and involving them in the planning of child-friendly green spaces. Asking children what they like to do when they play rather than what they would like in a playground and community involvement contributes to these guidelines used by Melville for developing child-friendly green spaces [17].

The Kadidjiny Park in Melville was specifically planned

and designed for children, in order to interact with the natural environment, and increase their abilities to think, learn and listen [10]. The park can accommodate residence and specifically children of all ages, and provides a unique multi-use landscape which has a positive effect on all residents of Melville. The park opened in November 2010 [1] and promotes imaginative and active play outdoors. The open space is much focused on the green aspects, such as the conservation and expansion of natural habitat, and improving the sustainability of the town where over 40 000 indigenous plants and trees have been planted to improve the "green" aspect of the park and the sustainability of the town in general [10].

The park situated in the suburb of Melville, also known as the Dr Seuss playground because of the red and white poles (see Fig. 2) that can be identified on the open space, making the space more attractive, was developed on the former Melville Primary School site, but when the school was relocated The City of Melville purchased it from the Education Department in 2006 and decided to develop it into a constructive space, bringing meaning and function to the area and contributing to the spatial needs of children of the area in a positive way. Many of the schools children were involved in the designing and planning of the park, resulting in comprehensive public participation processeswere the children were included in the developing process, and articulating their ideas of how a child-friendly park should look or feel like [10]. The execution of the proposed plan of the child-friendly park was successful, because of the participatory planning process and the principle of not planning for children, but with children.

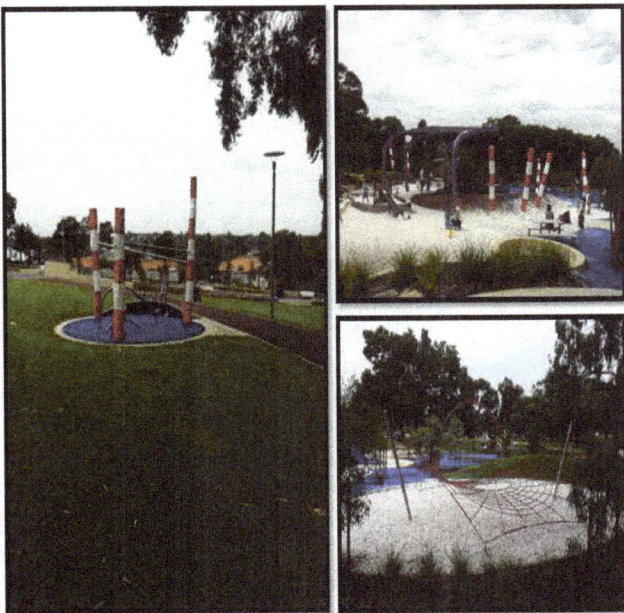

Figure 2. Kadidjiny Park in Melville, Australia.

The children of Melville have had a purpose in the development of the park, by being included in the planning process of this specific green space. Through this process, the children have witnessed the change it has brought and the positive influence it has on the town in terms of providing open green space for children and adults. But the main goal of providing a space that is appealing and where children can explore the environment with a feeling of safety and improve their natural abilities and the way they think and interact with people, has been achieved in this case. [1,10]. Fig. 2 illustrates Kadidjiny Park situated in Melville, Australia.

In terms of specific design guidelines, the Kadidjiny Park is fenced which improves the safety element for all children, distancing the children from the roads and other objects that can be a threat. The park is almost four hectares in size, which is quite large but the size contributes to the fact that it caters for all age groups to enjoy the outdoor space in Melville, with sufficient benches provided in the open space. The space promoted the idea of children to be creative in the space, where the name Kadidjiny comes from the Aboriginal Noongar word meaning "learning. thinking, listening" [10].

6.2. Mullerpier Public Playground, Rotterdam, Netherlands

Rotterdam situated in the Dutch province of South Holland, the Netherland, forms part of the Randstad capital, making it the second largest city in the Netherlands. The Port of Rotterdam is however one of the world's largest and busiest harbours [41]. With many immigrants, Rotterdam strives to be a clean, safe and healthy living environment for their residence living and working there [41].

Rotterdam has strived to become more child-friendly for the young citizens of this city [42,44]. It is believed that children form an intrinsic part of the city and therefore they should feel like they belong and fit in with the city. The child-friendly Rotterdam program started in 2007 with the goal of improving or incorporating child-friendly open spaces within the city for children to be able to enjoy the spaces developed for them. The program was developed to strengthen the economy of the city and improve the quality of life for the children of Rotterdam [44].

A method of "Building Blocks for a child-friendly Rotterdam" was developed [44] to identify areas where there is room for improvement in certain neighborhoods in Rotterdam. This method was used to improve the sustainability of the city, focusing on children and for them to be able to grow and learn how to interact with the environment and people, impacting their development skills [44]. These urban development tools of incorporating building blocks to improve child-friendliness in the Rotterdam city have succeeded to draw families back to the neighborhoods of the city [44]. Many families tended to leave the city and move out of the neighborhoods after the post-war era. This trend needed to be reversed, because a neighborhood that is not attractive for people, to live and work there, is destined to struggle [42] and the city of Rotterdam tried to achieve this through the development of the building blocks and was successful in doing so.

By attracting these families back to the neighborhoods of Rotterdam, the method of "Building Blocks for a child-friendly Rotterdam" promoted the idea of combining smaller apartments together to develop larger apartments with the end result of families having more living space and this idea is more striking to families to relocate themselves back in the neighborhoods of Rotterdam [42,44].

Rotterdam has made child-friendliness a key strategic goal. There are purposes and focus points on what has to be done or can be done to improve child-friendly spaces, because the City of Rotterdam believes a successful city for children is a successful city for all people [44]. This statement stresses the importance once again on how important it really is to have child-friendly open spaces in an urban setting that are welcoming for children to interact with the environment, feel safe and develop themselves.

For Rotterdam it was important to develop a city for children and adults that are lively and sustainable for one main reason, to keeps families in the city. The urban planning tool of "child-friendliness" that Rotterdam has developed, by focused on children and building a healthier city for them will lead to sustainability of the city and strengthen the economy. Different policies take children's needs and wants in consideration and these spaces are not only about developing playgrounds or providing an open space for children to run around in, it is rather a holistic way of thinking about urban planning, making the city more sustainable and attractive and specifically designed open green spaces for children to think, interact and learn on different levels with the end result of having a positive effect on their development and the way they react in different situations.

The city of Rotterdam believes that children form an intrinsic part of the city. By improving the city to become more child-friendly, resulting in children to become an asset to the city, there are physical and socio-economic requirements that Rotterdam should meet in order to cater for the needs of the children [44]. Rotterdam strives to ensure a child-friendly city, by incorporating the needs of children in their development frameworks.

Rotterdam developed an urban planning method, Building Blocks for a child-friendly Rotterdam [44]. This method consists of four building blocks, where the building blocks not only identify problems, but also opportunity for improvements [44].

Table 1. Rotterdam Building Blocks for a child-friendly Rotterdam.

building blocks	description
child-friendlyhousing	Single-family houses with a garden.
public space	Gear public space to the specific needs of children. Play areas, outdoor space, use of space between front door and street level (luminal space).
facilities	Providing opportunities for socializing.
safe traffic routes	Routes are made more child-friendly to encourage children to explore the city (sidewalks and no-through traffic zones).

Mullerpier public playground, situated on the Mullerpier in Rotterdam, was developed by Bekkering Adams Architects in 2007 [2]. This interesting squire was developed to serve as a playground for the nearby primary school, but currently also

meets the needs of the children in the surrounding area by serving as a public playground for each one. The square challenges children to play in different ways, interacting with one another and exploring their natural environment [2]. With sufficient activities provided in the space (Fig. 3), opportunities are provided for children to interact with strangers and improve their skills [2].

Figure 3. Mullerpier Public playground in Rotterdam,Netherlands.

Rotterdam truly seeks to develop a child-friendly city, including the children of the city when development takes place. Their needs and want are first priority with the child-friendly monitor developed by Rotterdam, to measure the results of efforts to make urban neighborhoods more child-friendly [44].

7. Local Evaluation of Child-Friendly Urban Green Spaces: Durban Case Study Area

Essenwood, Durban, South Africa

Durban is a large, growing city situated in the province of KwaZulu-Natal (KZN), on the east coast of South Africa, where the majority of the population is aged 15-34 years. The Child-Friendly City Campaign of Durban strives to develop the urban area to become more child-friendly by incorporating child-friendly green spaces [16].

The Child-Friendly City Campaign of Durban strives to develop the urban area to become more child-friendly by incorporating child-friendly green spaces [16]. Bulwer Park, situated in the suburb of Glenwood, in Durban is a pilot project of the eThekwini Municipality's Sustainable Public Spaces Programme and forms part of the IDP's Quality Living Environment Plan [16]. The aim for Bulwer Park was to serve as a valuable green space, offering opportunities for recreation and conservation and at the same time provide a creative space for the community (children and adults) [30].

What make Durban as a city liveable are its green spaces. As the population grows, the city itself expands, resulting in the importance of green spaces to increase [30]. Green spaces are nevertheless vital for the well being of the residents of the city. Durban has around 187 m² of green space per person [29], where Hawaan Forest and Burman Bush plays an important role when contributing to the environment.

The eThekwini Municipality of Durban has launched its very own Green Map, an interactive tool, displaying sustainable features in communities and is used all over the world [15,30]. Green Map provides opportunity for sustainable development within the city and different communities. Working with Durban's informal workers from the city's public spaces, AsiyeeTafuleni (AeT) an NGO promotes inclusive urban planning and design [15].

The Durban Metropolitan Open Space System (D'MOSS) is also an important system included in the Spatial Development Framework (SDF) of the eThekwini Municipality. D'MOSS was previously known as the eThekwini Environmental Services Management Plan (EESMP) and is a system of open spaces [15]. This includes high biodiversity open spaces, linked together in a viable network. D'MOSS covers a great deal of land, including nature reserves, large rural landscapes, coastal corridors and some privately owned land. Apart from contributing to the attainment of these areas, D'MOSS also provides recreational opportunities for the residents of Durban [15].

Durban has identified different project objectives, including creating an identity for the park with a catalyst and implement a strategy that should be managed. A high quality sustainable public space ought to be created when developing child-friendly/ open green spaces in the urban area [16].

The city of Durban further aims to implement specific planning elements such as footpaths, benches, bins and park lights to improve the current open spaces to become safe, inviting and child-friendly [16,29].

Figure 5. Current use of identified space in local case study.

Essenwood is a suburb of Durban, a few blocks to the Northwest of the city centre, located central, within reach of some of the main attractions in Durban. Currently green spaces situated in this area, play space, is neglected, with an uninviting and unsafe feeling, as identified in Fig. 5. With public inputs, these spaces have been described as unsafe and unappealing for children. The majority of the residents in the

Essenwood area live in apartment block, having no backyard space or access to open green space. Certain residents who do however live in houses have undersized backyards [30].

Based on evaluations of the concept that enhanced child-friendly green spaces, the Essenwood case study was found to be successful in terms of integration, traffic calming elements and access and linkages. Gaps were however identified in terms of safety, green space, access, separation, different surfaces, sufficient benches in space, comfort and image, uses and activities and sociability.

8. Results and Discussions

In comparison to international cases, it can be concluded that there is minimal provision made for green space planning in South Africa, and even less for the provision of child-friendly spaces in urban areas. International approaches and examples could guide and influence the local approach to planning in South Africa and aim to provide guidelines for the development and implementation of green spaces, more specifically child-friendly spaces. As identified in the international case studies, the key to success is to include public participation methods, ensuring that the public's needs are met. Neighborhoods should be developed to become more child-friendly, traffic calming elements, safety, open green spaces, and such can only be achieved by observing individual neighborhoods and identifying where there is room for improvement of development of child-friendly green spaces.

Improvements can be made on macro (surrounding area) and micro (specific green space) levels to ensure specific planning for child-friendly spaces.

8.1. Macro Environmental Recommendations

On a macro scale recommendations should be made to improve the safety around the identified space. These safety elements will contribute to improve the child-friendly component in and around the space.

Different planning tools, such as stop streets, separation elements and traffic calming elements (speed bumps) should be developed around the space to ensure optimal safety.

8.2. Micro Environmental Recommendations

The micro environment recommendations are made in terms of the different functions that will be provided in the open space identified. It is important that the space has an inviting feeling, provides different activities for children of different ages and where interaction can take place, improving sociability.

Important elements that can be developed in child-friendly green spaces. These elements ensure a safe environment for children where they interact with other children and improve their skill development:

(1)   Entrance to the child-friendly public park
(2)   Incorporation of benches
(3)   Introducing playing courts

(4)   Providing skipping ropes
(5)   Introducing a balancing pole
(6)   Providing a jungle gym

Different policies and legislations could be adapted to support child-friendly green spaces in urban areas. The majority of the policies and legislations do support open/public green spaces and strive to improve sustainability by improving the green element in urban areas. Unfortunately child-friendly green space development is not supported as much as it is necessary in specific policies and legislations.

The recommendations made on a macro and micro environment scale in are to improve the current open space and to enhance the planning and development of a child-friendly urban green space. These recommendations can be used for planning similar public spaces in all regions of South Africa. The main priority is to improve the safety element of the open space and secondly to improve the child-friendly element. The recommendations are set to provide an inviting, green and safe child-friendly public park for the community to enjoy, but specifically for the children to improve their physical and skill development as they enjoy and use the space.

9. Conclusion

Planning in urban areas in South Africa does address the provision and development of green spaces. Unfortunately there are not enough provision made for the development of child-friendly green spaces in urban areas, which contribute to their skill development and natural abilities. Policies and legislations should be amended to support open green spaces that are specifically designed for children. By incorporating child-friendly spaces, the quality of life of the children in the cities are improved, as they will have access to unpolluted environments and enjoy the outdoor natural environment in the area of their homes. The activities provided, through these child-friendly green spaces, improve children's social interactive skills and ultimately result in them being active outdoors and enjoying the natural environments.

Different policies and legislations could be adapted to support child-friendly green spaces in urban areas. The majority of the policies and legislations do support open/public green spaces and strive to improve sustainability by improving the green element in urban areas. Unfortunately child-friendly green space development is not supported as much as it is necessary in these specific policies and legislations.

The recommendations made on a macro and micro environment scale are to improve the current open space and to enhance the planning and development of a child-friendly urban green space. These recommendations can be used for planning similar public spaces in all regions of South Africa. The main priority is to improve the safety element of the open space and secondly to improve the child-friendly element. The recommendations are set to provide an inviting, green and safe child-friendly public park for the community to enjoy, but specifically for the children to improve their physical and skill development as they enjoy and use the space.

Acknowledgements

- This research (or parts thereof) was made possible as a result of a financial contribution from the NRF (National Research Foundation) South Africa.
- Any opinion, findings and conclusions or recommendations expressed in this material are those of the author(s) and therefore the NRF does not accept any liability in regard thereto.

References

[1]   A coffee in the Park. 2013. Kadidjiny Park Melville.www.acoffeeinthepark.com/kadidjiny-park-melville. Date of access: 26 Mar. 2014.

[2]   ArchiTravel. 2013. Public playground.www.architravel.com/architravel/building/public-playground. Date of access: 10 Jul. 2014.

[3]   Atiqul, H.A.Q. & Shah, M.D. 2011. Urban green spaces and an integrated approach to sustainable environment. *Journal of Environmental Protection*, (2):601-608. Date of access: 10 Jul. 2014.

[4]   Berthelsen, T.C. 2012. Seven Ideas from Tokyo for child-friendly spaces.http://thisbigcity.net/seven-ideas-from-tokyo-for-child-friendly-spaces/. Date of access: 17 Feb. 2014.

[5]   Birnbaum, S., Farrell, K. & Katz, A. 2012. New York family: The 9 best water parks near NYC.www.newyorkfamily.com/float-on/. Date of access: 11 Mar. 2014.

[6]   Brooker, L. &Woodhead, M. 2013. Early childhood in focus. The right to play. The Open University. http://www.google.co.za/url?sa=t&rct=j&q=&esrc=s&source=web&cd=1&ved=0CBwQFjAA&url=http%3A%2F%2Fwww.bernardvanleer.org%2FThe-Right-to-Play%3Fpubnr%3D1849%26download%3D1&ei=WwggVI2OEPKd7gbxkIHICw&usg=AFQjCNH-e30gbOSkuwcls9jqc3UbEP-1Rg&bvm=bv.75775273,d.ZWU. Date of access: 29 Mar. 2014.

[7]   Carmona, M., Heath, T., Oc, T. &Tiesdell, S. 2003.Public places-Urban spaces. Massachusetts: Architectural Press.

[8]   City of Tshwane. 2005. Proposed Tshwane open space framework, (1):1-135.http://www.google.co.za/url?sa=t&rct=j&q=&esrc=s&source=web&cd=2&ved=0CCIQFjAB&url=http%3A%2F%2Fwww.tshwane.gov.za%2FServices%2FOpenSpaceManagement%2FOpen%2520Space%2520Framework%2FOpen%2520Space%2520Framework%2520Vol%25201.pdf&ei=RQwgVMOMEYHW7QbC6oDACw&usg=AFQjCNHLFRnHder5JKScUTEjdh-y-H0N1g&bvm=bv.75775273,d.ZWU. Date of access: 8 Jul. 2014.

[9]   Coetzee, D. 2014. Child kinetics: Fine and gross motor activities for children [personal interview]. 24 Jul. 2014. Potchefstroom.

[10] Commissioner for Children and Young People. 2011.Caring for the future growing up today: Building spaces and places for children and young people. http://www.google.co.za/url?sa=t&rct=j&q=&esrc=s&source=web&cd=1&ved=0CBwQFjAA&url=http%3A%2F%2Fwww.ccyp.wa.gov.au%2Ffiles%2FBuilding%2520spaces%2520and%2520places%2520for%2520children%2520and%2520young%2520people.pdf&ei=2wwgVKDlA8XR7Qa2-4GIDA&usg=AFQjCNEgUTKo1DHIJT4zmEFLq_tbFiDyAw&bvm=bv.75775273,d.ZWU. Date of access: 17 Feb. 2014.

[11] Department of Cooperative Governance and Traditional Affairs and The Presidency in partnership with the South African Cities Network. 2009. National Urban Development Framework. http://www.google.co.za/url?sa=t&rct=j&q=&esrc=s&source=web&cd=1&ved=0CBwQFjAA&url=http%3A%2F%2Fwww.gov.za%2Fdocuments%2Fdownload.php%3Ff%3D107790&ei=YQ0gVJzeBseM7AbE_IHwCg&usg=AFQjCNEOldYujk5mAS23owkdnX8Bb3MMZw&bvm=bv.75775273,d.ZWU. Date of access: 13 Aug. 2014.

[12] Dewar, D. &Uytenbogaardt, S. R. 1995. Creating vibrant urban places to live: a primer. Cape Town: Headstart Developments.

[13] Diederichs, N. & Roberts, D. 2002. Durban's local agenda 21 programme: tackling sustainable development in a post-apartheid city. *Environment & Urbanization*, 14(1):189-201.

[14] eThekwini Municipality. 2013. Integrated Development Plan. http://www.google.co.za/url?sa=t&rct=j&q=&esrc=s&source=web&cd=3&ved=0CCkQFjAC&url=http%3A%2F%2Fwww.durban.gov.za%2FCity_Government%2FCity_Vision%2FIDP%2FDocuments%2FeThekwini%2520IDP%25202013_14.pdf&ei=bmMOVdyCIoGv7Aagj4BI&usg=AFQjCNEllPTsawaSR3tZ_b9cXXqBIuhYQ&sig2=BeH6ynpTq8sDbme-wK7GCw&bvm=bv.88528373,d.d24. Date of access: 17 Mar. 2015.

[15] eThekwini City Council. 2013. Spatial development framework of eThekwini.http://www.google.co.za/url?sa=t&rct=j&q=&esrc=s&source=web&cd=1&ved=0CBwQFjAA&url=http%3A%2F%2Fwww.durban.gov.za%2FResource_Centre%2Freports%2FFramework_Planning%2FDocuments%2FSDF_Executive_Summary_2013-2014%2520-%2520May%25202013.pdf&ei=dhEgVKrpDNSM7Abu5YD4Cg&usg=AFQjCNHnpweTXzrclswdlYoeNoINnaMZqQ&bvm=bv.75775273,d.ZWU. Date of access: 14 Mar. 2014.

[16] Galvin, M. & Payne, SE. 2010. Bulwer Park Revitalization Strategy. http://www.google.co.za/url?sa=t&rct=j&q=&esrc=s&source=web&cd=1&ved=0CBwQFjAA&url=http%3A%2F%2Fwww.imaginedurban.org%2FDocLibrary%2FBULWER%2520PARKsum11.doc&ei=bL07VNykNszg7QbG64GIBg&usg=AFQjCNGrWWXMQFBDsLU-tJNNXepfu5GvzA. Date of access: 20 Sep. 2014.

[17] Government of South Australia: Department of Education and Children's Services. 2012. Promoting Natural Outdoor Learning Environments. http://www.google.co.za/url?sa=t&rct=j&q=&esrc=s&source=web&cd=1&ved=0CBwQFjAA&url=http%3A%2F%2Fwww.lga.sa.gov.au%2Fwebdata%2Fresources%2Ffiles%2FDesigning_Outdoor_Learning_Environments_for_Young_Children.pdf&ei=tBEgVLiDIaSd7gb3xoDwCw&usg=AFQjCNFpKOg3m4dQ-wMdZkJc5k-yocnMAA&bvm=bv.75775273,d.ZWU. Date of access: 24 Jul. 2014.

[18] Howard, A. 2006. What constitutes child friendly communities and how are they build?*Australian Research Alliance for Children & Youth.* P1-57.http://www.google.co.za/url?sa=t&rct=j&q=&esrc=s&source=web&cd=1&ved=0CBwQFjAA&url=http%3A%2F%2Fwww.aracy.org.au%2Fpublications-resources%2Fcommand%2Fdownload_file%2Fid%2F165%2Ffilename%2FWhat_constitutes_child_friendly_communities_and_how_are_they_built.pdf&ei=9RIgVKPHJ6md7gbXpYG4Cw&usg=AFQjCNHpM6ZljCovRkGdSQydyIROzeZSQA&bvm=bv.75775273,d.ZWU. Date of access: 28 Mar. 2014.

[19] Levent, T.B., Vreeker, R. &Nijkamp, P. 2004.Multidimensional Evaluation of Green Spaces: a comparative study on European cities. dare.ubvu.vu.nl/bitstream/1871/8928/1/20040017.pdf.Date of access: 17 Feb. 2014.

[20] Lynch, K &Hach, G. 1984. Site Planning. 3rd ed.Cambridge: The MIT Press.

[21] Madanipour, A. 1996. Design of urban space: an inquiry into a socio-spatial process. England: John Wiley & Sons Ltd.

[22] McAllister, C. 2008. Child friendly cities and land use planning: Implications for children's health. *Environments Journal*, 35(3):45-56.https://www.google.co.za/?gws_rd=ssl#q=McAllister%2C+C.++2008.++Child+friendly+cities+and+land+use+planning%3A+. Date of access: 2 Feb. 2014.

[23] McConnachie, M.M. &Shackleton, C.M. 2010.Public green space inequality in small towns in South Africa.http://contentpro.seals.ac.za/iii/cpro/DigitalItemViewPage.external?sp=1006874. Date of access: 8 Jul. 2014.

[24] McDonald, L. 2012. Belfast healthy cities: child friendly spaces. http://www.google.co.za/url?sa=t&rct=j&q=&esrc=s&source=web&cd=1&ved=0CBwQFjAA&url=http%3A%2F%2Fwww.iphopenconference.com%2Fsites%2Fdefault%2Ffiles%2FLaura%2520McDonald%2520Final%2520iph%2520presentation.pdf&ei=oBYgVJ3gMYqd7gaxs4CgDA&usg=AFQjCNHLp45LYuN2oOMFrRXIAsXJIm5p4A&bvm=bv.75775273,d.ZWU. Date of access: 3 Apr. 2014.

[25] Metropolitan Planning Council. 2008. Place making Chicago: A neighborhood guide to placemaking in Chicago. Four key qualities of a successful place.http://www.placemakingchicago.com/about/qualities.asp. Date of access: 8 Mar. 2014.

[26] Ministry of Agriculture and Land Affairs: Republic of South Africa. 2001. White Paper on Spatial Planning and Land Use Management. http://www.google.co.za/url?sa=t&rct=j&q=&esrc=s&source=web&cd=1&ved=0CBwQFjAA&url=http%3A%2F%2Fwww.sapi.org.za%2Fsites%2Fdefault%2Ffiles%2Fdocument-library%2FWhite%2520Paper%2520on%2520Spatial%2520Planning%2520and%2520Land%2520Use%2520Management%2520 2001.pdf&ei=QGcOVaTNJafa7AaQkoD4Bw&usg=AFQjCNGOgDnMeUFEiZn5Djj763ToFdTwBg&sig2=Nztu1-PV0rS16I0YLGD1yA. Date of access: 17 Mar. 2015.

[27] Minister of Rural Development and Land Reform: Republic of South Africa. 2012. Spatial Planning and Land Use Managing Act.http://www.google.co.za/url?sa=t&rct=j&q=&esrc=s&source=web&cd=1&ved=0CBwQFjAA&url=http%3A%2F%2Fwww.ruraldevelopment.gov.za%2Fphocadownload%2FSPLUMB%2Fnationalspatplanning_landusemgmtbill.pdf&ei=6hYgVJKgNeud7gaVw4CYCw&usg=AFQjCNEgZQk4qxETZC094uedY2jXOwrhyA&bvm=bv.75775273,d.ZWU. Date of access: 15 Aug. 2014.

[28] Nordström, M. 2010. Children's views on child-friendly environments in different geographical, cultural and social neighborhoods. *Urban Studies*, 47(3):514-528.http://www.google.co.za/url?sa=t&rct=j&q=&e src=s&source=web&cd=1&ved=0CBwQFjAA&url=http%3A %2F%2Fwww.hphpcentral.com%2Fwp-content%2Fuploads% 2F2010%2F09%2FUrban-Studies-47-2010-Childrens-Views.p df&ei=rBcgVOH4BqyV7Aby_ICwCg&usg=AFQjCNFKmeK fy8_rZ2kQsS_qLAf-tbHFNg&bvm=bv.75775273,d.ZWU. Date of access: 3 Feb. 2014.

[29] Parker, E. 2014. Proposals for the development of child-friendly spaces in urban areas [personal interview]. 10 Mar. 2014. Durban.

[30] Prange, M. 2014. Urban Design tools to improve child-friendly green spaces [personal interview]. 14 Apr. 2014. Durban.

[31] South Africa. 1996. Constitution of the Republic of South Africa 108 of 1996. http://www.google.co.za/url?sa=t&rct=j&q=&esrc=s&source= web&cd=1&ved=0CBwQFjAA&url=http%3A%2F%2Fwww. gov.za%2Fdocuments%2Fconstitution%2F1996%2Fa108-96. pdf&ei=8hcgVNG3CaPP7gaplIDwCw&usg=AFQjCNGXsaiz QhB_lA7ZPMli7rI4w7YD_A. Date of access: 17 Aug. 2014.

[32] South Africa. 2005. Children's Act 38 of 2005. http://www.google.co.za/url?sa=t&rct=j&q=&esrc=s&source= web&cd=2&ved=0CCYQFjAB&url=http%3A%2F%2Fwww. justice.gov.za%2Flegislation%2Facts%2F2005-038%2520chil drensact.pdf&ei=KRggVML3Lcrd7QaguYHYCw&usg=AFQ jCNFnJ7-7G0wQvlt3smjV64hrxj_QLg. Date of access: 17 Aug. 2014.

[33] South Africa. 1998. National Sport and Recreation Act 110 0f 1998. http://www.srsa.gov.za/MediaLib/Home/DocumentLibrary/46 4863(3).pdf. Date of access: 17 Mar. 2015.

[34] Statistics South Africa. 2013. Mid-year population estimates 2013. http://www.google.co.za/url?sa=t&rct=j&q=&esrc=s&source= web&cd=1&ved=0CBwQFjAA&url=http%3A%2F%2Fwww. statssa.gov.za%2Fpublications%2FP0302%2FP03022013.pdf &ei=WRggVLD_LMGV7Aas9oHwCg&usg=AFQjCNE-6qi1 dtxUVaE5LUwMomT3E_jFpA. Date of access: 11 Feb. 2014.

[35] The Presidency Republic of South Africa. 2006. National Spatial Development Perspective.http://www.google.co.za/url?sa=t&rct=j&q=&esrc =s&source=web&cd=2&ved=0CCIQFjAB&url=http%3A%2 F%2Fwww.npconline.co.za%2FMediaLib%2FDownloads%2

FHome%2FTabs%2FDiagnostic%2FMaterialConditions2%2F The%2520Presidency-%2520National%2520spatial%2520dev elopment%2520perspective.pdf&ei=HhkgVIT4ENSS7Ab89Y HwCg&usg=AFQjCNHLTfTuT_XKs8JQhvKe-xXrlsx9NA. Date of access: 13 Aug. 2014.

[36] Thai Utsa, B., Puangchit, L., Kjelgren, R. &Arunpraparut, W. 2008. Urban green space, street tree and heritage large tree assessment in Bangkok. *Thailand, Forestry and Urban Greening*, 7(3):219-229. http://www.google.co.za/url?sa=t&rct=j&q=&esrc=s&source= web&cd=1&ved=0CBwQFjAA&url=http%3A%2F%2Ftreene tmedia.com%2Fup%2Fpdf%2F2013%2FTree-et13D1S11.pdf &ei=9RggVIuVKeaY7gbv1IH4Cw&usg=AFQjCNFLKMZm wqJxa2jeni_m6ry4rPUyag.Date of access: 17 Feb. 2014.

[37] Thomas, J. 2008. Child in the city.http://www.ombudsnet.org/resources/infoDetail.asp?ID=1 8893&flag=report. Date of access: 3 Feb. 2014.

[38] Trancik, R. 1986. Finding lost space: theories of urban design. Canada: John Wiley & Sons Inc.

[39] UNICEF. 2012. South Africa annual report: unite for children.http://www.google.co.za/url?sa=t&rct=j&q=&esrc=s &source=web&cd=1&ved=0CBwQFjAA&url=http%3A%2F %2Fwww.unicef.org%2Fsouthafrica%2FSAF_resources_annu al2012.pdf&ei=rBkgVN_4DaqV7Ab2qYGACQ&usg=AFQj CNEdmPBiufjnQ_k0xW-WlnJbYEnt7g. Date of access: 16 Aug. 2014.

[40] UrbSpace. 2010. Green spaces in urban areas. Date of access: 2 Apr. 2014.

[41] Van Boxtel, E. &Korenman, K. 2013. My public space.eu: How public is our public space these days?http://www.mypublicspace.eu/content/320132/rotterdam. Date of access: 29 Apr. 2014.

[42] Van den Berg. M. 2013. Rotterdam, city with a future: How to build a Child-friendly City. Date of access: 5 Mar. 2014.

[43] Van Heerden, I.V. 2011. Many SA kids obese.http://www.health24.com/Diet-and-nutrition/Weight-los s/Many-SA-kids-obese-20120721. Date of access: 12 Feb. 2014.

[44] Wapperom, R. 2010. Rotterdam, city with a future: How to build a child-friendly city.Paper presented at the Child in the City Conference, Florence, 28 October.http://www.rotterdam.nl/JOS/kindvriendelijk/Presenta tion%20Child%20in%20the%20City%20october%202010.pdf. Date of access: 26 Mar. 2014.

22

Bee-Keeping for Wealth Creation Among Rural Community Dwellers in Imo State, South-Eastern, Nigeria

Nwaihu E. C.[1], **Egbuche C. T.**[1], **Onuoha G. N.**[2], **Ibe A. E.**[1], **Umeojiakor A. O.**[1], **Chukwu, A. O.**[3]

[1]Department of Forestry and Wildlife Technology, Federal University of Technology Owerri, Imo State Nigeria
[2]Department of Chemistry, Federal University of Technology Owerri, Imo State Nigeria
[3]Department of Agric Economics, Extension and Rural Development, Imo State University Owerri, Nigeria

Email address:
ctoochi@yahoo.co.uk (Egbuche C. T.)

Abstract: This study was carried out in Imo State, South Eastern Agro-Ecological Zone of Nigeria. Five Local Government Areas and five communities were selected for the study. From the five communities, eight (8) Bee-Keepers were selected on purposive basis based on list of bee-keepers collected from Imo ADP field staff. This gave a total of 40 respondents for the study. Data for the study was collected using questionnaire, and oral interview schedule. Both primary and secondary data were used in addition to internet services. The information elicited from the respondents were based on the objectives of the study such as socio-economic characteristics, cost and return on beekeeping, constraints militating against beekeeping and the prospects of the enterprise. The data generated were analyzed using descriptive statistics such as percentages, frequency distribution tables, mean, Gross Margin and Net Farm Income. The result showed that the mean age of the respondents was 37 years, male respondents accounted for 72.5%, 40% had tertiary education, and family labor was the major source of labor. Personal savings (equity fund) was the major source of finance (85%), 55% had information from ADP Extension Agents, 80% use Kenyan Topbar, major Bee products processed were honey (60%) and Bee wax (40). It is profitable in the area as initial cost outlay was N15,900 and returns (Total revenue) is N42,000, thus getting N39,300 as gross marginal income with N26,100 as Net Farm Income (NFI). Lack of finance was the major constraint militating against the enterprise (39.6%) followed by Non-colonization of hives (18.7%). However, worthwhile recommendation on making fund available to Beekeepers by Commercial Banks, engaging the services of extension staff and use of appropriate attractants like sugar solution and sweet fresh palm wine were proffered as solution to some of the teething constraints. However, the enterprise of beekeeping has bright future prospects in the area, considering the number (40) already in practice. Therefore, beekeeping can create wealth in the area and beyond.

Keywords: Bee-Keeping, Wealth Creation, Rural Community Dwellers, Imo State Nigeria

1. Introduction

The Science of Bee-keeping is known as Apiculture. This technology has hive construction, honey harvesting and processing, wax extraction, disease and pest control as well as setting up the Hives (site selection). There are messages on absconding and control [1]. This Agro Forestry System can be integrated into many farming systems. It is best practiced where fallow system is still practiced. The use of non-timber forest products (NTFPs) is as old as human existence [2]. In subsistence economies, the roles and contribution of NTFPs are crucial because of their richness of varieties as source of food, fodder, fibre, fertilizers, herbal medicine and craft activities. Bees are traditionally an important part of small-scale integrated farming systems. Bees do well in natural forests and on integrated farms with abundant water and flowers [3]. There are altogether 20,000 species of honey bees, most of which are found in Asia. Only one of these species occurs in other parts of the world, either naturally or imported by man. Two species are domesticated and used in beekeeping. *Apis cerena* in South East Asia and *Apis mellifera* worldwide [4], [5]. Bees are vital contributors to pollination and crop production. Bee keeping as NTFPs resources provides raw materials to support processing enterprises which include internationally important commodities used in food products and beverages, construction materials, cosmetics and cultural products. They support village level confectionary, flowering, perfumes, medicines, paints, polishes Bees produce honey which contains about 80% of sugars that are readily absorbed by the body and it is extremely suitable food for children, sick people and those who perform heavy mammal Labor [6]. Honey as produced by bees is extensively used in pharmaceutical and beverage industries as pleasant-testing food component, sweetener for food and drinks, effective in the treatment of superficial wounds, treatment of sore or throat complaints. In many countries of the world, honey is used to make beer or wine [7].

It has high economic value, hence a good trade commodity. Both local and international markets are readily available. It requires a minimal amount of labor and no external inputs such as fertilizers or pesticides. Bee Keeping encourages minimum tillage in the acre (0.4ha) around the hive, since noisy equipment may disturb the bees [8]. Honey can also be used as a symbol of social prestige and is used to pay bridal dowries. In some cultures, an exchange of honey symbolizes the settling of major conflicts. Honey is used as a food preservative [9]. Bee wax is used in the manufacture of cosmetics, candles, foundation sheets for hives, polishes and cosmetics. Generally, it is used in shoes and cosmetics industries [10]. Bees use pollen to feed their larvae. The pollen can be collected using a simple trap placed at the flight entrance of the hive. Pollen contains 35% protein and can be eaten dry or added to other foods, as a good source of protein. Pollen is sold to the perfume industry, eaten or for medicinal purposes. Pollen is hygroscopic and must be protected against moisture, and it deteriorates quickly when attacked by fungi. Bee propolis is a resin that bees collect from the plant and they use it to cover the inside of the hive. Propolis has some therapeutic and antibiotic characteristics used for embalmment in early Egypt [11].

1.1. Problem Statement

In spite of the importance of NTFPs, their contributions to rural livelihood in many developing countries are yet to be acknowledged [12]. Timber was perceived as the dominant reason for forest management and hence no attention was paid to NTFPs by foresters, policy manners and economic planners. In Nigeria particularly, there is no clear-cut policy directed at NTFPs at Federal, State and Local Government or even communal levels. NTFPs has long been considered minor or secondary forest products. There was therefore general lack of appreciation of the value and roles of NTFPs in the livelihood of rural dwellers [13]. Modern bee-keeping technology, though not quite popular among the rural community dwellers can be used as a poverty reduction mechanism in Nigeria. The specie of bee, *Apis mellifera* commonly known as the honey bee is the most widely-spread and abundant insect on earth [14]. Due to ignorance of the profitability bee-keeping and fear of being stung by bees deter people from venturing into bee-keeping. Bee keeping for wealth creation has practically remained untapped in the

country. Those already involved in beekeeping in the rural communities are not utilizing all the bee products but are mostly interested only in honey and bee wax extraction.

Another set of problem for bee keeping is the incidence of pest attack, bush burning, indiscriminate pesticide use, and abscondment of bees, non colonization and inadequate information on the enterprise reduces the productivity of beekeeping [15] and [16]. Local beekeepers do not keep records of their activities making it difficult to determine the level of progress they make. Bees may sting the careless if they become aggressive on being disturbed by children.

1.2. Objectives

The objectives of the study are to;
1. Determine the socio-economic characteristics of the respondents.
2. Determine the cost and returns of beekeeping.
3. Determine the constraints of beekeeping.
4. Assess the prospects of beekeeping in the study area.

1.3. Justification

This study will create awareness on the profitability of beekeeping and encourage non-beekeepers to venture into it as means of wealth creation. Bee-keeping does not require a large piece of land or compete with crops or livestock for land space. Bee products provide farmers with an additional source of income as both local and international markets are readily available. Beekeeping can be used as a means of poverty alleviation, hunger reduction and job creation especially in the rural areas of the country where there is a high level of unemployment and the people are mostly engaged in subsistence agriculture for food production [17], [18]. Beekeeping can be undertaken by anyone who has the ability and determination to look after bees properly enough and courage to work with bees. Working with bees requires a gentle touch and calm disposition. It also requires basic understanding of the honey bees behavior during the various seasons and during handling and moving [19]. Depending on the country and environmental factors, a typical colony of bees can produce 80-120 pounds of surplus honey and 10 pounds of pollen in a year [20], [21] and [22]. Honey can be eaten or used in any type of cooking or to sweeten beverage. It can be used to make jams and marmalades [23]. It can be used as medicine alone and sometimes in combination with herbs for common colds, cough, gastric ulcer, restlessness, hypertension, infected surgical wounds, burns and sores, eye itching and taken as tranquilizer [24]. The bee wax is used as a water proof agent for wood and leather,, production of candles, ointments, soaps, polishes, battery cells, transformers, clothes (Attire) and used by dentists as an artificial denture, and used by shoe makers for strengthening shoemakers threads. The venom can be used to cure disease such as arthritis and for treatment of nervous system disorder [25].

The waxy substance, propolis contains enzymes which are believed to contain immunity factors which are used internally. They stimulate the body and give it a natural resistance to diseases. The bee venom has useful medicinal properties and can be collected by placing a clean polyether sheet at the entrance of the hive. The queen and the worker bee produce the venom [26]. The propolis otherwise known as bee glue differs in composition according to the plant from which the bees have been collecting. The bees bring the glues on their hind legs to the hive. They mix it with their own wax and saliva to produce propolis [27]. The propolis is used to prepare cough syrups, toothpastes, lotions, skin soaps, skin oils. Health care products and medical ointments containing propolis are used for wounds, scares, infections, muscle ailments, eczema, warts nail cuticle [28].

The enterprise needs relatively small investment capital and most of the equipment needed for both traditional and modern beekeeping can be sourced locally. *Apis mellifera*, commonly known as honey bee, is most widespread and abundant insects on earth [29]. It is used throughout history for the production of honey and for pollination of crops. *Apis mellifera* has proved to be an extremely useful species. It is estimated that the total value for worldwide crop pollination is 153 billion Euros [30]. Bees are social. Insects that live in colonies and are divided into three groups namely the queen, the drones and the worker bees. Honey has been harvested from, the wild nest for several years until it was discovered that honey crops can be obtained in more convenient and easier way if bees are encouraged to nest in hives. This led to the origin of bee keeping and management of bees in hives. It is widely practiced in Nigeria and other countries of the world as a result of magnificent importance of honey in the area of food and medicine [31]. Modern technologies of bees and keeping them were introduced into Nigeria in the early 1990's [32] [33]. [34] described the enterprise as means of empowering youth economically because of its many advantages over other types of agriculture enterprise. Men, women, youths and the elderly can participate in beekeeping within the homestead. Beekeeping can be practiced where fallow system is still practiced. The newness of the technology and demand from farmers as well as the potentials of this technology in poverty reduction makes it to be pursued with vigor and ease [35]. Honey is formed during alternate swallowing and regurgitation of nectar and pollen repeatedly, by which sucrose in the nectar and pollen are hydrolyzed to glucose and fructose by special gland in the bee. Honey is a complete food of its own and contains every vitamin and minerals needed by the body.

2. Materials and Methods

Bees normally collect nectar from various plants around its hive to produce honey. The process helps cross pollination of flowers, which is crucial for flowering plants. After flying long distances, bees find open flowers which they buzz to, hungry to drink the nectar; they gather the flowers sugary sap in their glands [36]. The bees suck in the juice with their long proboscis and pass it to their honey stomach where it is processed with saliva, transforming it from a complex

sucrose into simple fructose that is easy to digest [37]. Though honey contains sugar, it is not the same as white sugar or artificial sweetener. Its exact combination of fructose and glucose regulate blood sugar and it is readily assimilated and more acceptable to the stomach. This takes care of constipation, diabetes [38]. Honey is used for cooking, baking, and as sweetener in commercial beverage production, cosmetics industries for the production of body cream and soaps. This is due to the anti bacteria property of honey and when combined with other ingredients can be moisturizing and nourishing to the skin. In cases of barrenness, or infertility, the royal jelly contained in honey has been of immense solution. Royal jelly is used in the production of skin ointment, fertility drugs, etc and it is also believed to be the reason for the prolonged life of the queen bees [39]. Bee sting confers body immunity to malaria attack on the victim. According to [40], single sting is reported to be medicinal but multiple stings can be dangerous as many alarm pheromones might have been injected into the body.

Imo State is situated in Southeastern geographical zone of Nigeria. It lies between latitudes 5°10'N and longitudes 6°35'E and 7°28'E covering an area of 5,289.4 square kilometers. It has a population of 4.7 million people. The population density is over 500 persons per square kilometer [41]. It is bounded on the east by Abia State, on the South and South West by Rivers State, on the Southwest by Niger across which lies Delta State and on the North by Anambra and Enugu States. The area falls within the tropical rainforest zone characterized by high intensity of rainfall in most parts of the state which measures up to 2,550 millimeters. It has two distinct seasons; the rainy and the dry seasons with intervening cold dry spell called "August break". The rainy season extends from April to October and precipitation reaches its maximum towards the end of the season when there is an almost immediate change to dry month conditions. The mean temperature is 270°C while relative humidity ranges from 60-90 percent [42]. Five distinctive soils types have been identified in the area and include Lithosols, alluvial, ferralithic, medium fine altisol and clayey hydromorphic soils while the natural vegetation is the tropical rainforest.

The State has three agricultural zones namely: Okigwe, Orlu and Owerri and twenty seven Local Government Areas. The inhabitants of Imo State are predominantly farmers which are mainly on subsistent level under rain fed agriculture. The rainy season stretches from mid March to mid October and the dry season extends from late October to early March. However, there is no clear cut demarcation between the rainy season and dry season these years due to climatic change which has made the farmers to be unpredictive of the seasons. Though, the rainy periods are often more than the dry periods within the year. The area is covered with woodlots or open trees/shrub vegetation interspersed with short grasses. Agriculture is the major source of income of the rural populace producing mainly staple food crops such as yams, cassava, maize, cocoyam, vegetables, rice, bananas, pineapples, plantain, etc. Livestock rearing, fish farming, agro-forestry farming complement these food crops, cash and fruit tree crops [43].

The rural household economy revolves around two major sectors: Small holder farmers and nonagricultural sectors. Family labor is the major source of labor while land is acquired by inheritance and leasehold. The tropical disposition of the place favors beekeeping as there is abundant supply of nectar from flowering plants for bees' consumption (foraging) and production of honey and other bee products. Purposive sampling technique was used to select five Local Government Areas (LGAs) based on information gathered from Imo State Agricultural Development Programme, an apex agricultural institution in the state. From the five LGAs, five communities were selected and later, eight (8) beekeepers were selected from each of the selected communities to get a sample size of forty (40) beekeepers. The data used for the study was collected from both primary and secondary sources plus oral interview schedule for clarity of purpose. Questionnaires were administered to the forty (40) respondents. The information (data) elicited were on cost and return of beekeeping and the constraints militating against beekeeping in the study area. The secondary data were from relevant journals, textbooks, the internet, ADP and other relevant sources. The generated data were analyzed using Descriptive Statistics such as percentage, frequency counts, tables, mean, net farm income and Gross Margin analysis. The model specification is stated thus:

$$Percentage = \frac{x}{N} + \frac{100}{1}$$

Where
x = values of respondents
f = frequency of the respondents
n = total number of the respondents
x = the mean value of x
The Gross Margin Analysis
GM = TR-TVC
Profit = GM — TFC
Where TR = Total Revenue
TVC Total Variable Cost
TFC = Total Fixed Cost

3. Results and Discussion

3.1. Socio-Economic Characteristics of Respondents

Table 1 show that 72.5% of the respondents were male while 27.5% were female. This implies that both sexes participated in bee-keeping in the study area. The wide gap in the level of participation could be attributed to the fact that men are more courageous in keeping or working with bees. This is in keeping with the assertion of [44] that beekeeping can be undertaken by anyone who has the determination to look after the bee properly and courageously. The same table shows that the respondents fall within the age of 26-35 years which accounted for 35%, 2.5% were within 16-25 years, 22.5% fell into 36-45 years, 32.5% were within 46-55years

while 7.5% fell within 56 and above. The mean age of the respondents in the study area was 37 years old. Equally, in the same table, 72.5% of the respondents were married, 12.5% were single, 5% widows while 10% were widowers. This indicates that marital status has no strong influence and limitation for one to run beekeeping enterprise.

Table 1. Distribution of Respondents According to Socio-economic characteristics

	Frequency	Percentage (%)
Socio-Economic Characteristics Gender		
Male	29	72.5
Female	11	27.5
Total	40	100
Total	60	100.0
Age		
15 below	-	-
16-25	1	1
26-35	14	35.0
36-45	9	22.5
46-55	13	32.5
56 and below	3	7.5
Total	60	100.0
Mean = 37		
Marital Status		
Married single	29	72.5
Widowed	5	12.5
Widower	2	5.0
Divorced/separated	4	10.0
Total	40	100.0
Household		
1-2	6	15.0
3-4	3	7.5
5-6	9	22.5
7-8	22	55.0
9-10	-	-
Total	40	100.0
Mean = 5		
BEEKEEPING Experience (Years)		
1-2	2	5.0
3-4	5	12.5
5-6	15	37.5
7-8	18	45.0
9-10	-	-
Total	40	100.0
Mean = 3		
Education		
Tertiary	16	40
Secondary	12	30
Primary	7	17.5
No formal education	5	12.5
Total	40	100.0
Source of Labor		
Family	31	77.5
Hired/Exchange	3	7.5
Both	6	15
Total	40	100.0

Married life is an added responsibility and as such, the desire to engage in beekeeping may be to beef up sources of income for the household. Table 1 shows that 55% of the respondents had household size of 7- 8, 22.5% had 5-6, 15% had 1-2 while 7.5% had 3-4. Those with reasonable household size were sure of regular labor supply as family labor is the major source of labor in the study area. It is also interesting to know that 45% of the respondents had 7-8 years experience, 37.5% had 5-6 years, and 12.5% had 3-4 years while 5° h had 1-2 years beekeeping experience. This implies that the business is on course and is such a bright prospect for the industry/venture in the study area. It is equally worthy to point out that majority of the respondents attained tertiary level of education. This represents 40%, 30% has secondary education, 17.5% had primary education while 12.5% had no formal education. Finally, the table shows that 77.5% of the respondents depended on family labor as source of farm hand, 15% made use of both hired and family labor while 7.5% relied on hired labor.

Table 2. Distribution of Respondents According to Source of Finance for their Apiary

	Frequency	Percentage %
Source of Finance		
Personal fund (equity fund)	34	85
Relations/Well-wishers	4	10
Thrift Associations (Eludiegwu)	2	5
Banks	-	-
Total	40	100.0
Source of FINANCE		
Extension agents from ADP	22	55
Fellow beekeepers	8	20
Radio Farmer	6	15
Workshop	4	10
Bulleting/posters	-	-
Total	40	100.0
Types of Hives in Use		
Kenyan Topbar Hive	32	80
Langstroth Hive	4	10
Both Local and Modern Hives	4	10
Only local Hive	-	-
Total	40	100.0
Number of FINANCE		
1-2	1	2.5
3-4	6	15
5-6	18	45
7-8	15	37.5
9-10	-	-
11 and above	-	-
Total	40	100.0
Bee products		
Honey	24	60
Bee-wax	-	-
Pollen	-	-
Propolis	-	-
royal Jelly	-	-
Been Venom	-	-
Honey and Bee-wax	16	40
Total	40	100.0

Source: Field Survey Data 2014.

Table 2 shows that 85% of the respondents sourced their takeoff fund from personal savings (equity fund) and sometimes got financial assistance from relations/well-wishers at interest free which represents 10% of the respondents. 5% of them borrowed from thrift associations that operate at community level. The same table indicates that 55% of the respondents got their information on beekeeping from IMO ADP extension agents resident in their communities and local government, 20% got information from fellow beekeepers that started the business of beekeeping as early adopters. Others (15% and 10% respectively) got information through radio farmer and workshop attendance whenever organized. From the foregoing, 80% of the respondents made use of the Kenyan top bar, '0% made use of Langstroth, while 10% combined both local (drums, raffia palm log and clay pot) modern hives. In the same table, 45% of the respondents had 5-6 hives, 37.5% had 7-8 hives while 15% had 3-4 and 2.5% had 1-2 hives. This indicates that the respondents if given the right support will move the business of beekeeping into a professional venture rather than a hobby. Furthermore, 60% of the respondents were interested only in honey production while 40% were interested in both honey and bee wax production.

3.2. Evaluation of the Cost and Return from Beekeeping in the Study Area

From the result below, it can be observed that beekeepers in the study area are 'breaking even' from their investment into beekeeping business. The initial investment can be recouped from first harvest and at the same time were able to make substantial profit to cope with. This implies that beekeeping in the study area is a profitable business that one can venture into as a means of generating income. From the gross margin analysis, the result indicates that the major cost of beekeeping business comes from the fixed cost items such as bee hive, smoker, protective cloth, centrifuge, etc. This being the case, it can be predicted correctly that in the next harvest, the beekeepers are going to record more profit due to the fact that once the enterprise has been set up, little or no expenses are required to maintain the enterprise as the bees can fend for themselves without the intervention or involvement of the beekeeper. This result also shows that if one should start up with more than one hive, the person will make more profit as the quantity of products that will be gotten will increase, consequently• occasioning an increase in revenue. Another advantage of starting with more than one hive is that the only additional fixed cost required will be the purchase of the additional hives and the hive stands. In this case, other Fixed Cost (FC) will remain constant while there will be only a little increase in the Total Variable Cost (TVC). The harvest of other bee products will also increase the revenue gotten. The economic implications of this are that beekeeping industry is very lucrative, rewarding, good payback, and suitable means of wealth creation, especially in the rural areas. Beekeeping as a component technology in agriculture is not buying and selling, it is serious business [45].

Table 3. Average Gross Margin and Net Farm Income Analysis for one Hive Harvest in a Season

Items	Average Cost ₦
Variable Items	
Attractants	600
Transport	300
Packaging Materials (Bottles)	1,500
Label	300
Total Variable Cost (TVC)	2,700
Fixed Cost Items	
Bee Hive	5,000
Hive Stand	2,000
Protective Cloth	1,000
Rubber Hand Gloves	500
Boots and Scrapper	1,000
Smoker and Press	1,000
Knife (Extruder)	1,000
Sieve and Filter (Centrifuge)	800
Hat and Veil	400
Pin and Thread	100
Bucket (Plastic Bucket)	400
Total Fixed Cost (TFC)	13,200
Total Cost (TC)	15,900
Products from Hive	
Honey	36,000
Bee Wax	6,000
Total Revenue	42,000

Source: Field Survey Data, 2014.

Gross Margin and Net Farm Income

GM = TR - TVC

= ₦42,000 - ₦2,700

= ₦39,300

Net Farm Income (NFI) = GM - FTC

= ₦39,300 - N13,200

= 26,100

3.3. Constraints Militating Against Beekeeping

Table 4 indicates that there were identified constraints militating against the smooth activities of beekeeping in the study area. Beekeeping though less capital intensive, lack of finance is the most outstanding challenge against large scale beekeeping among the beekeepers. This represents 39.6% of the respondents. 3.3% accounted for pests and predator attack, 18.7% had problem of non-colonization of their hives, 8.8% had problem of theft and bush burning, 14.3% had problem of inadequate technical skill and information and 2.2% had problem of indiscriminate pesticide use. Furthermore, 13.2% of the respondents had problem of abscondment of bees.

Table 4. Distribution of Respondents According to Constraints Faced

Constraints	Frequency	Percentage
Lack of Finance	36	39.6
Pest and Predator Attack	3	3.3
Non-Colonization	17	18.7
Stealing and Bush Burning	8	8.8
Inadequate Technical Skill and Information	13	14.3
Indiscriminate Pesticide bee	2	2.2
Abscondment of bees	12	13.2
Total	91	100.0

Source: Field Survey Data, 2014

Multiple responses were recorded hence total frequency exceeds the sample size.

3.4. Prospects of Beekeeping in the Study Area

Considering the number of people already in the business, the low financial requirement/investment and low labor need, the business has a promising bright future for those already in the business and intending people who want to join. Nevertheless, the beekeepers expressed that they had problems like lack of fund, incidence of pests and predators, bee abscondment, non-colonization of hives, theft and bush burning, inadequate technical skill, inadequate information and indiscriminate pesticide use.

The study also showed that the beekeepers were more interested in honey and bee wax products probably due to low level of technical skills acquire. Above all the result showed that beekeeping has a bright future. Given the limited attention the business needs, it can be carried out on part-time basis even by Civil Servants in the non-agricultural sector. The use of more hives and extraction and processing of the bee products like propolis, pollen, bee venom, royal jelly apart from honey and bee wax will increase the profit margin and net farm income of the enterprise. There should be traces around areas where hives are sited, controlled application of pesticides be adopted, extension services should be extended to beekeepers to teach them how to handle other bee products other than honey and bee wax. Hives should be baited with appropriate attractant like sugar solution and sweet fresh palm wine. Intensive awareness campaign should be mounted to enlighten people on the benefits of beekeeping as this will lure more people into the business. Credit/loan should be provided by commercial banks to beekeepers at reasonable interest rate. If all these measures are taken, the prospects of beekeeping in the study area will be overwhelming.

References

[1] Udah, C. A. (2006). Overview of Forestry and Agro Forestry Systems with Adaptable Technologies for Extension to Imo Farmers. An unpublished paper presented at the Intensive Training Workshop for Newly Employed Local Government Agricultural Officers in Imo State. 6th April, 2006.

[2] Okafor, 3. C. (1980). Edible Indigenous Woody Plants in the Rural Economy of Nigeria Forest Zone. Forest Ecology and Management. 3: 45-65.

[3] IIRR (1998). Sustainable Agriculture Extension Manual for Eastern and Southern Africa. International Institute for Rural Reconstruction Nairobi, Kenya pp. 7-10.

[4] ADP (2007). Beekeeping Handbook. Extension Guideline. Pp. 3-5.

[5] Seeley, Thomas D. (2010). Honey Bee Democracy. Princeton: Princeton Up Print.

[6] ADP, (2007) Beekeeping Handbook. Extension Guideline. Pp. 3-5 .

[7] ADP, (2007). Beekeeping Handbook. Extension Guideline. Pp. 3-9

[8] IIRR (1998). Sustainable Agriculture Extension Manual for Eastern and Southern Africa. International Institute for Rural Reconstruction Nairobi, Kenya pp. 7-10.

[9] IIRR (1998). Sustainable Agriculture Extension Manual for Eastern and Southern Africa. International Institute for Rural Reconstruction Nairobi, Kenya pp. 7-10.

[10] ADP, 2007 Beekeeping Handbook. Extension Guideline. Pp. 3-5.

[11] ADP, 2007 Beekeeping Handbook. Extension Guideline. Pp. 3-5.

[12] Shackleton, C. and Shena, S. (2004). The Importance of Non-Timber Products in Ruraommunity Livelihood-Security and a Safety Nets. A Review of Evidence from South Africa. South Africa Journal.

[13] Oyun, N. B. (2009). The Role of Non-Timber Forest Products on the Livelihood of Fringe Communities of Idanre Forest Reserve, Nigeria. Journal of Forest and Forest Products Vol. 3, No. 5.

[14] Goulson D. (2003). Effects of Introduced Bees on Natural Ecosystem. Annual Reviews of Ecology and Evolution System 34 (2003); 1-26 print.

[15] Gutierrez, E.G. (1999). Guide to Natural Remedies for Health and Well Being. Orvil Publishing, Mexico pp 263-283.

[16] Caruthers, I. and Rodriquez, M. (1992). Tools for Agriculture. Intermediate Technology Publication, Nottingham, United Kingdom. Pp. 288.

[17] Nlemchi, R. (2003). Beekeeping Managements. Unpublished Lecture Delivered to Imo ADP Staff. Preseason Training, April, 2003.

[18] Nicolas, B. (2004). Beekeeping and sustainable Livelihood. IBRA U.K. Obi, V. (2009). Management of Beehive. NRCRI, Umudike

[19] Morse Roger, A. (2007). Encyclopedia of Science and Technology, 10th Edition. Mcgraw Hill, United States. Pp. 674-678.

[20] Issa, Y. A. (1999). Economic Benefits of Beekeeping. A paper presented at the bee training in Ibadan, Oyo State, Nigeria; Held at Premier Hotel Ibadan. Organized by Beekeepers Association on 4th July, 1999.

[21] Standifer, C. N. (2007). Honey Bee Nutrition and Supplementary Feeding. Except from Beekeeping in the United States.

[22] Adjare, S. (1989). The Golden Insect: A handbook for beginners. Intermediate Technology Publishes, London, U.K.

[23] Adrain and Claire, W. (2006). Teach yourself Beekeeping. Pp. 133-240. Graw Hill Publishing Company, New Delhi, India.

[24] Eddy, J. (2007). Tropical Honey as a Treatment of Diabetic Ulcers. University of Wisconsin Study Test, University of Wisconsin, Madison.

[25] Ubeh, E. 0. and Nwajiuba, C. U. (2005). Economics of Apiculture. A Case Study of Federal University of Technology, Owerri (FUTO), Apiary. In: Ekenyem, B. U. and Madubuike,

F. N. (2005). Issues in Tropical Animal Science for Rural Development, Fegro Press, pp. 164-169.

[26] Wilipedia.org./wiki/beekeeping. Last Modified on 16th December, 2013.

[27] Ubeh, E. 0. (2011). Beekeeping.Unpublished Lecture Delivered at Federal University Technology, Owerri (FUTO).

[28] Nicolas, B. (2004). Beekeeping and sustainable Livelihood. IBRA U.K. Obi, V. (2009). Management of Beehive. NRCRI, Umudike.

[29] Goulson D. (2003). Effects of Introduced Bees on Natural Ecosystem. Annual Reviews of Ecology and Evolution System 34 (2003); 1-26 print.

[30] Gallai, N,Michael S. and Bernard, F. V. (2009). Economic Evaluation of the Vulnerability of World Agriculture Confronted with Pollinator Decline. Ecological Economics 6.8 (2009): 810-821 print.

[31] Ojeleye, B. (1999). Chemical Composition of Honey. The Bee Keeper. Journal of Beeking Vol. 1, pp 4-5.

[32] Olagunju, F. I. and Ajefomobi, J. 0. (2003). Profitability of Honey Production under Improved Method of Beekeeping in Oyo State, Nigeria. International Journal of Economic Development Issues 341; 148-151.

[33] Chinaka, C. (1995). Beekeeping Technology for Nigeria Farmers. Extension Bulletin No. 3, National Agricultural Extension and Research Liaison Services, ABU, Zaria, Nigeria.

[34] Ojo, S. 0. (2004). Improving Labour Productivity and Technical Efficiency in Food Crop Production. A Panacca for Poverty Reduction in Nigeria. Food, Agricultural and Environmental. Pp 2(2) 222-231.

[35] Udah, C. A. (2006). Overview of Forestry and Agro Forestry Systems with Adaptable Technologies for Extension to Imo Farmers. An unpublished paper presented at the Intensive Training Workshop for Newly Employed Local Government Agricultural Officers in Imo State. 6th April, 2006.

[36] Jayeola, O.A. Meduna, A.J. and Oluoku, N. S. (2009). Forest and Forest Products. Journal of Forestry. Vol. 2, No: 6.

[37] Gutierrez, E.G. (1999). Guide to Natural Remedies for Health and Well Being. Orvil Publishing, Mexico pp 263-283.

[38] Keystone, R. C. (2001). Marketing of Agricultural Products. Macmillan Company. 3rd Edition, Ibadan.

[39] Ubeh, E. 0. (2011). Beekeeping.Unpublished Lecture Delivered at Federal University Technology, Owerri (FUTO).

[40] Udah, C. A. (2006). Overview of Forestry and Agro Forestry Systems with Adaptable Technologies for Extension to Imo Farmers. An unpublished paper presented at the Intensive Training Workshop for Newly Employed Local Government Agricultural Officers in Imo State. 6th April, 2006.

[41] Federal Republic of Nigeria: FGN (2006). Census Result Official Gazzette Vol. 96, No. 2. Geological Survey of Imo State (1994).

[42] Meteological Department, (1993). Ministry of Agriculture and Natural Resources, Owerri.

[43] Akinyosoye, V.0. (1996). An Introduction to Senior Tropical Agriculture for West Africa MacMillan press, Lagos, Nigeria.

[44] Morse Roger, A. (2007). Encyclopedia of Science and Technology, 10th Edition. Mcgraw Hill, United States. Pp. 674-678.

[45] Ezeagu, D. (2010). Agriculture is not buying and selling. In: New Face Agriculture NewA Publication of Imo State Ministry of Agriculture and Natural Resources. Vol. 4 No. 1. Pp 13-14.

23

Addressing Water Concerns Through Spatial Planning Initiatives For Rural Communities

Hildegard E. Rohr

Faculty of Natural Sciences, North-West University, Potchefstroom, South Africa

Email address:

21082197@nwu.ac.za

Abstract: Planning in South Africa operates within a legal framework, which strives to ensure that municipalities deliver their developmental duties (in terms of Section 153 of the Constitution). South Africa's approach to Spatial Planning and Land Use Management is undergoing major changes in order to escape from the legacy of apartheid planning, as well as to ensure sustainable development and better management of municipal land. Developing countries such as South Africa do not have the luxury of centuries of learning to adapt to growth. There is no better time than the present to introduce innovative, multidimensional and effective evidence based planning practices by improving the connection between research and professional work to support sustainable development and to overcome the urban and rural challenges presented by rapid population growth. For the first time in the history of South Africa it has been legislated that municipalities must include previously secluded rural areas into their planning strategies. These strategies must take place with a full understanding of current and future challenges such as demographic; environmental; economic; social-spatial; and institutional challenges in order to foster sustainable development. This research will present opportunities in terms of planning for sustainable water management in rural areas (which also includes previously secluded townships) through the use of spatial planning tools such as a Spatial Development Framework and a Municipal Land Use Scheme based on the context of the Spatial Planning and Land Use Management Act No. 16 of 2013.

Keywords: Spatial Planning, Water, Rural Communities, Spluma, Sdf, Lus

1. Introduction

Among the greatest challenges of the twenty-first century is the increasing population growth [1]. South Africa stands as the second most populous country in the Southern African Development Community, after the Democratic Republic of Congo [2]. From an environmental perspective, as countries continue to grow, they place growing pressure on land, energy and natural resources (especially water), which can lead to greater environmental threats. South Africa is already rated as being the 30th driest country in the world [3]. This is due to low levels of rainfall, 450 mm per annum relative to the world average of 860 mm per annum, [3] and high level of evaporation due to the temperatures of annual mean above 17 °C [4]. With a population growth rate of 1 percent per year, key demographic observations indicate that South Africa's population will reach an estimated population total of 83,6 million in 2050 [5]. Population and economic growth inevitably creates more demand for water. How and where that growth takes place affects how much additional water is needed and how much it will cost to deliver.

The Department of Water Affairs (DWA) has embarked on a nationwide water-reconciliation study. The DWA found that surface water availability and its remaining development potential will be insufficient to support the growing economy and associated needs in full. Water development potential only exists in a limited few water management areas, whilst serious challenges remain in the majority of water management areas [6].

Consequently, countries around the world, including South Africa, have been actively seeking environmentally sustainable solutions through sustainable development planning practices, with water sustainability being one of the major priorities. The USA formulated the concept of

Low-Impact Development (LID), the UK's approach was Sustainable Urban Drainage System (SUDS), and New Zealand formulated their approach as Low Impact Urban Design and Development (LIUDD) [7].

Aiming to integrate all of the above mentioned approaches, Australia developed the concept of Water Sensitive Urban Design (WSUD). WSUD refers to the interdisciplinary cooperation of water management, urban planning, urban design and landscape planning which considers all parts of the urban water cycle, combines water management functions and urban design approaches and facilitates synergies between ecological, economic, social and cultural sustainability [8]. It bridges gaps between various sectors and provides a platform for trans-disciplinary planning, which is a challenge for sustainable water resources management in municipalities. The South African Water Research Commotion (WRC) recently published "*Water Sensitive Urban Design for South Africa: Framework and Guidelines*" based on the concept of Australia's WSUD, as a first attempt at guiding South Africa's settlements in becoming more water sensitive. Brown, et al., (2009) raised concerns that even though international best practices will be fit for purpose in formally-developed areas in South Africa, rural settlements are not being catered for [9]. South Africa cannot rely on purely international guidelines. South Africa's history and geographic makeup, linked with its unique climate, requires guidelines that are fit for local circumstances.

2. Rural South Africa

2.1. Rural Neglect

Rural areas present a number of challenges for the modern society as they are generally associated with the legacy of apartheid. In the 1910's South Africa consisted of four provinces. The segregation of the Black population through the concept of "apartheid" started in 1913, with ownership of land by the Black majority being restricted to certain areas known as "homelands" or "Bantustans" (today, better known as rural or traditional areas) [9]. In essence, the Apartheid State refused to acknowledge "Africans as permanent urban inhabitants...investment in housing, infrastructure, education and other essential services in the townships was pared back from an already low level, in order to eliminate any such attractions the cities might offer to people from rural areas" [10]. This resulted in significant backlogs in infrastructure.

After the first non-racial elections in 1994, the country's adoption of the first Interim Constitution made it possible for Provinces and homelands to merge, thus giving birth to the current nine provinces of South Africa. In 2000 the re-demarcation of the country's "wall-to-wall" municipalities were created thereby including all intervening land between the towns or former transitional local councils [9]. Even though this meant that rural areas were now included in the municipal boundaries, town planning practices (spatial development frameworks and land use schemes) were still limited to former town boundaries (excluding the rural or

traditional areas).

2.2. Changing Legislative Perspective

South Africa's planning history has seen a considerable array of legislation. Much of the legislation responsible for managing land uses in a municipality pre-dates 1994. Settlements in these rural areas was governed by Legislations such as the Development Facilitation Act No 67 of 1995 (DFA) and the Less Formal Township Establishment Act No 113 of 1991 [9]. In theory, these pieces of legislation directed planning authorisation in rural areas to Provincial Development Tribunals. Constitutional Court found sections of post 1995 legislation (DFA, 1995) invalid based on unconstitutionality. As a result, The Department of Rural Development and Land Reform published the Spatial Planning and Land Use Management Bill in 2012, which was enacted in 2013.

The Spatial Planning and Land Use Management Act No. 16 of 2013 (SPLUMA) repealed the DFA in its entirety. One of the key principles of the Act is the active pursuit of sustainability through the promotion of integrated planning and the involvement of various sector departments in formulating planning documents at various levels of government. Furthermore, the Act states that municipalities must adopt a single "wall-to-wall" Land Use Scheme and Spatial Development Framework including the previously secluded rural areas [12].

2.3. The Rural Reality

Statistics South Africa defined rural areas as "Any area that is not classified urban". Census 2011 indicated that 63 percent of South Africa's population resides in urban areas, 32 percent in rural areas and 5 percent of the population in commercial farming areas [13]. Map 1 below presents the geographic extent by South Africa's geo-type. The red areas in the Map 1 represents the country's rural areas, yellow represent urban and the light green areas are commercial farmland.

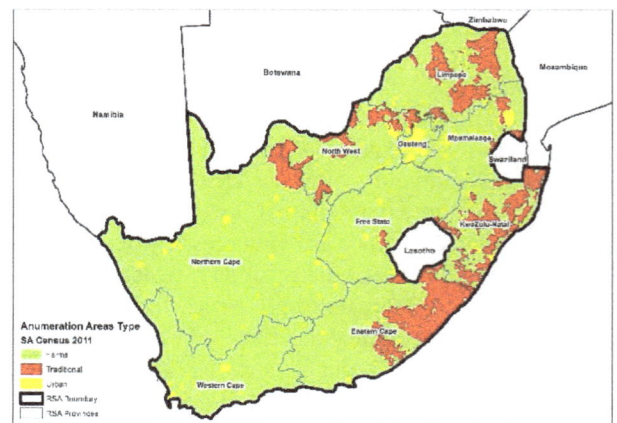

Map. 1. Geographic extent of enumerated area type.

Even though the majority of the population resides in urban areas, the geographic extent of these populated areas paints a different story. Table 1 sets out the geographic extent of these areas in hectare and percentage of land coverage. Rural areas in South Africa covers 13 percent in comparison to the urban areas which is only a small 3 percent and commercial farming 63 percent. Placing this information in perspective, spatial planning practices in South Africa has ignored 13 percent of its vitally important populated geographic area.

Table 1. South Africa Geo-type.

Geographic type – Enumerated Area	Hectare	% of cover-age
Farms	103 345 006	84%
Rural / Traditional	15 740 885	13%
Urban	3 746 130	3%
South Africa Total Areas	122 832 021	100%

Municipalities cannot afford to ignore the population that resides in the previously unplanned rural areas. The South African Constitution states that municipalities have the responsibility to make sure that all citizens are provided with services to satisfy their basic needs, e.g. water, electricity, refuse removal, etc. These services have a direct and immediate positive impact on the quality of the lives of the people in that community. It is no secret that underlying tension between rural/traditional and municipal leadership exists. This is mainly due to the ideological paradigms sometimes depicted as "African versus western" or "customary versus individual civil rights" frameworks. That said, rural areas presents major challenges for town planners, some of which includes the following:

- There is a lack of spatial information and data necessary for planning exercises.
- Geographically these communities are inaccessible and far from urban centers.
- Communication presents a barrier as little English is spoken or understood in the majority of these areas.

Issues of land management and ownership are viewed with suspicion by traditional leaders, making any form of spatial planning rather difficult.

Adding to the strain of planning for rural areas is the fact that rural areas are faced with many spatial structuring elements which cannot be changed, generally referred to as the inefficient rural form.

2.4. An Inefficient Rural Form

Rural settlements which are commonly associated with large plots, low density, and dispersed development all increase the cost of delivering conventional water infrastructure. Other factors in terms of planning for infrastructure and the increasing cost thereof include:

- Reconstruction and Development Programme – RDP - (though well intended) places a burden on existing water sources as this is often accompanied by policies such as free basis service delivery or an unaffordable high level of service.
- Low-density, dispersed development requires longer pipes, which lose more water through leakage and raise transmission costs (6 – 25 percent) [12].

- Infrastructure investment that supports water system expansion instead of upgrading and maintaining existing networks can lead to increasingly inefficient systems, greater waste, higher capital and operating costs [13].
- Location of development beyond urban development boundary can reduce return on investment in infrastructure - Adding new developments to the existing network spreads the system's capital costs over a larger customer base, lowering the cost of water service per customer [14].
- Inappropriate land use near water sources could potentially lower downstream water quality (for example locating landfills near aquifers) [13].
- Service costs associated with water infrastructure are strongly influenced by a development's location and density. Therefore, any new system should be located within existing, growing communities (in the form of infill development) as opposed to the creation of new infrastructure on the periphery of the urban boundaries [14].

There is a need to use a different approach when dealing with rural areas. Rural areas have resources that are unique form the urban areas and that gives them a comparative advantage, however the exclusion of rural areas into municipal planning and incoherent development approach has made rural areas seen as unworthy, hopeless areas where growth and development can never happen. Land Use Management System will be incomplete if rural areas are not included in the municipal planning processes.

3. Current Institutional and Legislative Context

The management of water sources (dams) currently happens at a national level through the DWA. The distribution of water from its source to its clients is the constitutional responsibility of local government (metros, local or district municipalities) who act as the water services authorities and often also water service provides for all communities in their areas of jurisdiction. Plans at this level includes water services development plans (WSDP) which must be updated on an ongoing bases. The WSDP integrates technical, social, institutional, financial and environmental planning and feeds into the integrated development plan process [6].

Spatial planning (to date) happens at a municipal or district municipality level through the Spatial Development Framework. In theory the SDF must be aligned to the integrated development plan (same as the WSDP). Land uses are managed on a municipal level through a land use scheme (also called zoning code, zoning scheme or town planning scheme) [15].

Reviewing the above, it would seem as if all basis are covered. In practice, very little (if any) of the above plans are in fact not aligned with each other. The result thereof is that planning for water supply and planning for water demand (as expressed through population and land use growth) is disconnected [16]. An example of this can be seen in the Greater Tubatse Municipality in South Africa, where large planned developments (to accommodate an urbanising population) never happens due to the unavailability of bulk water [15].

4. Changing legislation Spatial Planning and Land Use Management Act

Recent court cases involving land use management and urban planning in South Africa led to the development of the new and recently enacted SPLUMA. As stated earlier, one of the key principles of the Act is the active pursuit of sustainability through the promotion of integrated planning and the involvement of various sector departments in formulating planning documents at various levels of government [11].

- A National Spatial Development Framework (NSDF) must take into account policies, plans and programs of public and private bodies that impact on spatial planning, land development and land use management (for e.g. the national water resource strategy).
- Provincial Spatial Development Framework (PSDF) which must be consistent with the NSDF. The PSDF must coordinate, integrate and align provincial plans and development strategies with policies of national government, provincial departments and municipalities (this could also imply, policies and plans developed by water service authorities whose boundaries overlaps with local municipal boundaries).
- The Municipal Spatial Development Framework (MSDF) must give effect to the development principles and applicable norms and standards set out in Chapter 2 (this include the principle of sustainable development as mentioned earlier in this document). The MSDF must identify, quantify and provide location requirements of engineering infrastructure and services provision for existing and future development needs for the next five years (water provision for sustainable development). The MSDF must provide the spatial expression of the coordination, alignment and integration of sectoral policies of all municipal departments (most importantly water management).
- Municipalities must also adopt a land use scheme that gives effect to and be consistent with the municipal spatial development framework and determine the use and development of land within the entire municipal area.

It is critical that sector departments from different spheres of government as well as utilities and agencies communicate with one another.

4.1. Planning for the Future Water Sensitive Planning Exploring the Options

Spatial planning and land use management must value water as a natural resource playing a critical role in the overall sustainability of the municipality. South Africa has been given a golden opportunity to adopt a water sensitive planning approach, which includes rethinking the role of strategic spatial planning and land use management and its obligations towards water in the built environment.

Water sensitive planning ensures that development interacts with the hydrological cycle in ways that: provide the water security essential for economic prosperity through efficient use of the diversity of water resources available; enhance and protect the health of watercourses and wetlands; mitigate flood risk and damage; and create public spaces that harvest, clean and recycle water [17]. Given the new planning environment as stipulated by SPLUMA, this section will specifically focus on the proposed content of SDF's and a land use scheme and how this can contribute the sustainable planning for the management of water.

4.2. Integrating Water Sustainability Through Innovative Spatial Planning Management Tools

A Spatial Development Framework (SDF) is the principal strategic planning instrument which guides and informs all planning and development, and all decisions with regard to planning, management and development in the municipality. An SDF is a document accompanied by a set of plans that illustrates the future spatial form of the municipality. It uses tools such as nodes and corridors and concepts such as densification, containment, protection and growth areas to indicate how land uses in the municipality must be managed to arrive at this future spatial form. Keeping in mind that rural areas have always been excluded from this planning instrument, SPLUMA now legislates that a municipal SDF must include preciously secluded rural areas.

In the section below, the legal requirements for a municipal SDF (highlighted in bold) will be discussed with specific reference to water sensitive planning. SPLUMA, in section 21 specifies the content of a municipal spatial development framework as consisting of the following [11].

Section 21(A) Give Effect to the Development Principles and Applicable Norms and Standards Set out in Chapter 2

One of these principles includes sustainability. Climate change is affecting the sustainability of South Africa's natural resources, especially the availability of water sources. Climate change is not only about the change in the earth's system, it is also about the impact of these changes on vulnerable communities. The impact on both rural and urban communities, particularly in the absence of effective risk reduction strategies, are expected to be significant in changing climate scenario and require an effective response. In communities where access to clean water is already a problem, a slight decrease in rainfall has an amplified effect. Water being a scare natural resource in South Africa is one of the major reasons why municipalities are currently unsustainable. Planning should therefore address this at a policy level.

Section 21(B) Include A Written and Spatial Representation of A Five-Year Spatial Development Plan for the Spatial Form of the Municipality

The spatial development pattern consists of land uses.

Urban and rural land users consume water. Therefore the decisions regarding the spatial patterns should also consider the availability of water and the impact on water resources. Pro-active planning and water management can contribute to sustainable allocation of water and optimal development of water infrastructure. By aligning development priorities with the availability of water in the municipal area, municipalities will be able to make provision for new infrastructure of redirect development to better allocated locations.

Section 21(C) Include a Longer Term Spatial Development Vision Statement for the Municipal Area Which Indicates a Desired Spatial Growth and Development Pattern for the Next 10 to 20 Years

See the point made above. This longer term spatial vision should, however, also incorporate the longer term outlook of water availability as well as changing climate considerations such as global warming and its impact on water sources.

Section 21(D) Identify Current and Future Significant Structuring and Restructuring Elements of the Spatial Form of the Municipality, Including Development Corridors, Activity Spines and Economic Nodes Where Public and Private Investment Will Be Prioritised and Facilitated

Corridors, activity spines and economic nodes can either mean future development or redevelopment and is not only limited to urban areas. Water sensitive planning includes the development of blue and green corridors which integrates the different land uses and stakeholders through the development of surface water mitigation measures. Blue-green corridors imply retaining stormwater from small, frequent rainfall events as the source of public-, residential- and commercial buildings, through the use of source and site control measures such as rainwater harvesting, green roofs or rain gardens.

Municipalities and National Departments are responsible for delivering community facilities e.g. Department of Education is responsible for the provision of schools. These community facilities can be designed as a rainwater harvesting structure, providing water for users within the facility. Many will argue that this approach will be too expensive, and that budgets are not capable of catering for these request. Fact is, research has proven that the long-term benefits of water harvesting systems overwrite the short-term construction cost [20].

Section 21(E) (F) Include Population Growth Estimates as Well as Estimates of Economic Activity and Employment Trends and Locations for the Next Five Years

This is most important in order to quantify future water demand. If the population growth and water allocations do not match, serious considerations must be given to innovative rainwater harvesting technologies. Municipalities must identify the development pressures presented by local economies. The vision must be supported and aligned with the water service delivery plan of the municipality. Growing rural population also increases the need for food security, which in essence means more pressure on water sources. Today, an estimated 14 million people of about 35 percent of the country's population are vulnerable to food insecurity. Alternative approach such as improving food security with

less water should be explored in rural areas.

Section 21(H) Identify, Quantify and Provide Location Requirements of Engineering Infrastructure and Services Provision for Existing and Future Development Needs for the Next Five Years; Include an Implementation Plan Comprising of Sectoral Requirements, Including Budgets and Resources for Implementation

Knowing the cost of water utilities and the cost of water infrastructure can have a significant impact on the size, location and density of development. Often land use planning or proposals are done without taking into account the availability of infrastructure or the cost of establishing adequate bulk infrastructure and reticulation. This should in fact be an iterative process. The planning scenario should be measured against the infrastructure cost, if the answer exceeds available infrastructure or the cost of establishing capacity, the planning scenarios should be adapted to be in line with capacity constraints.

Section 21(J) Include A Strategic Assessment of the Environmental Pressures and Opportunities Within the Municipal Area

Preserve open space, farmland, natural beauty, and critical environmental areas. Address water resource issues and conservation of biodiversity at the catchment and sub-catchment level especially in rural areas. Households in rural areas are often located near rivers as it is used as their primary water source. Legally, there should be no development within the 1:100 floodline and where the integrity of a river bank may be compromised. There is no data that exists on the calculations of the floodlines in most rural areas. Spatial development framework and land use schemes can assist Traditional authorities in developing a buffer zone around these river banks, in order to protect households from flooding and also protecting rivers from pollution. Protecting the environment and water sources should be prioritized.

Section 21(M) Provide the Spatial Expression of the Coordination, Alignment and Integration of Sectoral Policies of All Municipal Departments

This very important principle should enforce cooperation between the planning departments and the department responsible for water. Often this is not the case – the spatial development framework (compiled by the planners) and the water services delivery plan contradict each other.

4.3. Alignment of Spatial Development Framework and a Land Use Scheme

While an SDF provides an indication of acceptable land uses or intensity of land uses in certain geographical areas, land use rights are managed though a land use scheme. This scheme can be amended to reflect new or additional land uses or land use rights. Because the aim of the SDF is to provide an overview of the future spatial form of the municipality, it is the primary tool used to decide if a change in land use rights (through the amendment of the land use scheme) should be allowed. In the past, this link between planning tools has been tenuous at best, but SPLUMA specifically calls for a stronger linkage by insisting that an SDF:

- determine the purpose, desired impact and structure of the land use scheme to apply in that municipal area;
- propose (as part of the implementation plan) a list of amendments to the land use scheme that is necessary to achieve the aims of the SDF; and
- propose geographical areas where the normal processes and procedures of changing land use/rights may be shortened as a way to ensure that the spatial objectives of the SDF will be met.

In the section below, the legal requirements for a municipal land use scheme (highlighted in bold) will be discussed with specific reference to water sensitive planning. SPLUMA, in Section 24 specifies the content of a municipal land use scheme as consisting of the following [11].

Section 24(2) (A) Include Appropriate Categories of Land Use Zoning and Regulations for the Entire Municipal Area

Zoning for groundwater protection directs development away from groundwater-sensitive or aquifer recharge areas and prohibits potentially polluting uses. A land use scheme must be aligned with environmental policies and plans, therefore, zoning codes and scheme clauses can and must address site-specific ecological conditions. "Overlay zones" can protect stream corridors, lakeshores, and watersheds thereby maintaining and improving the water quality - even as the community becomes more developed. An overlay zone can protect water quality by setting additional standards for development and by incorporating site-specific review procedures. These site specifics include – building lines, boundaries, densities [18].

Zoning can regulate development by directing development to appropriate locations, requiring development to be setback from riparian areas, limiting the total impermeable site coverage, establishing appropriate lot sizes, limiting or enhancing density, requiring appropriate drainage, and prohibiting potentially polluting uses in areas where aquifers must be protected. This might not be fully accepted in rural areas as these urban-planning related terms are typically associated with "western thinking" but an incremental introduction of these land use management guidelines will benefit all.

Protect ecological and hydrological integrity - Use natural channel design and landscaping to ensure that the drainage network mimics the natural ecosystem. Control sediment-laden runoff from disturbed areas, in particular during construction of developments. The land use scheme must include groundwater recharge and conservation regulations where environmental management authorities aim to limit impermeable areas, drainage, and keeping permitted uses to those that are non-polluting. Natural channel design and landscaping can include the planting of edible and nutritious food. Opportunities for job creating can be introduces, as this is a familiar activity for farmers in rural areas.

Section 24(2)(C) Include Provisions that Permit the Incremental Introduction of Land use Management and Regulation in Areas Under Traditional Leadership, Rural Areas, Informal Settlements, Slums and Areas Not Previously

Subject to A Land Use Scheme

The high cost of providing services and infrastructure in rural areas, and especially in places that are remote and have low population densities requires innovative solutions. For example installation of rainwater tanks to collect rainwater to supply toilet flushing and outdoor uses. Many rural municipalities lack the financial and technical capacity to manage water services adequately. Planners must recognise the need for site-specific solutions and implement appropriate non-structural and structural solutions. Minimise the use of hard engineered structures.

Section 24(2) (E) Include Land use and Development Incentives to Promote the Effective Implementation of the Spatial Development Framework and other Development Policies

Provide incentives through the use of rebates for implementation of on-site measures which may reduce the need for drainage infrastructure upgrade. Ensure developments incorporate water efficient appliances (community facilities). Ensure fit for purpose re-use is incorporated on site or in the catchment. More relevant to urban areas than rural areas, Land Use Scheme and development controls have the ability to incentivise developers on a financial or non-financial manner. Typically, in urban areas incentives are linked to existing dwellings or retrofit situations where opportunities for implementing of water sensitive planning and site design elements. Non-financial incentives include for example, increasing the allowable floor-space ration, or increasing the impervious area of a residential allotment, provided the development includes sustainability initiatives such as rain water harvesting, living green walls, water reuse and water efficient appliances.

Section 24(2) (F) Include Land Use and Development Provisions Specifically to Promote the Effective Implementation of National and Provincial Policies and Give Effect to Municipal Spatial Development Frameworks and Integrated Development Plans

Ensure water management planning is precautionary and recognises intergenerational equity, conservation of biodiversity and ecological integrity. Protect waterways by providing a buffer of natural vegetation. Use of native vegetation in all runoff management measures and all landscaping to maximise habitat values.

By developing effective, adaptive and innovative water sensitive spatial development frameworks and land use schemes, sustainable management of the total water cycle in rural and urban areas can be accomplished.

5. Conclusion

The link between the land use and provision of infrastructure is evident in the development principles of spatial planning highlighted in the SPLUMA. In order to plan sustainably, spatial planning and land use management systems must consider all current and future water resources, and costs thereof, to all parties for the provision of infrastructure and social services specifically in rural areas.

SPLUMA also requires that Municipal Spatial Development Frameworks and land use schemes identify, quantify and provide location requirements of engineering infrastructure and services provision for existing and future development in both urban and rural areas. Land development must optimise the use of existing infrastructure and needs to introduce innovative surface- and groundwater protecting tools to fight against the looming water scarcity reality.

The focus of spatial planning in South Africa has been redirected adopt a national, provincial and local municipal planning system which is integrated with all sectors impacting on development. If implemented according to law spatial planning and land use management can and will provide water security for rural and urban communities in near future.

References

[1] B. Ruble, "The Challenges of the 21st Century City," Wilson Center, Washington, DC, 2012.

[2] Central Intelligence Agency, "The World Factbook SADC region.," 2009. [Online]. Available: https://www.cia.gov/library/publications/the-world-factbook/index.html. [Accessed 02 March 2015].

[3] Department of Water Affairs, National Water Resource Strategy, Republic of South Africa , 2012.

[4] T. McMahon and M. Peel, "Global stream flow Part 3: Country and climate zone characteristics," pp. 347: 272-291.

[5] G. Ara, M. Jonathan, R. Mickey and S. Julia, "Population Futures: Revisiting South Africa's National Development Plan 2030,," Hanns Seidel Foundation, 2013.

[6] Department of Water Afairs, "Strategic Overview of the Water Sector in South Africa 2013," Department of Water Affairs, Pretoria, 2013.

[7] H. E. Rohr, "Water Sensitive Planning - An integrated approach towards sustainable urban water system planning in South Africa," North-West University Potchefstroom, Potchemstroom, 2012.

[8] I. Wagner, "Water Sensitive Urban Design Task Group," SWITCH, 2010.

[9] R. Brown, N. Keath and T. Wong, "Urban Water Management in Cities: Historical, Current and Future Regimes.," in Water Science & Technology, vol. 59(5), 2009, pp. 847-855.

[10] F. John, "An Introduction to Municipal Planning within South Africa," South Africa Planning Instituete, Durban, 2011.

[11] I. Turok, "Urban Planning in the Transformation from Apartheid: Part 1: The Legacy of Social Control.," in The Town Planning Review, vol. 65(3), 1994, pp. 243-259.

[12] Rural Development and Land Reform, "Spatial Planning and Land Use Management Act No. 16 of 2013," Republic of South Africa, Pretoria, 2013.

[13] Statistics South Africa, "Statistics South Africa," 2010. [Online]. Available: www.statssa.gov.za. [Accessed 27 March 2015].

[14] Calgary Region Focus, "Let's talk density," Calgary Region Focus, 23 Febuary 2013. [Online]. Available: http://www.calgaryregionfocus.com/2013/02/20/lets-talk-density/. [Accessed 4 Febuary 2014].

[15] The World Bank Institute, "Suasainable Urban Land Use Planning- Module 03: How to Integrate Land Use and," The World Bank Institute, Unknown, 2014.

[16] P. Van Lare and D. Arigoni, "Growing Towards More Efficient Water Use: Linking Development, Infrastructure and Drinking Water Policies," United Sates Enviromental Protection Agentcy, Washington DC , 2006.

[17] W. Fourie, Interviewee, Spatial Planning and Land use management in South Africa. [Interview]. 18 05 2014.

[18] Glen Steyn & Associates, "Greater Tubatse Local Economic Development Strategy," Greater Tubatse Local Municipality, Greater Tubatse, 2007.

[19] CRC, "Water Sensitive Cities," Australian Goverment Inisiatives, Melbourne, 2014.

[20] A. Niel, V. Michael, F.-J. Lloyd, W. Keven, S. Andrew and D. Jessica, "The South African Guidelines for Sustainable Drainage Systems.," Water Research Commission, Cape Town, 2013.

[21] J. Russell, "Overlay Zoning to Protect Surface Waters," 2014. [Online]. Available: http://plannersweb.com/2004/04/overlay-zoning-to-protect-surface-waters/. [Accessed 02 06 2014].

[22] Brundtland Commission, "Report of the World Commision on Enviroment and Development," United Nations , 1987.

[23] Robeco & RobecoSAM, "Measuring Country Intangibles," 2013[Online].Available: www.robecosam.com/images/CS_Ranking_E_Rel.FINAL.pdf. [Accessed 03 04 2014].

[24] UN-HABITAT, Urban Planning for city leaders, 2nd Edition ed., Germany: Swedish International Development Cooperation Agency, 2014.

[25] J. Verbeek, "Developing Countries Need to Harness Urbanization to Achieve the MDGs: IMF-World Bank report," The World Bank, Washington, DC, 2013.

[26] K. Kevin, F. Ann and S. Carissa S, "Is There a Role for Evidence-Based Practice in Urban Planning and Policy?," Planning Theory & Practice, vol. 10, no. 4, pp. 459-478, 2009.

[27] [National Planning Commission, "National Developemt Plan - Vison of 2030," National Panning Commicion , Pretoria, 2011.

[28] American Water Works Association, "Fact sheets," 2004. [Online]. Available: <http://www.awwa.org/pressroom>. [Accessed 10 June 2014].

[29] S. Rhimes, "GoodReads," 2014. [Online]. Available: https://www.goodreads.com/author/quotes/3888197.Shonda_Rhimes. [Accessed 19 May 2014].

[30] N. Taylor, Urban Planning Theory since 1945, 1st ed., London: SAGE, 1998.

24

Planning Child-Friendly Spaces for Rural Areas in South-Africa

Ma-Rene' Kriel

Unit for Environmental Sciences and Management, North-West University, Potchefstroom, South Africa

Email address:

marenekriel@gmail.com

Abstract: Child-friendly space are not successfully implemented in South Africa due to problems such as urbanization, development pressure, lack of qualitative open spaces and lack of policy and legislation guiding the planning and protection of such spaces. This study explores the possibility of creating qualitative, playful, educational and environmentally preserving open spaces through the creation and provision of child-friendly spaces for children within their surrounding neighbourhood. The priority within rural areas is usually focussed on providing basic facilities and infrastructure, and the provision of qualitative open spaces is often neglected. In this sense, there is no qualitative child-friendly space currently documented or successfully implemented in rural areas in South Africa. This study provides an overview on what is considered as a child-friendly space and the importance to create such spaces. The main challenges faced by rural areas in South Africa in creating child-friendly spaces are explored. Furthermore current planning approaches in providing child-friendly space in rural areas are identified and evaluated. Lastly green guidelines in creating child-friendly spaces are established.

Keywords: Child-Friendly Spaces, Open Spaces, Rural Area, Qualitative Green

1. Introduction

Open spaces within the South African urban planning context include areas such as parks, boulevards, green belts, buffer strips, lagoons, escarpments and trials [1]. All of these examples, including outdoor play spaces, are components that create an open space system and provide numerous benefits for the public and community in terms of social cohesion, recreational opportunities, health and aesthetic enjoyment [1, 2]. Open spaces such as parks are crucial in developing healthy-communities as it contributes to quality of life by improving, protecting and preserving the quality of the urban environment. Benefits of open spaces and parks include, but are not limited to, visual and aesthetic appealing; places for social interaction, physical and spiritual activity; increase property value; provide shade and protection from natural elements; offers habitat for wildlife and form the image of the local community [1, 2]. In this sense, open spaces provide a qualitative function within the urban planning context.

According to [1] parks and open spaces are classified in to three levels namely, Level 1: Neighbourhood level, Level 2: Community level and Level 3: Regional level. Neighbourhood level, include playgrounds and tot lots defined as *"soft*

landscape of grass, trees, and planting areas, usually located in a residential setting and detailed and furnished for a variety of active and passive uses". Tot lots and playgrounds are typical neighbourhood level open spaces and serve a population of approximately 2000 residents. Neighbourhood level parks provide both active (sports, play, waling) and passive (sitting, sunbathing, resting) recreation opportunities, and in this sense provide a basis for the development of child-friendly spaces on a neighbourhood level [3].

Community level open spaces serve two to three neighbourhoods and include a broad choice of amenities. Regional level includes open spaces such as Nature Reserves, regional athletic parks, golf courses and campgrounds [3].

Numerous of literature confirms that outdoor play spaces are vital for children's learning and developing stages throughout life [4, 5, and 6]. [3] confirm this statement by emphasizing the importance of play spaces for normal child development. Development includes 1) *Physical development* (large-muscle or gross motor activities) such as climbing, running and jumping and 2) *Intellectual development* (manipulative play) where children begin to formulate concepts of action and relationship by energetically manipulating the elements of the environment. Active interaction with the environment is furthermore important for children to learn to conserve and respect the natural environment [3, 5].

[7] and [5] stated that outdoor play spaces have the ability to inspire children's imaginations and exploration as well as improve their confidence and connection with friends, family and have a positive effect on community cohesion. It is thus important to ensure adequate child-friendly play spaces within neighbourhoods.

Furthermore [3], [5] and [6] confirms that play is a child's way of learning. Play is complicated, intimate processes which develops and teach children to become socialized. Play is essential for the healthy development of children for their physical, social and cognitive development. It allows children to develop a sense of well-being, improves their interpersonal abilities, develops language skills, establishes creating thinking and involves exploring and problem solving skills.

[5] and [8] states that today's children have fewer opportunities for outdoor play than previous generations. Reasons include urbanization and development pressure where open spaces are used for the development of businesses and housing. Children are a powerful icon of the future. Youth is considered the most critical periods in life in forming an individual's unique relationship with the environment. If children are not able to create their personal relationship with the environment through actively participating with the world around them, their ability to address environmental problems in the future can be threatened [9]. They provide us with a captivating reason to protect the environment, and provide adequate open spaces for outdoor play, thus, through creating child-friendly spaces within open spaces.

From above statements the importance of child-friendly spaces are emphasized. Child-friendly spaces, in context of this research, implying qualitative open spaces developed primarily to be used by children. The concept of child-friendly spaces is thus defined as "a complex multi-dimensional and multi-level concept, referring to settings and environmental structures that provide support to the participation of children and youth in the shaping of their setting, consequently playing a central role in the creation of child-friendly environments in spatial planning" [10].

Literature offers an abundance of definitions defining child-friendly spaces but criteria for creating such spaces are often relatively broad, vague and not easy to implement and only deals with the immediate surrounding environments of children without considering the impact on social, political and historic factors and furthermore do not approach the issue from the child's perspective [10, 11]. The lack of participation of children throughout the planning proses is the main problem when creating child-friendly spaces. Children's perspective differs from adults and they perceive the natural environment more intensely [10, 12 and 13].

An abundance of literature and authors confirms that numerous problems become major factors in determining the quality of children's outdoor play environments, such as safety and security issues [4, 5]; children's restricted independent mobility [6, 13, 14 and 15]; and child obesity [13, 15 and 16].

This study is the first step in developing guidelines on planning for and creating improved child-friendly spaces within open areas, especially in rural areas in South Africa. It aims to identify and define essential elements of qualitative child-friendly spaces that support the physical and emotional growth of a child, as well as contribute to sustaining the ecological benefits of the environment.

2. Child-Friendly Spaces

2.1. Defining Child-Friendly Spaces

There is growing awareness of the importance and benefits of designing healthy, safe places for children [4]. Many aspects define what is considered as child-friendly space and the importance of creating such spaces. A child-friendly space is a safe space created for children where they can actively and passively interact with the environment and socialize with friends through playing and learning simultaneously [4].

According to [10] a child-friendly space can be defined as "a community product developed from local structures beyond the individual level. It comprises a network of places with meaningful activities, where young and old can experiences a sense of belonging whether individually or collectively. The participation of children and youth in the shaping of their setting plays a central role in the creation of child-friendly environments."

The objectives in providing a well-designed child-friendly space are indicated in the following section. The space should provide opportunities for children's physical, cognitive and social development through a wide range of 1) play settings, 2) cultural and racial groups and 3) natural setting and the need for human contact.

1) Play settings [3] implies: Firstly, motor skill development which includes a range of opportunities for children to test their limits and abilities through providing them with a wide range of activities. Secondly, children should be able to make their own decisions about their activities. They should be in control of most or the entire environment. The play space should provide a wide range of decision points that is appropriate to different age and skill levels for the continuing of a present activity, ceasing it, or instigating a new one [3 ,4]. Thirdly, the environments should provide opportunities for learning where children learn to solve problems, manipulate the environment, redesign it and develop their own viewpoints towards the environment. Furthermore the space must provide opportunities for fantasy play where children stimulate their imaginations. The space must not restrict children's imaginative play through being too literal or too abstract.

2) Cultural and racial groups implies that the space must support social development where positive interaction and socializing can take place between different cultural and racial groups and the most important of all a child-friendly space should be fun. Smiling and laughing children are the purest indicator of an effective play space [3, 4].

Child-friendly spaces have two main purposes to provide a

3) natural setting and the need for human contact. The natural setting must be able to create a rich aesthetic environment where children can enjoy nature, feel comfortable and peaceful. The main motive for a children space it to have direct interaction with the environment where they can observe and socialise when favoured [3].

In conclusion the following characteristics are used to create a successful child-friendly space according to [7] and [5]: Are well located; enables active and healthy lifestyles; make use of natural elements; designing green and promotes sustainability; providing child-friendly transportation options; provides a safe place; provide a wide range of play experiences, risk and challenge opportunities; multi-use and accessible to both disabled and non-disabled children; allow children of different ages to play together.

When considering all above statements the quality of a children-friendly space is more than a piece of play equipment; it's only as rich as the supporting physical and social environments [4].

A child-friendly space should adhere to four main characterisations in order to be successful. This include safety, open space or natural setting, access and sociability and integration [12]. Each of these themes is thoroughly explained in the following division in the context of urban planning provision, focussing on rural areas.

2.1.1. Safety and Comfort

A safe environment fosters feeling of security and makes people more willing to engage in outdoor activities according to [12]. A child-friendly space is where children can play safely without fearing the surrounding environment. Safety is a main aspect in creating a child-friendly space and determines whether a play space will be used successfully by the surrounding community. Parental fear are reduced in safe child-friendly spaces and the need for constant supervision is decreased which enables children to explore and discover independently. Children need to be able to play willing in the space without any dangerous hazards and risk but also not totally eliminate the ability to stimulate risk taking opportunities [17].Safety in a child-friendly space especially in rural areas can be improve in terms of planning through providing adequate lightning in and around the space, enhancing the visibility of the play space in all directions for easy supervision, locating emergency public telephones near entrances and providing sufficient drop-off and pick up points away from traffic.

2.1.2. Natural Open Space

Research has shown that children prefer to play in natural areas and need access to rich stimulating environments [18]; therefore child-friendly spaces are directly connected and created within the surrounding natural environment. Natural spaces offer sensory stimulation and physical diversity which is critical for childhood experiences outdoors. Child-friendly spaces support greening which refers to the integration of natural elements and processes in a play space. Children's direct social and individual involvement in nature has a positive effect on children's motor skill development, social

development, attentiveness and activity level. Integrating the natural environment is a crucial element in creating child-friendly spaces because is forms children's environmental identity and guide their future environmental actions [12, 18, 19 and 20]. The natural environment can be integrated into a play space through providing a wide range of vegetation, trees, shrubs and opportunities for water and sand play.

2.1.3. Access

Access is an important factor in creating child-friendly spaces. Children need access to rich appealing environments that are free from unacceptable risk, such as parks to create the opportunity to explore and discover. A children-friendly space must be accessible for all ages and cultures of children. Furthermore it should be accessible for children with disabilities where they can also play and explore freely without limiting their abilities [12, 18 and 19]. Accessible child-friendly spaces can be provided in terms of planning, through locating the space within walking distance and in close proximity of residential areas and schools. Furthermore entrances must be visible, through adequate signage, and accessible for all children disabled and non-disabled.

2.1.4. Integration and Sociability

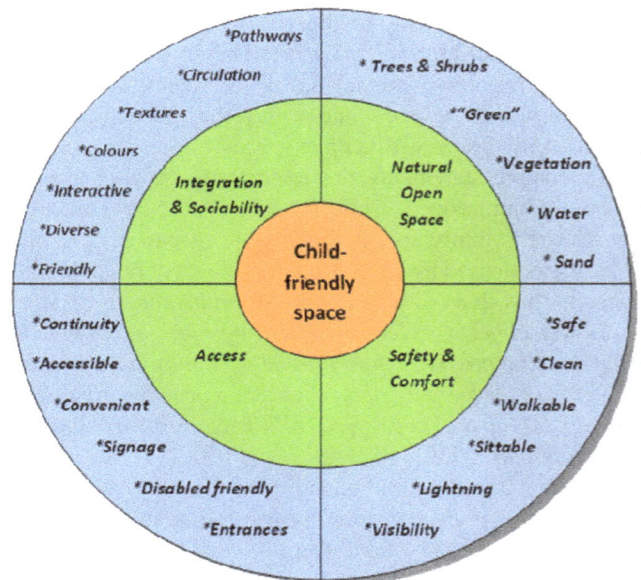

Figure 1. Child-friendly Model.

Source: Own Creation based on [12, 18, 19 and 20]

Child-friendly space need to be created to support different ethnic groups and improve integration in communities. Boys and girls must be able to play freely without discrimination as well as children with disabilities. Integration help children learn gender differences and the ability to see all children as equals no matter the age, size, gender or disabilities [17, 18 and 19]. Sociability and integration can be improved in terms of planning, through integrating different textures, colors and facilities in the play areas which develops their physical, emotional and mental skills. Providing a variety of pathways and seating in the space improves circulation, integration and creates opportunities to socialize. Incorporating water and sand setting in the play area enhances opportunities for socializing and integration. "Fig. 1", illustrates that child-friendly spaces are created according to 4 main themes namely, Natural open spaces, Safety & comfort, Access and Integration & sociability. Each of these themes is divided in to sub-themes that integrate the planning of a child-friendly space in rural areas.

2.2. The Importance of Child-Friendly Spaces

The importance of child-friendly spaces is examined through the term *"play"* by many authors such as [3]; [7]; [18]; [20] and [21],thus, consequently the broad definition for the term *"play"* can be defined as *"fun or serious"*. Through play children explore social, material and imaginary worlds and their connection with them, expanding all the while a supple range of responses to the challenges they stumble upon. By playing, children learn and develop as individuals and as members of the community [18]. Furthermore play is the foundation in forming a child's intellectual, social, physical and emotional skills. For example sand and water play develops logical mathematical thinking, scientific reasoning and rational problem solving. Playing in outdoor environments help children learn through direct concrete material that inspires exploring, manipulation and active engagement [21].

[3] and [7] state that the response of a child to his environment is far more direct and active than an adult's. Children are constantly making discoveries through physically exploring concepts (high and low, near and far, hard and soft, light and dark) that stimulate their imagination and learning abilities. The physical surroundings in which children grow, influence and shape their interactions, development and experience of life into adulthood.

Not only is play important for imaginative skills but risk taking are crucial for children to develop confidence and abilities during childhood. Play stimulated children's minds and help them overcome trauma, fear and stress [7, 18]. Consequently play leads to creating strong supportive communities and helps reduce children and young people in anti-social behavior that may lead to cults and vandalism [3]. Children obtain the following benefits through play according to [18]: Opportunities to enjoy freedom, and exercise choice and control over their actions; Opportunities for testing boundaries and exploring risk; Offers a very wide range of physical, social and intellectual experiences for children; Improves children's independence and self-esteem; Develops social interaction and respect towards others; Improves quality of life; Ensures healthy growth and development; Promotes creativity; Increase knowledge and capacity to learn.

In conclusion children can benefit from play opportunities as highlighted above especially children from stressful circumstances in rural areas. A play area is where children can escape from their fears, poverty and family and experienced freedom and emotional healing. They can improve their communication skills, attitudes, problem solving approaches and even their circumstances in the long term [20].

2.3. Challenges for Rural Areas and Provision of Child-Friendly Spaces

South Africa faces many problems due to the apartheid and post-apartheid eras especially in rural areas. The apartheid spatial planning had a few consequences namely cities that are undersized, but sprawling, marginalising, decentralising, planned to obstruct movement and under serviced areas. The post-apartheid settlement planning had its own consequences namely housing-driven settlement planning, de-densification, on-going decentralisation and worsening service levels in sprawling new informal settlements [22]. The main problems include urbanization, poverty, political structure, health, lack of open spaces and environment derogation. The underlying causes of these problems are lack of public awareness, crime, insufficient governance, poor policies, and the lack of knowledge [23].

2.3.1. Challenges 1: Opportunities
In South Africa urbanization is mainly caused by the huge supply of job opportunities in cities and the huge demand of job opportunities in rural areas. According to [24] urbanization can be defined as ''the shift from rural to an urban society, and involves an increase in the number of people in urban areas during a particular year." Due to the rapid urbanization taking place all over the world, the concept of child-friendly spaces has emerged disputes about urban development from an environmental perspective. Additionally there are no guidelines or frameworks to guide urbanization and results into insufficient open spaces [13]. Rural areas lack open space because the main focus is on housing provision and not open spaces, consequently limiting their opportunities to sufficient open spaces.

2.3.2. Challenge 2: Independent Mobility
Rural areas are safety hazards due to poverty, poor street lighting, and degraded urban environments, lack of food, housing and education leading to an increased restriction on children's independent mobility.

The term *"children's independent mobility"* refers to their freedom to move around without out adult supervision, that is critical for their physical, social, cognitive and emotional development. In modern society, the active, independent mobility of children and young people is becoming increasingly restricted for various reasons such as a lack of safe environments and support bases, limited facilities and development opportunities, increased road traffic have negative consequences for children such as obesity due to lack of exercise, lack of risk taking opportunities, lack of environmental contact and the sense of environmental preservation [6, 7 and 13].

These factors can restrict a child to discover their internal abilities and environmental possibilities as well as fall behind in their social and personal development. Independent mobility helps promote children bonding with their peers, how to preserve and interact with the natural and built environments, thus, creates a stronger sense of community and responsibility for the environment, a reduced fear of crime, and increases feelings of isolation during puberty [6, 13, 14 and 15].

Children who are limited in their independent mobility fall behind in the following aspects according to [6]: (1) *Social development:* Children learn to socialise with friends through playing and meeting new friends in outdoor play spaces as well as learn the ability to adapt to new situations.

(2) *Physical development:* Sufficient outdoor exercise is not only healthy but a necessity to reduce obesity and health problems in adulthood as well as healthy bone development, posture and balancing skills. (3) *Cognitive development:* Discovering new things develop children's spatial awareness and their understanding of how the world is structured.

2.3.3. Challenge 3: Participation

The improvement of existing park layout and facilities is important to address if repeating past mistakes want to be avoided. Public participation in a specific park design or redesign is essential especially the involvement of children because their needs are not always well represented [3, 19]. The Commissioner for Children and Young People has developed participation guidelines, *"Involving Children and Young People: Participation Guidelines"*, to support the encouragement of practitioners to involve children in the planning and designing of spaces [7]. Children are mostly neglected in land use planning in one of two ways. Firstly, they are given little consideration when it comes to design and secondly, there is lack of planning for children [12].

In rural areas the lack of participation leads to an increase unawareness of environmental preservation and the desire to protect the local environment. According to [12] involving children in the planning process improves integration and gives them a sense of self-worth and a more connected feeling towards the created space. A child-friendly spaces offer recreational opportunities to rural families and provides a liveable place where the people can experience a sense of community [25].

2.3.4. Challenge 4: Crime and Safety

The fear of crime and concern for personal safety is consistently within the top issues in South Africa especially in rural areas. Consequently, leading to a change in the ways in which people use public spaces within their communities. According to [3] *"fear of crime keeps people off the streets, especially after dark, and out of parks."* Safety and security are major factors in determining the quality of children's outdoor play environments. Without being able to take risk children cannot grow to their full potential. Risk taking and challenge have an especially important role in children's play development [4, 5].

The fear of crime limits a child opportunity to play in the outdoors. Safe space must be created to enable children to participate in activities with some independence [7]. An approach to address the fear of crime is to consider what the

root causes of crime are such as disadvantages, neglected and lack of open spaces and recreation opportunities and discrimination. These causes can be addresses through providing education, job opportunities, economic development, community involvement and efficient planned open spaces in rural areas.

2.3.5. Challenge 5: Obesity

Numerous researchers such as [13, 15 and 16] emphases that childhood obesity, due to a lack of exercise, is becoming a significant public health issue all over the world including South Africa. The main reason for above statement is the lack of children's independent mobility and safe, playful child spaces such as parks. In play areas children usually find various play equipment's containing different activities, but rarely opportunities for imaginative play and environmental contact. According to [25] obese people are likely to suffer from high blood pressure, high cholesterol, heart disease, strokes, osteoarthritis and emotional problems such as depression, eating disorders and low self-esteem. Rural areas are restricted to safe play environments and these areas are especially prone to health problems such as obesity.

Children with low-income backgrounds have less if any access to play spaces and are therefore at a disadvantage. Creating child-friendly spaces in rural areas will eliminate all above challenges and help develop healthy and educated children which can improve their quality of life and grant them the opportunity to become more successful in life in adulthood and improve their own circumstance [18].

2.4. Planning Approaches in Providing Child-Friendly Spaces

Different approaches to child-friendly space development have the tendency to lead to one of two outcomes 1) a man-made manufactured design or 2) a natural rugged approach. Child-friendly spaces can be divided into two paradigms namely conventional paradigm and conservational paradigm [26]. These paradigms will be discussed and compared to determine which one is more likely to create a successful child-friendly space.

2.4.1. Conventional Paradigm

Conventional paradigm focuses on a formal design approach where the equipment is manufactured and firmly designed play equipment. The play area is ordered in a logical manner that can be understood by adults and is not a freely open-minded design where children can explore and fantasise. People are individuals of their one experience and this usually shapes conventional wisdom methods in planning. Adults see playgrounds as asphalts areas were play equipment such as swings and jungle gyms are placed due to the image of their childhood memories, as a result, this perspective is understood as the ideal model of how a children's playground ought to be [5, 26].

Play equipment are selected according to catalogues, which appears decent trough an adults perspective and are place in an outdoor space with no effort and ease. The conventional paradigm planning approach follows a structured order that is fixed and precise without opportunity for flexibility, consequently leading to an isolated process rather than an integrated whole [27]. However, limiting outdoor spaces with manufactured play equipment is not the ideal planning approach but rather incorporating it with the natural environment which include vegetation, water, sand and wilderness [26].

2.4.2. Conservational Paradigm

Conservational paradigm focuses on a more natural and informal design method and can be seen as a discovery play garden rather than a formal structured play area. Children see and experience the natural environment different than adults. Through a child's perspective beauty is seen as rough wilderness rather than an orderly design. In a conservational design approach the space provides openness, variety and openings for manipulation, discovering and experimentation. The conservational paradigm is the shift from a fix structure design approach, as can be seen in the conventional paradigm, to a looser design that includes loose parts such as sand, water and natural manipulative materials. The play space ought to flow from one area to the next, be as flexible and simple as possible and stimulate children's senses and curiosity [26].

2.5. Summary

The four main themes contributing in creating child-friendly spaces where identified as, 1) safety and comfort, 2) natural open space, 3) integration and sociality and 4) access. The importance in creating such spaces where linked with physical, emotional and social benefits through the term "play". Furthermore the challenge faced by rural areas where identified such as lack of opportunities, independent mobility, participation, obesity, crime and safety, consequently children in rural areas do not have adequate outdoor play space. Two planning paradigm where identified and discussed namely conventional paradigm which refers to manufacture play equipment, and conservational paradigm which refers to the integration of natural surroundings in the play equipment. In rural areas there are many obstacles to overcome in creating child-friendly spaces these obstacles include the following; lack of children participation in the planning process; parks are not uniquely design according to the community's needs; a quantitative planning approach are followed and not a qualitative planning approach; and lastly the planning process is isolated and not an integrated approach.

3. Conceptual Framework

3.1. Origin of Parks

The roots of open spaces can be traced back to the 1830's in America where burial grounds were set in pleasant landscaped environments, mostly at the edge of the town. This encouraged the creations of public parks. The New-York's Central Park was the first planned park in America and was designed by the nation's first landscape architect Frederick Law Olmstead in 1858 [2, 3]. According to [28] Olmstead initiated the first concentrated park and recreation movement in the United States. He reasoned that the entire population could not flee to the countryside, but rather bring the rural landscape to the heart of the city and create a pleasing natural setting where people can escape the city. Concurrently, parks were defined as *"naturalized passive retreats"* and recreation areas as *"active sport oriented facilities-playgrounds, hard-surface court areas, team sport fields."* [28]

Eventually open spaces became part of a regional open space system in America and spread to Europe such as Holland, Scandinavia and Germany. The *"garden city"* movement during 1904 in England, conceived by Sir Ebenezer Howard, was the main objective was to promote the concept of greenbelt, open spaces and parks [2].

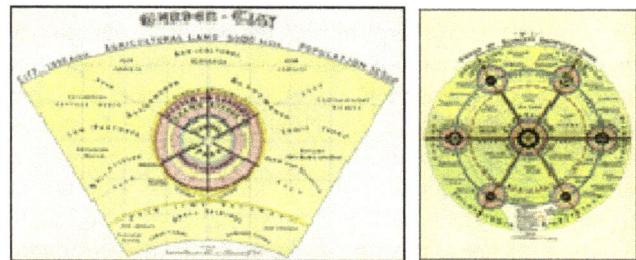

***Figure 2.** Garden City Model.*

Source: [2]

"Fig. 2", illustrates the Garden City proposed by Ebenezer Howard in the 1890's. Howard's concept was for a town of limited size surrounded by agriculture lands. The benefits of rural and urban would be balanced in a self-contained, self-sufficient community. His goal was to devise the ideal plan that would bring urban and rural into harmony with each other. Each garden city will be surrounded with 1,000 acres of urban land and 5,000 acres of agriculture land. The area would support a population of 32,000 people.

From 1946 to 1950 around London 14 New Towns were designated to rehouse the overloud population from the capital. Each of these New Towns had an unusual high proportion of green space. The green spaces were designed as a public amenity and became central features in cities. In the 1970's the public open spaces spread to older towns as well to improve the environmental quality [2].

Today parks are known as "people places" where people actively engage with the environment through observing and socializing in safe, aesthetic spaces that is supported by furnishing and other supporting amenities [3]. Open spaces is ideal in creating child-friendly spaces where children can actively participated in their surrounding natural environment and improve their physical, mental and cognitive development. Open spaces create the opportunity to integrate these spaces into neighborhood settings that is easy accessible for children and supervision is within appropriate distance from surrounding houses.

3.2. Green Space Guidelines for Creating Child-Friendly Spaces

The golden rule in creating an efficient child-friendly space in rural areas is to design the space for the specific location through integrating the surrounding landscape and vegetation as the play settings and nature as possible play materials, consequently it creates a space that can been seen as a discovery play garden according to [5] and [26]. The space must be aesthetic appealing and functional to satisfy the senses of the children. Creating child-friendly spaces is a multidisciplinary and complex task according to [5]; [7] and [28]; therefore guidelines are established in providing successful child-friendly spaces in rural areas. These guidelines are discusses accordingly:

3.2.1. Location and Size

The most important factor in a successful play space is the location of the area where children feel safe and want to play. The location of the child-friendly space ought to be away from dangerous roads, noise and pollution and located in an area that are easy accessible for children and visual from all directions [5]. The child-friendly space must be walking distance from surrounding neighborhoods and schools that are easy accessible for all and grant easy supervision for adults. The proposed space must be designed to fit the surrounding attractions and enhance the local character of the environment [28]. Primary features need to be identified on the site and be incorporated into the design such as a structure of an old building, a tree with character or an old sculpture [29]. Inspiration in creating a unique space can be accomplished through historical background on the area or any occurring materials and geographic features [5].

The size of the space depends on the available space in the area and the function of the space. Smaller parks should be small enough to maintain a sense of intimacy, and enable easy visibility and perception, ± 25 m maximum, consequently small parks should be between 450 m2 and 1 000 m2 in size, with widths of between 15 m and 25 m, and lengths of between 30 m and 40 m. The area and dimensions of a play space vary according to the nature of the play equipment, and whether or not the play space is part of a larger soft open space. Play spaces should be small enough to enable easy supervision, ± 25 m maximum, consequently play space should be between 450 m2 and 1 000 m2 in size, with widths of between 15 m and 25 m, and lengths of between 30m and 40m [28, 30].

3.2.2. Creating of a Safe Space

The creation of a safe space refers to the health and well-being of all children under all circumstances through ensuring that all hazard conditions are removed from the child-friendly spaces [4, 7]. Adequate lightning should be provided throughout the space especially on all pathways and entrances. Fences, barriers and lightning can be used to prevent vandalism. Emergency telephones must be located at entrances and appropriate fencing which surrounds and protects the space [28]. Circulation in the space should be clear with no entrapment zones and clear signage must be provided throughout the space to inform the user of all possible entrances and exits [3]. Furthermore the child-friendly space should be connected to accessible streets and roads that provide routes for walking and cycling and ensure the safety of the children as well as promote their independent mobility [7, 31].

3.2.3. Creating Accessible Entrances

The space need to ensure that there are clearly defined entrances which orientate, informs and introduce the users to the specific site. The entrances must be located between transportation areas and where it is easy accessible for children and people with disabilities. The creation of useful accessible entrances is where people can gather, talk and have easy access to the space [4, 7]. The main aspects of quality entrances in a child-friendly space include the following, according to [4]:

(1) *Functionality:* The entrances must be located near accessible pathways, drop-off zones, streets and parking areas.

(2) *Access:* The entrance must encourage a welcoming feeling and provide nonslip walking access pathways that are wide enough for people and children with disabilities.

(3) *Drop-off Zones:* Entrances is where parents drop-off and pick up their children.

(4) *Waiting zones:* Entrances are mostly used as waiting zones therefore seating, bicycle stands and shelter against the weather should be provided.

(5) *Communication images:* The entrances must be visual and attract the user as well as inform them about the information concerning the space and provide them with a map and direction boards [3, 4].

3.2.4. Creating a Variety of Pathways

Pathways are multi-purpose and provide accessibility to the space and separate different uses within the space. They help users to move between different elements located in the space and improve circulation [28, 29 and 32].The main aspects in creating pathways in a child-friendly space in rural areas are listed below:

(1) Dimensions: Pathways must be wide enough (primary routes 9-10 feet and secondary routes no less than 3 feet) for the appropriate use of the specific pathway and the surface must be accessible, even and non-slip. The slope must be easy accessible for children with disabilities and handrails must be provided on both sides if the slope is steep [4, 33].

(2) Variety: A separate bikeway path should be provided on primary paths. The space should provide variety types of pathways away from the primary path to accommodate different uses such as running, walking and biking [4].

(3) Intersecting and connectivity: Pathways must intersect at some points to support continuity of movement. The hierarchy of pathways can promote movement and help children to understand the play space [28, 29].

(4) Surfaces: Pathway surfaces should be accessible, stable, firm, flat, and slip resistant and raised edges must be provided at hazard areas [4, 32].

3.2.5. Creating Appropriate Signage

Signs are a form of communication and provide the users with important information concerning the space as well as provide directions and support of traffic flow within the space [29]. A similar style amongst signs creates familiarity and eliminates confusion, thus, standardization of all signs is proposed [28]. The following aspects are considered when providing signage in a child-friendly space in rural areas:

(1) Types of signage: Signage can be divided into 3 main types: (1) *Informational signs:* Present general information of the space in words and graphics and is located at the entrances such as rules, closing times and background information. (2) *Directional signs:* Present directions to different facilities, routes and play areas. (3) *Identification signs:* Present information indicating specific features such as water or bathroom facilities [4].

(2) Design considerations: Signs must be placed logically and free of obstruction. They must be appropriate heights for children with interesting colors and symbols. Signs can also be used as a learning objective through providing buttons that provide a verbal response. Furthermore signs should support children with all disabilities thus providing words, raised letters, pictures and the appropriate languages spoken in the area (Herrington, 2006:37 & Moore et al., 1987:57).

3.2.6. Creating a Variety of Seating Options

Seating can be used to encourage interaction between people. Seating arrangements can either support or preclude social interaction. Grouped benches facing each other provide opportunity for conversation-making and social interaction while back-to back benches provide opportunities for a more private setting [3]. In a child-friendly space a variety of comfortable seating is essentials for different tasks such as observing, privacy, interaction and waiting. Providing different types of seating also create an interesting aesthetic atmosphere [4, 29].

3.2.7. Boundaries and Fencing Considerations

Fences are used in a child-friendly space to define, protect, separate and create activity settings. Fences can also be used to direct pedestrian movement and protect the surrounding vegetation. According to [4] and [5] the following design considerations must be taken into account when providing fencing in a child-friendly space in rural areas:

(1) Barriers: Fencing can be designed as barriers to protect vegetation and provide play elements. The main objective of the barrier is not to keep the children out the vegetation but rather reduce the impact so that the plants can recover and survive. Fencing is also used to define intimate social areas and provide privacy away from activities [4].

(2) Barriers against weather conditions: Barriers can be used to protect children against harsh sunlight and strong, cold winds [29].

(3) Aesthetic appearance: Fences and barriers must be attractive and not obstruct sight. Vegetation, groundcovers and vines can be used to make fencing attractive.

(4) Play setting: Providing fencing with peep holes and interesting colors and textures an attractive play settings for children [4].

3.2.8. Creating Child-Friendly Play Equipment

Play settings stimulate large muscle development and supports movement, social interaction as well as fantasy play which stimulate children's mental development. Equipment settings should be multipurpose and support creativity and coordination development [3, 4 and 33]. A well designed play space ensures play equipment where disabled and non-disabled children can play together. The play space should have a stimulating layout and be aesthetic appealing through providing imaginative equipment [29]. The following criteria are used in choosing and creating play equipment for children:

(1) Hazard versus challenge: Children use equipment in all possible ways and the equipment must be designed to incorporate safety to eliminate dangerous hazards and provide different levels of challenges. A good play setting allows children to take risk and challenges them through swinging, jumping and climbing [4, 20 and 33].

(2) Separate play areas: Separate play areas should be provided for different age groups especially for children less than three years of age. Barriers can be provided to protect smaller and younger children. Well-designed play areas have different degrees of challenges, which enabled integration of different age groups and prevent physical obstructions [4].

(3) Options: Providing children with a variety range of play settings stimulates a wide range of activities and ensures that that play area are used frequently such as climbing, swinging, crawling, bouncing, jumping, balancing and sliding [20, 29 and 33].

(4) Sensory variety: Play equipment need to stimulate all the senses of a child through providing opportunities for touch such as different textures sand, water and vegetation, fragrant plant materials for smell and colorful play equipment that are visually stimulating. Children usually respond best to bright cheerful primary colors [4, 5, 28 and 29].

(5) Movement, linkage and flow: Play equipment should support movement and linkage between different play equipment and provide different levels of play. Play equipment must be visually understandable for children with many ways to get on and off. Orientation can be supported through different colors, textures and shapes [3, 5 and 28].

(6) Disabilities: Children in wheelchairs must be able to access the play equipment through providing wheelchair entrances or raised play settings as well as handle bars to support body weight [3, 4 and 33].

(7) Themes: Thematic elements provides opportunities and encourage fantasy play through creating slides that look like rocket ships, climbing areas as castles and fixed in pieces such as a steering wheel [4, 29].

(8) Slides and swings: The play area should include different heights of slides to accommodate all ages of children. Access for children with disabilities to the slides can be created through providing stairs alongside the slide. Slides can contribute in creating a play area with variety and challenge such as slide designs that includes waves, spirals and tunnels

[3]. Swings should be situated away from other equipment and tire swings make it accessible for children with disabilities. Furthermore swings ought to include safety straps to accommodate all ages of children as well as children in wheelchairs [3].

(9) Climbers and balancing equipment: Play structures should include a form of climbing to develop upper body strength and create challenge opportunities. Balance equipment such as rocks, logs and chains can be used to link different play equipment and areas with each other and is vital important for balancing and coordination development in children [3, 5 and 29].

(10) Surfacing under play equipment: The best impact absorbing surfacing under high zone fall areas include the following: (1) *Rubber:* Chopped compact tire is the best impact absorbing surface to use under play equipment due to its ability to spread outside its containment barrier, thus, children are less likely to obtain injuries. (2) *Sand:* Sand is the second best impact absorbing surface and is most frequently used as fall cushioning due to the ability to deform to the shape of the falling child. Sand is also must cheaper than rubber absorbing products.(3) *Gras:* Gras is not recommend as surfacing under equipment with a high fall zone risk because grass is prone to minor injuries and is preferred on open areas for chasing and rolling games [3, 5].

3.2.9. Supporting Amenities and Lightning

In a child-friendly space the provision of supporting amenities are essential, such as toilet facilities, and coverage facilities against rainfall and strong winds. Other facilities may include storage area for play equipment and management of the space [3]. Adequate lightning are an essential factor in creating an efficient child-friendly space in rural areas, consequently providing protection and peace of mind. Lightning can be used for the following:

(1) Safety lightning: Provides necessary lightning for people during night time to provide safety against crime.

(2) Security lightning: Provides a degree of protection to the property against vandalism and crime.

(3) Aesthetic lightning: Is used to enhance the beauty of a visual element such as a fountain or statue [32].

3.2.10. Supporting Vegetation and Trees in Child-Friendly Areas

People-plant interaction is important in creating a rich child-friendly space where users can make close contact with vegetation such as groundcovers, shrubs and trees. Vegetation and trees create a wide range of play activities such as tree climbing, hide and seek games, exploring, discovering, and manipulative play, collecting and touching of plants. Leaves, flowers, fruits, seeds, sticks and nuts stimulate a variety of senses and imaginative responses in children [5, 20 and 29]. Trees are used as barriers against cold winds, harsh sunlight and rainy days, as well as provides a shady play area and attracts wildlife and birds, consequently creates opportunities for people-wildlife interaction [3, 5].

Vegetation marks the passing of seasons and develops children sense of time as well as creates a pleasant atmosphere

of textures, smells and colors'. Children in wheelchairs can enjoy vegetation through creating raised planting areas. Plants chosen must be fast growing, easy to maintain, resilient and comfortable to touch and does not irritate the skin [5]. Child-friendly spaces include plants that eliminate poisoning and injuries and trees that can endure tree climbing activities which creates a sense of achievement amongst children [3]. Trees are chosen according to their rooting pattern, water requirements, climbing durability and growing behavior [2].

Vegetation can be used for the following in a child-friendly space in rural areas [28]: To form spaces and shapes; To direct circulation; To provide detail interest; To supply shade and protection against the weather; To buffer odors and noise; To provide sensory stimulation; To provide opportunities for learning.

3.2.11. Creating a Garden or Vegetable Setting

Garden or vegetable settings are a best way of enabling children to interact with each other and nature. They learn about the ecological cycle, how to preserve the environment and it stimulates the cooperative work between children. Raised beds create easy access for children in wheelchairs and protect plants against direct impact and pests. Children can experience different taste and learn more about different kinds of fruit and vegetables, thus, encouraging them to eat healthy. Vegetable gardens are a great way in involving the community and promote interaction between people; furthermore it also provides the community with a source of income if the fruits and vegetables are harvest and sold [4, 5 and 31].

3.2.12. Promoting Environmental Sustainability

A good play space can be designed through using recycled materials and incorporating the natural environment as far as possible to ensure sustainability. As mentioned above the integration of trees and vegetation support environmental sustainability as well as the provision of a vegetable garden setting. Fallen leaves, twigs and grass cuttings can be reworked back into the environment to preserve the ecological cycle and it provides additional play opportunities [3]. Furthermore the play space should provide decent recycle bins which sort's plastic, cans and paper to create awareness of sustainability amongst the children and this is a fun meaningful way to educate them. The recycle bins can be designed in an interesting way with capturing color's to gain children's attention [32]. Not only will the play space support sustainability but the desire to protect the environment is established amongst children and they will preserve the environment over the long-term [5].

3.2.13. Creating a Sand Setting

Sand is multi-functional and can serve as play material and safety surfacing as well as improve the quality of the play space. Sand areas in child-friendly spaces should be located near water play and paths where it is easy accessible for children with disabilities [5, 29]. Furthermore the sand play areas are best located near or under trees for shade in the summer and protection against cold winds during the winter [3]. The following criteria area important to take in consideration when creating sand plays areas in child -friendly spaces in rural areas:

(1) Playability: The type of sand should contain a small grain size which makes it easier to mold and sculpture with as well as have a low dust content to prevent unwanted allergies.

(2) Design considerations: The sand areas must be located near water play and separated from active play equipment, furthermore, have multi-level sand tables and access points to accommodate disabled and nondisabled children. The sand pits should be between 18-24 inches and have a 2-3% slope to prevent sand from falling out. Providing a form of table setting in the sand play area provides children the opportunity to sculpt and mold forms easier [3, 4].

(3) Vegetation and water: Incorporating vegetation and water play with sand play settings enhances the range of fantasy play and is essential for good sand play settings [3, 29].

3.2.14. Creating a Water Setting

Water play creates a multi-sensory function such as sounds, textures and a substantial aesthetic dimension. Children are excited and relaxed through water play and it enables them to have physical contact with water which can be incorporated with sand play to provide a more appealing play setting [5, 29]. Shallow pools and sprays are used in child-friendly spaces for children to cool-off during the hot summer months. Children can interact and experiment with water play through observing materials that sink and float, thus, supporting the development of their intellectual skills. Water play can be integrated in a child-friendly space through forms such as streams, drinking fountains, spray pools, sprinklers and water tables [3, 4]. The following criteria are used in creating a water play setting:

(1) Design considerations: Water and sand play areas should be located next to each other away from active play equipment and under trees for shelter against wind and the hot summer days. Seating should be provided for adult supervision and the water setting must be easy visible from all directions. Water depth must be carefully considered to eliminate drowning risks therefore spray pools, drinking fountains and water tables are best recommended. Shallow ponds can be integrated with stepping stones and bridges for close contact with the water and is fascinating for children [3, 5].

(2) Circulation: Water must circulate to eliminate health risk and provide a relaxing atmosphere this can be accomplish through running water tables, water pumps and wandering streams. The water play setting should be connected to the public water supply if natural rainfall is insufficient in the area [3, 32].

(3) Sprays: Spray areas are preferable to standing water and uses less water supply. A fine spray is best comfortable for all weather conditions. Spray areas should contain non-slip surfaces and different spray types can be used to create an attractive atmosphere such as sprinkles, hoses and nozzles [3].

(4) Fountains: Drinking fountains can be integrated as primary aesthetic features in a child-friendly space and provide drinkable water for all. The fountain should be located where it is convenient for most children and the appropriate height. Each fountain should be equipped to provide clean water in a sanitary manner [32].

(5) Water tables: Water tables make access to water possible for children in wheelchairs and create an appealing feature [3].

Table 1. Comparative summary of guidelines in creating a child-friendly space.

	Play value	Accessibility	Integration	Safety
Location & Size	Children must feel safe and want to play in the space.	Walking distance from surrounding neighborhood.	Enhance the local character of the environment.	Visual from all directions.
Safety	Improves health and well-being.	Promotes children's independent mobility.	Connected to accessible streets and roads.	Fences, barriers & lightning improve safety.
Entrances	Social meeting space.	Located near transportation.	Arriving & leaving points which ensure people interaction.	Visual with signage & direction boards.
Pathways	Different types of paths support different types of activity.	Slip proved & user friendly pathways. (disabilities included)	Pathways improve circulation.	Pathways located near drop-off zones & adequate lightning.
Signage	Can be designed as play objectives.	Appropriate heights, colors, pictures & tactical qualities for all types of users.	Supports an image of "All users are welcome".	Improves user's sense of security & provide information & safety tips of equipment.
Seating	Encourages interaction between children.	Seating options for disabled and non- disabled users.	Seating integrates different tasks such as observing, privacy and interaction.	Comfortable and appropriate height.
Fences	Provides a sense of security, enclosure & support for activities.	Define open spaces for entrances.	Supports integration between disabled and non-disabled users.	Important safety devices. Protects space & users against vandalism & crime.
Equipment	Support large muscle development, social interaction and fantasy play.	Accessible for children of all ages & disabilities.	Supports integration between disabled and non-disabled users.	Designed to protect children as far as possible.
Amenities & lightning	Provides safety and aesthetic atmosphere.	Access to toilet and coverage facilities.	Enhances beauty of environment and focus points in play space.	Safety during night time and ensure peace of mind.
Trees & Vegetation	Stimulates exploring &	Low branching & weeping trees	Trees create opportunities for	Appropriate trees must be

	Play value	Accessibility	Integration	Safety
	discovery behavior. Encourages fantasy & imaginative play. Tree climbing develops upper body strength.	are accessible for children in wheelchairs. Vegetation can be integrated into accessible settings for all.	interactive play between children. Vegetation is the most important element for integration as it can be shared and loved by all.	selected which support tree climbing activities. All harmful and poisonous vegetation must be eliminated.
Garden or vegetable setting	Improves social interaction, fine motor skills development & sensory stimulation.	Raised beds area accessible for all users.	Gardening is a group activity that improves integration and community involvement.	Gardens need to be raised or enclosed where direct impact is minimized.
Sand play setting	Excellent medium for creative play & social interaction.	Multi- level sand areas make it accessible for disabled & non-disabled users.	Supports & motivates interactive play.	Is used as safety fall surface material.
Water play setting	Multisensory character includes sounds & textures. Excites & relaxes children.	Multi- level sand areas make it accessible for disabled & non-disabled users.	Supports & motivates interactive play.	Shallow water and sprays are easy accessible and safer & circulation improves the water quality.

Source: Own Creation

4. Conclusion

Creating child-friendly spaces is a multidisciplinary and complex task. The golden rule in creating an efficient child-friendly space in rural areas is to design the space for the specific location through integrating the surrounding landscape and vegetation as the play settings and nature as possible play materials. The space should be aesthetic appealing and functional to satisfy the senses of the children. These issues should be incorporated in the spatial planning process. The main challenges, cross reference to Section 2.3, faced by rural areas in South Africa in creating child-friendly spaces were explored and was identified as a lack of opportunities, independent mobility, participation, obesity, crime and safety, consequently children in rural areas do not have adequate outdoor play space. The current planning approaches were identified and describe namely as, conventional paradigm which refers to manufacture play equipment, and conservational paradigm which refers to the integration of natural surroundings in the play equipment.

The conclusion was made that a conservational paradigm approach in spatial planning needs to be pursued to create an effective and qualitative well-designed child-friendly space in rural areas. The importance in creating such spaces where linked with physical, emotional and social benefits through the term "play". A child-friendly space should adhere to four main characterizations in order to be successful, namely 1) safety and comfort, 2) natural open space, 3) integration and sociality and 4) access, cross reference to Figure 1. Children can benefit from play opportunities especially children from stressful circumstances in rural areas. A play area is where children can escape from their fears, poverty and family and experienced freedom and emotional healing. Children with low-income backgrounds have less if any access to play spaces and are therefore at a disadvantage. Creating child-friendly spaces in rural areas will eliminate challenges and help develop healthy and educated children which can improve their quality of life and grant them the opportunity to become more successful in life. Children don't realize they need recreation and this leads to inappropriate behavior such as crime, vandalism, cults and drug abuse due to being bored and not using their free time healthy.

Green guidelines in creating child-friendly spaces were established and the guidelines were summarized into a guiding framework, Table 1, which can be used to evaluated future child-friendly spaces in rural areas. This study was the first step in developing guidelines on planning for and creating improved child-friendly spaces within open areas, especially in rural areas in South Africa

Acknowledgements

I would like to thank the God for giving me the knowledge and perseverance to complete this project and Dr. E. J. Cilliers, for her mentorship, support and guidance.

Nomenclature

CCYP- Commissioner for Children and Young People

References

[1] J. Harper, "Planning for Recreation and Parks Facilities: Predesign Process, Principles, and Strategies," Pennsylvania: Venture, pp. 2-180, 2009.

[2] B. Clouston and K. Stansfield, "Trees in Towns: Maintenance and management," London: Architectural Press, pp.8-133, 1981.

[3] C.C. Marcus and C. Francis, "People Places: Design Guidelines for Urban Open Spaces," 2nd ed., New York: Wiley, pp.5-293, 1998.

[4] R. Moore, S. Goltsman, and D. Iacofano, "Play for all: Guidelines," Berkeley: MIG Communications, pp.6-155, 1987.

[5] A. Shackell, "Design for Play: A guide to creating successful play spaces," England, pp.1-130, 2008.

[6] J. Zomervrucht, "Gradually grow to cycle: Experiences with a child-friendly public space," Netherlands, pp.8-17, 2005.

[7] CCYP, "Building spaces and places for children and young people," Australia, pp.4-19, 2011.

[8] G. Woolcock, B. Gleeson, and B. Randolph, "Urban research and child-friendly cities: A new Australian outline," 2nd ed., pp.177-192, 2010.

[9] G. Thomas and G. Thompson, "A child's place: Why environment matters to children," Green Alliance / Demos report, pp.3-23, 2004.

[10] L. Horelli, "Constructing a theoretical framework for environmental child-friendliness. Children, Youth and Environments," 4th ed., 2007.

[11] S. Schulze and Moneti, "The child friendly cities initiative. Proceedings of the Institution of Civil Engineers," 2nd ed., pp.77-81, 2007.

[12] C. McAllister, "Child Friendly Cities and Land Use Planning: Implications for children's health," 3rd ed., pp.47-56, 2008.

[13] M. Nordström, "Child-friendly cities – sustainable cities," Sweden, pp.44-45, 2004.

[14] A. Carver, A.F. Timperio, and D.A. Crawford, "Young and free? A study of independent mobility among urban and rural dwelling Australian children," 2012.

[15] M. Huby and J. Bradshaw, "A Review of the environmental dimension of children and young people's well-being," Heslington, pp.4-39, 2008.

[16] A. Abolahrar, "Freedom Play Sticks for Play Spaces," Gothenburg, pp.2-24, 2011.

[17] S. Munoz, "Children in the Outdoors," Scotland: SDRC, pp.5-24, 2009.

[18] PLAYLINK, "Best Play: What play provision should do for children," Scotland: Sutcliffe Play, pp.1-9, 2000.

[19] DTLR, "Improving Urban Parks, Play Areas and Open Spaces," London, pp.8-189, 2002.

[20] A. Parsons, "Young Children and Nature: Outdoor Play and Development, Experiences Fostering Environmental Consciousness, And the Implications on Playground Design," Virginia, pp.4-74, 2011.

[21] J. Hewes, "Let the children play: Nature's answer to early learning," Canada, pp.2-6, 2006.

[22] S. Campbell, "Green Cities, Growing Cities, Just Cities? Urban Planning and the Contradictions of Sustainable Development," Journal of the American Planning Association, pp.1-7, 1996.

[23] L.B. Sohn, "Stockholm Declaration on the Human Environment, 1973."

[24] K. Nsiah-Gyabaah, "Urbanisation Processes- Environmental and Health effects in Africa," pp.1-2, 2004.

[25] P.M. Sherer, "The Benefits of Parks: Why America Needs More City Parks and Open Space," San Francisco: The Trust for Public land, pp.6-22, 2003.

[26] R. White and V. Stoecklin, "Children's Outdoor Play & Learning Environments: Returning to Nature," White Hutchinson Leisure & Learning Group: Kansas, pp.1-8, 1998.

[27] M. Jansson, "Management and Use of Public Outdoor Playgrounds,"Alnarp: SLU Repro, pp.9-78, 2009

[28] A.B. Rutledge, "Anatomy of a Park," NewYork:McGraw-Hill, pp.1-146, 1971

[29] Department of Housing, "Guidelines for Human Settlement Planning and Design," 1 ste ed., vol. 1. Pretoria: CSIR Building and Construction Technology, pp.13-278, 2000.

[30] DEMOS, "A child's place: why environment matters to children," London, pp.3-21, 2004.

[31] M.L. Christiansen, "Park Planning handbook," New York: Wiley, pp. 1-219, 1977.

[32] KOMPAN, "Play for All: The newest developments in universal design, accessibility and inclusion in playgrounds," pp.2-11, 2007.

[33] S.Herrington, "CHILD: Seven Cs an Informational guide to young children's outdoor play spaces," 2nd ed., pp.1-59, 2006.

25

Planning for Sustainable Communities: Evaluating Place-Making Approaches

Sanmarie Schlebusch

Unit for Environmental Sciences and Management, North-West University, Potchefstroom Campus, South Africa

Email address:
24schlebusch@gmail.com

Abstract: People's survival and their quality of life, are irrefutable dependent on the natural environment in which they reside. To ensure that people's quality of life be maintained in any specific area, it is therefore imperative to find a sustainable equilibrium between the social and economic needs of people and the capacity of the natural resources in their environment. The composition of communities is, however, complex and diverse. The multiplicity of culture, gender and age in any particular community, give rise to needs that is unique regarding to that community as well as the impact thereof on the natural environment. For this reason, it will require an ingenious planning approach be followed, whereby the unique needs of people in an specific area as well as the protection of the natural environment simultaneously be addressed. Characteristics of a sustainable community are typically, a healthier, safer, greener, economically independent community which is well managed. Furthermore it has lower transportation costs and less traffic, is more economic in terms of housing and market demands, shows decreased costs of infrastructure and also has low level of air pollution. Place-making, through layout and design, is an integrative planning approach in creating sustainable communities. Place-making is fundamentally a strategy aimed at creating one or more places in an area which, serve as focal points for economic and social activities of people in the community. Such places will contribute to the quality of life in a community and will also encourage more people to visit the area.

Keywords: Sustainability, Sustainable Development, Sustainable Communities, Place-Making, Liveability

1. Introduction

Sustainability is essential in the process of community planning and plays an important role in the long-term success of communities. Planning for sustainable communities is primarily based on addressing the needs of the people in the community and ensuring a better quality of life. Public participation plays a critical part throughout the process of planning for sustainable communities and in this sense, a sustainable community is created through balancing the environmental, social and economic activities within the community.

Place-making is an integrative approach to the planning and sustainable development of communities. People are attracted to good places with high quality of life, which consist of effective place-making principles that are implemented through layout and design. Good places are a focal point of economic and social activity, thus place-making approaches can contribute to planning and creating attractive, focal points by including various functions within one space.

According to Giradet a sustainable city [in this instance a community] is planned to enable all of its citizens to meet their own needs and to enhance their wellbeing without damaging the natural world or endangering the living conditions of other people, now or in the future. Planning for sustainable communities is challenging and thus it is essential to compile a framework wherein detailed practical guidelines for implementation of sustainable solutions are described. [1]

Power states that the heart of sustainable development encompasses the simple idea of ensuring a better quality of life for everyone, now and for future generations. It implies meeting the following four objectives simultaneously:

- Social progress which recognizes the needs of everyone;
- Effective protection of the environment;
- Prudent use of natural resources;
- Maintenance of high and stable levels of economic growth and employment; and considering the long-term implications of decisions. [2]

There is a need for an integrated place-making approach

that will contribute to the planning for sustainability communities of rural and urban areas

2. Concept of Sustainable Communities

2.1. The Language of Sustainability

Einstein said: "We shall require a substantially new manner of thinking if mankind is to survive." [3]

Jaber, states: "…to create a world that works for all, we need to change the language we use to frame our mindset [4]. Language has real power. It communicates the concepts that shape thought, and, as such, we need to be vigilant about the terms we use". The term "sustainability" in the context of, to provide for, means "the ability to continue into the indefinite future by respecting the Earth's ecosystems, its limits, and providing space for the other beings on the planet to exist [4].

Sustainability is founded on the fundamental principle: "Everything that we need for our survival and wellbeing depends, either directly or indirectly, on our natural environment" [5]. Sustainability creates and maintains the conditions under which humans and nature can exist in productive harmony that permits fulfilling the social, economic and other requirements of present and future generations" [5].

According to Geis & Kutzmark sustainability is primarily determined by human traditions and practices. The latter, however, are influenced by external factors and may therefore, often change or completely disappear due to the variability of these factors [3]. Geis & Kutzmark asserts that, limited resources, urbanization, scientific knowledge, technology, social awareness, health and safety imperatives and new economics, are in this regard the main external factors of the twenty-first century. [3]

According to Filho there are many misconceptions regarding the concept of sustainability [6]. Filho is of the opinion that individuals and/or communities opposing sustainability or sustainable development often do not fully comprehend the all-inclusive value and significance of sustainability. These misconceptions usually have a negative impact on the community or society and affect their efforts to work towards a more sustainable future [6]. Perceptions that Filho asserts have a negative influence on society's attitude towards sustainability include:

Sustainability is not a subject per se and it is too theoretical. Notwithstanding sustainability being a high priority in virtually all scientific fields, many continue to view the concept as being vague, without scientific base and expensive to implement. The result is that many see the concept as indistinct and theoretical. Sustainability is too broad and people and institutions are intimidated and discouraged by the scope of the concept, believe the implementation of sustainability difficult to manage and it is too recent a field. Poor knowledge of the significant value of sustainability leads to the unfounded criticism that sustainability merely represents a fashionable trend that only the minority can afford [6].

To eradicate these misconceptions and reservations, Filho suggests that an aggressive informative effort which educates the community extensively on the importance and long-term advantages versus short-term economic sacrifices, simultaneously supported by practical pilot projects and initiatives that illustrate the feasibility of sustainability, will result in individual and collective resolve to pursue sustainable objectives and solutions [6].

2.2. Sustainable Development

Berke asserts that the history of the process that was followed in the development as applicable to urban planning of towns, cities and regions has been dominated by the physical design model and the rational planning model, both distinctively representative of a top down approach [7]. This approach permitted government and other major role players the opportunity to manipulate the planning and development process, thereby promoting subjective political and economic objectives and at times overlooking the aspirations and needs of people in specific communities. Since 1960, denunciation of the aforementioned development models mounted as critics progressively exposed the fundamental weaknesses in these models and displeased citizens [7].

The necessity to devise an alternative approach embracing a common goal that would serve the interests of all the people and at the same time protect the environment became noticeably essential [7].

The World Commission on Environment and Development (WCED) of the United Nations was commissioned to conceive a philosophy that will be instrumental to reverse environmental degradation, reduce over-consumption and grind poverty. In their report, Our Common Future, that was published in 1987, portraying the common goal as equity to future generations, the WCED defined the hypothesis of sustainable development as follows: "Sustainable development is development that meets the needs of the present generation without compromising the ability of future generations to meet their own needs" [7].

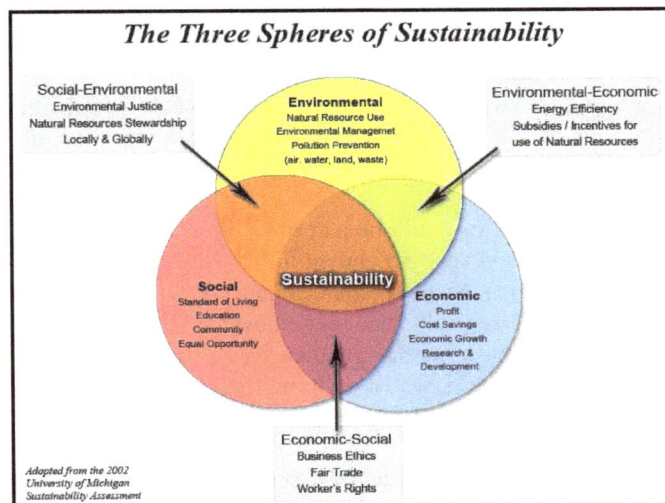

Figure 1. The Three Spheres of Sustainability [8]

An illustration that examines the three primary values of sustainable development is presented in Figure 1. The expanses where the circles transcend display the core characteristics of sustainable development.

Grasping the aim of the WCED's definition of sustainable development is undemanding, however, translating the concept into procedure shows a diverse interpreting

methodology, which is clearly evident in the seven sample definitions of sustainable development below.

Table 1 describes the definitions as provided by The World Commission on Environment and Development (WCED) to capture the essence of sustainable development on a variety of levels.

Table 1. Five examples of definitions for sustainable development [7]

Level	Definition
International	"Sustainable development respects and defines traditional livelihoods and indigenous culture and societies. It recognizes that communities must define and develop their own solutions to environmental and development problems. It also works toward shared power and participation, at the local, national, and international levels" (Canadian University Students Organization, 1989: 3).
National	"Our vision is of a life-sustaining Earth. We are committed to the achievement of a dignified, peaceful, and equitable existence. A sustainable United States will have a growing economy that provides equitable opportunities for satisfying livelihoods and a safe, healthy, high quality of life for current and future generations. Our nation will protect its environment, its natural resource base, and the functions and viability of natural systems on which all life depends" (President's Council on Sustainable Development, 1996: i).
State	"Sustainable development links the environment, economy and social equity into practices that benefit present and future generations" (North Carolina Environmental Resource Program, 1997: 1).
Regional	Sustainable development involves "achieving positive change that enhances the ecological, economic, and social systems upon which South Florida and its communities depend. Once implemented these strategies will bolster the regional economy, promote quality communities, secure healthy South Florida ecosystems, and assure today's progress is not achieved at tomorrow's expense" (Governor's Commission for a Sustainable South Florida, 1996).
Local	Sustainability is "long-term cultural, economic, and environmental health and vitality" (Seattle Planning Department, 1994). "As a community, we need to create the basis for a more sustainable way of life both locally and globally through the safeguarding and enhancing of our resources and by preventing harm to the natural environment and human health" (Santa Monica Planning Department, 1995: 1). Sustainable development is "the ability of [the] community to utilize its natural, human and technological resources to ensure that all members of present and future generations can attain high degrees of health and wellbeing, economic security, and a say in shaping their future while maintaining the integrity of the ecological systems on which all life and production depends" (Cambridge Planning Board, 1993: 43).

Deduced from the multiplicity of the above definitions, it is reasonable to argue that the people and the prevailing circumstances in a specific community, town, region, state or country will determine in what manner the definition of sustainable development should be paraphrased.

The Sustainable Development Commission an independent advisor to the United Kingdom Government on sustainable development, founded six core principles that, from their perspective, determine what sustainable development is and ought to be [2].

Putting sustainable development at the centre Sustainable development must be the organising principle of all democratic societies, underpinning all other goals, policies and processes.

We are and always will be part of nature, embedded in the natural world and very dependent for our own economic and social well-being on the resources and systems that sustain life on earth. Sustainable economic development means 'fair shares for all', ensuring that people's basic needs are properly met across the world, whilst securing constant improvements in the quality of peoples' lives through efficient, inclusive economies [2].

There is no one blueprint for delivering sustainable development. It requires different strategies in different societies. However, all strategies will depend on effective,

participative systems of governance and institutions, engaging the interest, creativity and energy of all citizens [2].

Adopting a precautionary approach Scientists, innovators and wealth creators have a crucial part to play in creating genuinely sustainable economic progress. However, human ingenuity and technological power is now so great that we are capable of causing serious damage to the environment or to peoples' health through unsustainable development that pays insufficient regard to wider impacts [2].

Although strategies for the sustainable development of any one community may differ due to different circumstances in the community, strategies should primarily be founded on the principles for sustainable development.

2.3. Sustainable Communities

"The sustainable community is a model, an ideal set of goals to work toward. But it also is a philosophy for envisioning those goals and a practical problem-solving process for achieving them" [3]. "A sustainable community seeks to maintain and improve the economic, environmental and social characteristics of an area so its members can continue to lead healthy, productive, enjoyable lives"[5]. The real challenge of creating a sustainable community lies in the process of harmonizing the expectations and needs of the community with the values of sustainability. A sustainable community is a community that is economically, environmentally, and socially healthy and strong [10]

To accomplish the process successfully, the fundamental values of sustainability have to form the nucleus of the development and planning approach. [9] A sustainable community underwrites objectives that reflect respect for both the natural environment and human nature [3] A sustainable community should essentially strive to achieve the following characteristics and goals:

Place a high value on quality of life. A sustainable community accepts that communities are first and foremost for people and that the primary objective of the planning and development process is to improve the quality of life of its residents socially, economically, psychologically, and spiritually. It implements policies to achieve quality of life and does so in a fair, open, and democratic manner. [3]

Respect the natural environment. A sustainable community recognizes its relationship to nature and sees nature's systems and components as essential to its wellbeing. It provides access to nature through metropolitan parks, open-space zones, and urban gardens. It understands the sensitive interface between the natural and built environment, develops in a way that will support and complement – not interfere with – nature, and avoids ecological disasters. [3]

Infuse technology with purpose. A sustainable community uses appropriate technology, while ensuring that technology in the built environment is a means to an end, rather than an end unto itself. It emphasizes learning and understanding how existing and new technology can serve and improve

communities, not vice versa. It sets clear and measurable goals for what it wants technology to achieve. [3]

Optimize key resources. A sustainable community takes an inventory of its human, natural, and economic resources and understands their finite quality. It ensures that forests are not overused, people are not underemployed, and the places of the built environment are not stagnant and empty. It reduces waste and reuses resources; it creates conditions in which all these resources can be used to their fullest and best potential, without harming or diminishing them. [3]

Maintain scale and capacity. A sustainable community recognizes the importance of scale and capacity with regard to the natural and human environment. It ensures that the environment is not overdeveloped, overbuilt, overused, or overpopulated. It recognizes the signs of tension that indicate when the environment is overstressed and can adjust its demands on the environment to avoid pollution, natural disaster, and social disintegration [3].

"[A] sustainable community reflects the interdependence of economic, environmental, and social issues by growing and prospering without diminishing the land, water, air, natural and cultural resources on which communities depend. Housing, transportation and resource conservation are managed in ways that protect economic, ecological and scenic values" [11].

The Institute for Sustainable Communities views the concept of a sustainable community as a framework to guide action; the following table offers some examples from their experience:

Table 2. The concept of a sustainable community as a framework to guide action [10]

Example:	Explanation
A Healthy Climate and Environment	Protection and enhancement of local and regional ecosystems and biological diversity. Conservation of water, land, energy, and non-renewable resources. Utilization of prevention strategies and appropriate technology to minimize pollution. Use of renewable resources no faster than their rate of renewal. Infrastructure that improves access to services and markets without damaging the environment.
Social Wellbeing	Satisfaction of basic human needs for clean air and water and locally sourced nutritious, uncontaminated food. Affordable provision of quality health prevention, care, and treatment services for all community members. Safe and healthy housing accessible to all. Equitable access to quality education services, formal and informal. The basic human rights of all community members are respected and defended against injustices including exploitation and psychological and physical harm. Protection, enhancement and appreciation of community manifestations of cultural diversity, treasures, customs, and traditions.
Economic Security	Community members equitably benefit from a strong and healthy community-centred economy. Diverse and financially viable economic base. Reinvestment of resources in the local economy. Maximisation of local ownership of businesses. Meaningful employment opportunities for all citizens. Responsive and accessible job training and education programs that enable the workforce to adjust to future needs. Businesses that enhance community sustainability.

2.4. Planning for Sustainable Communities

"The kind of change required by sustainability implicates each community, each household, and each individual. Successful solutions to problems at this level of society will need to be rooted in the cultural specificity of the town or

region if the people are to be supportive of and involved in such change" [12].

Following an eighteen-month investigation, during which contributions of over seventy national, regional and local organizations were scrutinized, the Sustainable Development Commission (SDC) concluded that fundamentally there are three aims which, should dominate the development or regeneration approach regarding a sustainable community. These aims are, a healthy environment, a prosperous economy and the social wellbeing of the inhabitants [2].

"Everyone has the right to an environment that is not harmful to their health or wellbeing; and to have the environment protected for the benefit of present and future generations through reasonable legislative and other measures that prevent pollution and ecological degradation as well as promote conservation and secure ecologically sustainable development and use of natural resources while promoting justifiable economic and social development" [13].

2.5. Conclusion

The concept of sustainability is more than only a theory. Fundamentally it is a long-term practical solution through which the quality of life of people is prolonged, improved and protected. This outcome can however only be realized, when the basic values, principles and objectives which are intrinsic to the concept, are entrenched in a clearly defined policy which is applicable to all facets of life and that are supported by the community and individuals.

Although the social composition and environmental characteristics, of any one community differs from the other, various place-making approaches had been devised, had been implemented, and could be integrated as building blocks for the planning sustainable communities. Place-making can be managed as it is an approach through which "The Three Spheres of Sustainability" are used to create a better place for people.

3. Place-Making Approaches

3.1. Understanding the Concept of Place-Making

In his post-World War II speech, Winston Churchill considering the reconstruction of neighbourhoods, communities and buildings, said, "[w]e shape our buildings and then they shape us." [3].

There are many descriptions of the concept of place-making, such as "both an overarching idea and a hands-on tool for improving a neighbourhood, city or region" Project for Public Spaces, [14] or, according to Placemaking Chicago [15] "the art of creating public 'places of the soul', that uplift and help us connect to each other." It is therefore evident that the concept of place-making cannot be encapsulated by one

specific definition, but should rather be understood as a wide range of community strategies and initiatives aimed at the improvement of the community's environment and their quality of [16].

"An effective Placemaking process capitalizes on a local community's assets, inspiration, and potential, ultimately creating good public spaces that promote people's health, happiness, and wellbeing". Thus, place-making is a continuous process, which encapsulate peoples' ideas and through which their needs in terms of the liveability and quality of life are fulfilled by using effective planning, layout and design or redesign of their environment [14].

The Project for Public Spaces asserts that perspectives that were presented by futurists Jacobs and Whyte [17] , were the inspiration which eventually gave way to place-making concepts. In her treatise, The Death and Life of Great American Cities, Jacobs proposes ideas, which irrevocably altered planners and activists' approach regarding urban planning. Jacobs) underlines five important perspectives regarding place-making, namely: Cities as Ecosystems, where cities should be viewed as living beings and ecosystems wherein the dynamics of streets, buildings and functions can change in response to human use patterns and related interactions; Mixed Use Development, a diversity of buildings in a city that are used at different times of the day by different genders and age groups to ensure liveliness in the city; Bottom-Up Community Planning – planning for the development of the community, is guided by the community itself and not by other external agendas; The Case for Higher Density, where a high concentration of people is imperative for city life, however, the difference between overcrowding and high density lies in the critical mass of people that are needed to stimulate the community's vitality; and Local Economies, in which case, a city's economy is not dependent on large corporation business, the growth of a city's economy is stimulated by more innovative small business entrepreneurs [17].

A rudimentary objective of the place-making approach is to discover the needs and ambitions of the local community by observing, listening and communicating with the community and subsequently drawing on this knowledge to devise and implement a strategy that effectively fulfils these needs [15] Project for Public formulated eleven (11) principles to direct a strategy towards efficient place-making. These principles are portrayed in Table 3.

Table 3. Key principles of place-making [15]

Principles	Explanation
The community is the expert	The community members themselves best communicate the community's needs.
Creating a place and not just a design	The place-making concept should be fundamental to the layout and design approach. The layout and design are only the tools.
You can't do it alone	Identify partners who can contribute in terms of management and innovative ideas and can provide political and financial support.
They will always say, "it can't be done"	"We've never done things that way before." Identify and engage people in the community that share the same vision. Use the positives and if possible elderly people to help influence the rest of the community.
You can see a lot just by observing	Observing a space enables you to absorb and understand how that specific space is used on a daily basis.

Principles	Explanation
Develop a vision	The people that use the space effectively should define the vision and character of a space.
Form supports function	Existing trends and habits of a specific area should guide the place-making process
Triangulate	Identifying elements that are situated next to each other to use in a way that promotes activity.
Start with the petunias	Render small changes and implement progressively.
Money is not the issue	Combining the location and the level of activity of the public space, with the involvement and willingness of the partners and local community members can elicit resources from those involved to improve these spaces.
You are never finished	Management is the key, because uses of places change constantly and effective responses thereto can only be achieved through good management.

Thus, place-making as an approach is on-going and driven by the community for the community, facilitated by planners and experts, and takes form in practice through a well-managed and effective layout and design approach, which will transform the community and their environment progressively into a place with good living conditions.

3.2. Criticism on the Place-Making Approach

There are critics who doubt the conclusive role of place-making in layout and design approaches for sustainable communities. Grant declares: "We can predict planners to continue to look for the one big theory that can explain all, predict all, and offer guidance for practice to create good communities. We can also safely predict that we are not likely to find such a model" [18].

Critics claims that place-making does not contribute to the development of local economies in previously disadvantaged communities. They assert that place-making only accelerates the gentrification of an area and thereby succeeds in reducing the pressure exerted by the local community and the general public in this regard. If gentrification is primarily project-driven, development-driven, design-driven or artist-led, this criticism is justified and transformation will only be superficial and limit in terms of the long-term outcome. However, in most instances, this criticism is largely due to ignorance regarding the value and objectives of place-making and confusion as to whom the stakeholders and beneficiaries of the process are [19].

3.3. Concepts Intrinsic to Place-Making

From the above definition and description of place-making, it is evident that two concepts are an inextricable part of the place-making approach, namely (1) liveability and (2) lively public spaces, which will be described accordingly.

3.3.1. Liveability

The theory of liveability maintains that a person's subjective appreciation of life primarily depends on the objective quality of life. In other words, the better the living conditions in an area or community, the more contented the people living in the area or community will be (Veenhoven & Ehrhardt, 1995). In turn, the comparison-theory advocates that people in a specific place will be contented if their living conditions are good, irrespective of the knowledge that people living in a different place may experience even better living conditions [20]. People have widespread needs; liveability is the collective arrangement to fulfil these needs. To regard a place as liveable, the collective requirements and demands have to comply with the needs and capacities of individuals. Hence, citizen-centred initiatives should be the principal angle of incidence in conceiving an approach intending to make a place more liveable [20]

Cilliers et al. states that "liveability reflects the wellbeing of a community and comprises the many characteristics that make a location a place where people want to live now and in the future, such as: employment and incomes, community strength, environment, amenity and place, planning, participation, and infrastructure. Economic and community strength are critical to liveability." [21].

Pacione asserts that the relation of people to their everyday environment or living space determines the living conditions in the area and that the prevailing living conditions are a measure of the liveability of the area [22]. Pacione delineates the two fundamental measures of liveability as the cost of living and the quality of life [22].

Economic, social and environmental factors are used when either liveability or quality of life is measured, however, the purpose and the results are different. When the liveability of a place is measured, the objective would be to gauge the liveability characteristics as well as the quality and incidence of services and facilities of a place in terms of these factors. Conversely, when the quality of life is measured, the focus would be to gauge the liveability characteristics and the wellbeing of the inhabitants of a place in terms of these factors. As opposed to quality of life that is primarily being dictated by the subjective experience of people, the liveability of a specific area can be manipulated and influenced through devised policies and layout and design [23].

Although indices of liveability and quality of life is derive from a weighted list of mostly locational characteristics that supposedly contribute to liveability, they are currently used as a benchmarking tool in the evaluation of towns and cities in terms of liveability and quality of life [23].

Table 4: encapsulates fundamental liveability indices currently employed to measure liveability in a city.

Table 4. Summary of the core liveability indices [21]

Indices	Measurement
Economist Intelligence Unit Ranks 127 cities on liveability as part of the Worldwide Cost of Living Survey, based on five weighted categories (VCEC, 2011:6):	1. Stability (25%) – crime and conflict 2. Healthcare (20%) – availability, quality 3. Culture and environment (25%) – climate, recreation, services 4. Education (10%) – availability, quality 5. Infrastructure (20%) – transport, links, housing, utilities, services
Mercer human resource survey The quality of living study has 39 factors that are grouped into 9 key categories (VCEC, 2011:6):	1. Political and social environment 2. Economic environment 3. Socio-cultural environment 4. Health and sanitation 5. Schools and education 6. Public services and transportation 7. Recreation, natural environment 8. Consumer goods 9. Housing
Anholt city brand index Assesses how people perceive the images of cities, using a survey of nearly 20,000 consumers in 18-20 countries. Cities are evaluated in terms of:	1. Presence (city's international status and standing) 2. Place (beauty, climate and other physical attributes) 3. Potential (economic and educational opportunities) 4. Pulse (urban appeal and lifestyle) 5. People (friendliness, openness, cultural diversification and safety) 6. Prerequisites (basic facilities: hotels, schools, transport, sports)
EU Urban Audit Benchmarking of quality of life in 58 European cities. Represents the most comprehensive attempt to assess the liveability and competitiveness of cities and regions (VCEC, 2011:7). The core issues include:	1. Population, nationality, household structure 2. Labour market, employment, income disparities, poverty 3. Housing 4. Health 5. Crime 6. Economic activity, civic involvement 7. Education and training, level of educational qualifications 8. Air quality, noise, water, waste management 9. Land use, travel patterns, energy use 10. Climate and geography, culture, recreation
Global competitiveness index Developed in 2004, measures national competitiveness in using a weighted average of factors that contribute to countries competitiveness. The factors are grouped into twelve categories (VCEC, 2011:13):	1. Institutions 2. Infrastructure 3. Macro economy 4. Health and primary education 5. Further education and training 6. Goods market efficiency 7. Labour market efficiency 8. Financial market sophistication 9. Technological readiness 10. Market size 11. Business sophistication 12. Innovation
Creativity index Indicator for 'overall standing in creative economy, economic potential' (Florida 2002), based on four factors:	1. Creative share of workforce (proportion in creative occupations) 2. High tech industries 3. Innovation (measured as the number of patents per capita) 4. Diversity (measured by the number of gay people per capita)

Although the combination liveability measures encompassed in the different surveys varies, common factors such as access to infrastructure and services, social equity and cohesion and climatic conditions are included. Notwithstanding fact that the weighting given to each factor is subjective and therefore differs according to the survey [23], it suffices to find the most common used issues of liveability.

It is important to understand that there is a definitive difference between liveability and liveliness.

While liveability is defined in terms of the quality and incidence of services in a place, the liveliness of a place is measured in terms of the frequency and way in which the community takes advantage of the services and facilities available in one place; "… liveliness is entirely associated with people and activities and it can be assessed by measuring pedestrian flows and movements, the uptake of facilities and the existence or otherwise of 'things to do'" [16]. The following section will elaborate on lively public spaces.

3.3.2. Lively Public Spaces

The people living in a specific place are the so-called "public", therefore, the focus throughout the process of creating lively public places should primarily be to ensure that the public grounds are accessible and open for a wide range of user groups [24].

The place-making approach is rooted in the principle that a successful public space is also a lively place with distinctive functions that attract a wide variety people. In these lively public spaces, the many functions and activities about community life that take place induce a feeling of ownership and connectedness that therefore influence people to stay or return to the place. Lively places can be regarded as spaces with a function [21].

A space signifies the physical and geometrical characteristics of an environment, which, when occupied by people and enhanced by lively elements, are transformed into a place [25].

Great public places have four main key attributes: access and linkages, comfort and image, uses and activities, and sociability. These are evaluated in terms of specific factors within each key attribute that are needed for the space to be regarded as a successful public place [21].

An ordinary place can be transformed into a lively place by augmenting certain key attributes in intangible measurements.

To effectively accomplish the transformation of a space into a lively public place, the public place should be made highly attractive [26] which can be done by including various initiatives such as residential development, educational institutions, open spaces and other facilities [24]. Paul Bevan notes that living, working and playing are ideally much closer together than often found and that, when an area is unlivable, it may be owing to the loss of this proximity [27].

Cilliers et al. assert that norms by which places are evaluated are wide-ranging and common factors that are representative of successful public spaces are not limited to the physical dimensions of a place [21].

Historically, public spaces were places with streets, marketplaces, boulevards, gardens, squares, courtyards, etc., where residents spent a great deal of their time (Loudier & Dubois, 2001). Present-day traditional planning schemes that are implemented have proved to be somewhat unsuitable to new lifestyles; public places are mostly rather dysfunctional and dehumanized places lacking quality and proper use, and the absence of on-site managers contributes to ineffective public spaces [28].

To create lively public spaces, efforts in the area should focus to render services and opportunities that are versatile, accessible and attractive to a wider range of user-groups and that encourage them to stay. To accomplish this, initiatives such as more residential development, more education institutions in the city centre and attractive facilities and open spaces can be developed [24]. Initiatives such as public transport and roads, arts, entertainment and sporting, social and cultural events may be added [27]. To achieve versatility in an area, alternative uses of the city space should be encouraged [24].

The current physical structure of cities provides for public life, but further opportunities should be developed to strengthen a range of activities within one space in order to create lively city spaces with many benefits.

Place-making is therefore a socially constructed process that shapes spaces by including different functions, aspects and even capital investment, designed to generate economic growth and promote cultural tourism in order to create a place worth living and working in [29].

3.4. Place-Making Approaches in Urban Planning Context

"You have to turn everything upside down to get it right side up" [30]. The aforementioned articulates, in essence, the basic principle underlying the bottom-up approach and community-scale of planning.

Place-making was introduced in the urban planning sphere to address community-scale planning with the objective to create qualitative, liveable environments that adhere to the principles of sustainability and progress in the transformation of areas from merely being places that people occupy, so-called ordinary places, into lively places that are good places to live in [16].

The place-making approach can be employed, within the urban planning context, to realize liveability by planning for, and implementing various functions within one space. This can entail the transformation of areas from solely being places that people occupy, into vibrant lively places, by focusing on current public spaces that have potential, and developing these spaces according to the place-making objectives in order to create places with function in which people can socialize and interact [21].

Five place-making approaches include: the livelihoods approach; Power of 10 approach: community participation approach: New Urbanism and Green planning approach. These approaches will be discussed accordingly, as these approaches enable place-making within the rural planning context.

3.4.1. Livelihoods Approach

Understanding the diverse needs and activities of people, ingrained in the different ways that different people live in different places, is known as the so-called livelihoods approach [31].

Livelihood can be explained as a prevailing condition that involves capabilities, material and social assets, and activities as a means of living. For this livelihood to become sustainable, it needs to cope with and recover from stresses and shocks; maintain or enhance its capabilities and assets while focusing not to undermine the natural resource base [32].

The sustainable livelihood approach is indicated as a framework that provides an understanding of survival strategies in poor communities and can help to put pro-poor tourism in a better position and guide it towards successful implementation. [33].

The collective term "livelihoods" is considered flexible because of its possible attachment to a number of other phrases such as "…locales (rural or urban livelihoods), occupations (farming pastoral or fishing livelihoods), social difference, (gendered, age-defined livelihoods), directions (livelihood pathways, trajectories), dynamic patterns (sustainable or silent livelihoods) and many more". [32]. Therefore, it is a widely applicable term and is especially relevant when planning for rural areas. The perspectives of livelihoods have been central to rural development thinking and practice in the past decade, and the perspective is rooted in the different ways that different people live in different places [32].

Within the livelihoods approach, the focus is on "diversity". Fundamental, single-sector approaches to livelihoods and liveability, like that of the comparison [20], have been challenged by this approach in order to address complex rural development problems in a more hands-on and adequate manner [32]. This approach is a simple and straightforward one, as it purely focuses on understanding things (needs, activities, people, etc.) from a local perspective [32]. In order to implement or promote the liveability theory and expand livelihoods accordingly, different aspects should be addressed, including knowledge, politics and scale and dynamics. Portrayed in Table 5 below explains the four above-mentioned perspectives.

Table 5. Perspectives to address

Knowledge	Livelihoods can be expanded by focusing on inclusive debates about livelihood frameworks and proposed directions of change, rather than relying on a bland listing of principles or by keeping questions of values and politics away.
Politics	Within these communities, a need for municipal and government services were identified. These needs include an explicit, theoretically based concern and knowledge of how class, gender and capitalist relations operate. They need to be given the opportunity and right to actively participate in politics and political discussions by being allowed to ask up front questions regarding gains and losses based on theories of power and the political economy.
Scale	Scale is an important element to take into consideration when expanding a community's livelihood. Therefore, a livelihood analysis needs to be developed and implemented. This analysis will examine networks, linkages, connections, flows and chains across different scales yet will remain in its specific place and context – i.e. rural communities.
Dynamics	The improvement of livelihoods in terms of dynamics requires local people, policymakers, outsiders, etc., to think about long-term change. This shift in mind-set can be ensured by providing future strategies and pathways for development and growth.

3.4.2. Power of 10 Approach

The Power of 10 place-making approach endorses the concept that an authentic, lively city has at least 10 great public places throughout the city that attract a wide range of user-groups. In these great public places, people are offered many mixed-use opportunities to take pleasure in public life. "And, it's not enough to have one liveable city or town in a region; you need a collection of interesting communities" [15].

A great place offers people opportunities of at least ten (10) things to do or ten (10) reasons to visit the place. For example, a place to sit, art to touch, music to hear, food to purchase, historic information to learn about, and books to read [21]. The opportunities, however, should give expression to the people's experience of the city [15]. "The concept also provides people something tangible to strive for and helps them visualize what it takes to make their community great" [34].

The concept of mixed use and multiple functions in these ten (10) great places should also be dynamic enough to stimulate continuous development and inspire people to come back to the place [15].

Cowan et al. are of the opinion that this type of public place will create lively neighborhoods where interaction arises between people, social gatherings are held and where people simply enjoy spending time [35]. An example of this approach is found in the Canadian city, Toronto, where the focus is placed on combining the rich cultural heritage with creativity. These activities include the Toronto International Film Festival (the largest and arguably the most influential festival in the world); Ontario College of Art and Design; The Young Centre; Wychwood Car Barns and numerous other similar examples [36].

3.4.3. Community Participation Approach

"When citizens are effectively engaged in the design process, designers and planners can be at their most effective in facilitating a process that synthesises local experience and wisdom with design principles and technical expertise. Designers can help people uncover their common interests and work towards practical and creative solutions that build local character and assets" [37].

Irrespective of the environmental attributes of an area, the community should be the primary source of information when planning and designing a specific place. Community participation can be seen as an approach to lively planning, or as an indispensable element needed to create a lively place.

However, the composition and dynamics of communities, especially in the urban environment, have become increasingly complex. Cultural diversity, in particular, offers an enormous challenge to public participation; the more diverse the group, the more needs that need to be taken into consideration and therefore the more complex the participation process and input will be [38].

Even though it is difficult to implement, participation remains a critical part of planning for sustainable communities and public places, and the participation of all residents along with supervision, reviews and awareness are important for effective place-making [28]. This qualitative participation approach is needed to address and successfully implement a bottom-up approach, as well as to ensure the planning of functional and usable spaces that can be regarded as lively. To create this type of situation, where active participation is present, it is crucial for the community to play a bigger role in deliberations with authorities, policy formalization and the devising of solutions [21].

3.4.4. New Urbanism Approach

The planning concept of New Urbanism has been known for some time, however, the implementation thereof only progressively increased since US Congress adopted The Charter of New Urbanism in 1993 that reads as follows: "We advocate the restructuring of public policy and development practices to support the following principals: Neighborhoods should be diverse in use and population. Communities should be designed for the pedestrian and transit, as well as the car. Cities and towns should be shaped by physically defined and universally accessible public spaces and community institutions. Urban spaces should be framed by architecture and landscape design that celebrate local history, climate, ecology and building practices" [39].

The invention and rapid development of the automobile has had a distinct impact on development of cities and towns that was noticeable in the decentralization from the central city. In the 1970s, while attempting to design a pedestrian based town that is sustainable, USA planners and designers started converting streets into pedestrian walkways as an experiment [40]. Craven declares: "New Urbanist town planners, developers, architects, and designers try to reduce traffic and eliminate sprawl" [40].

"In simplistic, layman's terms, New Urbanism might be defined as taking the most desirable land use and architectural features of communities from the past and adapting them to the technological needs of the present" [39]. basic principles of New Urbanism are explained in Table 6 below.

Table 6. *Principles of New Urbanism [41]*

Principle	Explanation
Walkability	Most things within a 10-minute walk of home and work. Pedestrian friendly street design (buildings close to street; porches, windows and doors; tree-lined streets; on-street parking; hidden parking lots; garages in rear lane; narrow, slow speed streets). Pedestrian streets free of cars in special cases.
Connectivity	Interconnected street grid network disperses traffic and eases walking. A hierarchy of narrow streets, boulevards, and alleys. High quality pedestrian network and public realm make walking pleasurable.
Mixed-Use & Diversity	A mix of shops, offices, apartments, and homes on-site. Mixed-use within neighbourhoods, within blocks, and within buildings. Diversity of people – of ages, income levels, cultures, and races.
Mixed Housing	A range of types, sizes and prices in closer proximity.
Quality Architecture & Urban Design	Emphasis on beauty, aesthetics, human comfort, and creating a sense of place; special placement of civic uses and sites within community. Human scale architecture and beautiful surroundings nourish the human spirit.
Traditional Neighbourhood Structure	Discernible centre and edge. Public space at centre. Importance of quality public realm; public open space designed as civic art. Contains a range of uses and densities within a 10-minute walk. Transect planning: highest densities at town centre; progressively less dense towards the edge. The Transect is an analytical system that conceptualises mutually reinforcing elements, creating a series of specific natural habitats and/or urban lifestyle settings. The Transect integrates environmental methodology for habitat assessment with zoning methodology for community design. The professional boundary between the natural and manmade disappears, enabling environmentalists to assess the design of the human habitat and the urbanists to support the viability of nature. This urban-to-rural transect hierarchy has appropriate building and street types for each area along the continuum.
Increased Density	More buildings, residences, shops, and services closer together for ease of walking, to enable a more efficient use of services and resources, and to create a more convenient, enjoyable place to live. New Urbanism design principles are applied to the full range of densities from small towns to large cities.
Green Transportation	A network of high-quality trains connecting cities, towns, and neighbourhoods together. Pedestrian-friendly design that encourages a greater use of bicycles, rollerblades, scooters, and walking as daily transportation.
Sustainability	Minimal environmental impact of development and its operations. Eco-friendly technologies, respect for ecology and value of natural systems. Energy efficiency. Less use of finite fuels. More local production. More walking, less driving.
Quality of Life	Taken together, these add up to high quality of life well worth living, and create places that enrich, uplift and inspire the human spirit.

Thus, New Urbanism is an urban planning approach, which provides for the implementation of place-making principles through which the urban environment is transformed into an integrated, compact, walkable, mixed-use, vibrant and sustainable community where people experience high quality of life.

3.4.5. Green Planning Approach

The widely accepted definition of urban green spaces is that they are "public and private open spaces in urban areas, primarily covered by vegetation, which are directly (e.g. active or passive recreation) or indirectly (e.g. positive influence on the urban environment) available for the users" [42].

Unplanned development and urbanization patterns, especially in cities, have had a negative influence on green spaces that consequently resulted in a significant decrease in the environmental benefits of green [43]. "If green spaces are so important for human wellbeing, how is it possible to increase these areas and maximise the positive aspects for humans, while at the same time decrease the negative aspects of cities for the environment?" [44].

Green spaces play a key role in the sustainable development of communities and likewise contribute decisively to the liveability of the built-up environment. Green spaces have a direct link to place-making and add quality to a place. The character of a community is often identified and labeled by the quality of its green spaces. Well designed, efficiently managed and maintained green spaces enhance living and working conditions, has social and visual value and, equally importantly, attract people and investment into an area [45].

Development of green spaces is an integrated approach to sustainable environments and plays an important role in terms of social, economic, cultural and environmental aspects of sustainable development [42]. A strategy for green spaces has to effectively and concurrently address a variety of (ecological) environmental, social, economic and sustainable development issues [46].

Green planning approaches of countries, cities and communities may be at variance, however, the central focus should underwrite the place-making concept and conclusively achieve transformation of a space into a lively public place. Therefore, a Green planning approach should include objectives such as: to safeguard the future of green spaces; to enhance the quality of urban areas; to render urban areas more attractive and thereby attract more resources; and to enhance the wellbeing of the user-group [46].

Benefits derived from an effective Green planning approach can be categorized according to three main groups, including: environmental benefits, economic and aesthetic benefits and social and psychological benefits [42]. These are discussed briefly in the table below.

Table 7. Environmental Benefits of Urban Green Spaces [42]

	Environmental Benefits
Ecological Benefits	Urban green spaces supply cities with ecosystem services ranging from maintenance of biodiversity to the regulation of urban climate.
Pollution Control	Pollution in cities is due to pollutants which include chemicals, particulate matter and biological materials, which occur in the form of solid particles, liquid droplets or gases. Air and noise pollution is common phenomenon in urban areas. The presence of many motor vehicles in urban areas produces noise and air pollutants such as carbon dioxide and carbon monoxide. Emissions from factories, such as sulphur dioxide and nitrogen oxides, are very toxic to both human beings and the environment.
Biodiversity and Nature Conservation	Green spaces function as protection centres for the reproduction of species and conservation of plants, soil and water quality. Urban green spaces provide the linkage of the urban and rural areas. They provide visual relief, seasonal change and a link with the natural world. A functional network of green spaces is important for the maintenance of ecological aspects of a sustainable urban landscape, with greenways and use of plant species adapted to the local condition with low maintenance cost, self-sufficiency and sustainability

Table 8. Economic and Aesthetic Benefits of Urban Green Spaces [42]

	Economic and Aesthetic Benefits
Energy Savings	Using vegetation to reduce the energy costs of cooling buildings has increasingly been recognised as a cost effective reason for increasing green space and tree planting in temperate climate cities. Plants improve air circulation, provide shade and they evapotranspire. This provides a cooling effect and helps lower air temperatures. A park of 1.2 km by 1.0 km can produce an air temperature between the park and the surrounding city that is detectable up to 4 km away. A study in Chicago has shown that increasing tree cover in the city by 10% may reduce the total energy for heating and cooling by 5% to 10%.
Property Value	Areas of the city with enough greenery are aesthetically pleasing and attractive to both residents and investors. The beautification of Singapore and Kuala Lumpur, Malaysia, was one of the factors that attracted significant foreign investments that assisted rapid economic growth. Still, indicators are very strong that green spaces and landscaping increase property values and financial returns for land developers of between 5% and 15% depending on the type of project.

Table 9. Social and Psychological Benefits of Urban Green Spaces [42]

	Social and Psychological Benefits
Recreation and Wellbeing	People satisfy most of their recreational needs within the locality where they live. Findings by Nicol and Blake (2000) show that over 80% of the UK's population live in urban areas, and thus green spaces within urban areas provide a sustainable proportion of the total outdoor leisure opportunities.
Human Health	The level of stress in people who were exposed to natural environments decreased rapidly compared to people who were exposed to urban environments, whose stress levels remained high. In the same review, hospital patients whose rooms were facing a park had a 10% faster recovery rate and needed 50% less strong pain relieving medication compared to patients whose rooms were facing a building wall. This is a clear indication that urban green spaces can increase the physical and psychological wellbeing of urban citizens. In other research conducted in Swedish cities, people who spent more time outdoors in urban green spaces were less affected by stress. Certainly, improvements in air quality due to vegetation have a positive impact on physical health, with such obvious benefits as a decrease in respiratory illnesses. The connection between people and nature is important for everyday enjoyment, work productivity and general mental health.

It is thus evident that, depending on the dominant conditions of a place, development of green spaces may present many challenges. Nevertheless, through careful planning and site-responsive design, urban green spaces can make a meaningful contribution to sustainable development at regional, district and local levels. The planning of layout and design approaches for rural green spaces should strive to meet the needs of the community, optimize opportunities in the community to grow towards sustainability, and furthermore contribute to the specific character and image of a place and the community. Uncomplicated access to green spaces will benefit these efforts and stimulate physical activity.

3.4.6. Conclusion of Place-Making Approaches in Planning Context

The main objective of the place-making concept is the improvement of the community's environment and their quality of life [16].

In order to evaluate the contribution of the three place-making approaches in planning sustainable communities, it is necessary to evaluate these approaches in terms of the Three Spheres of Sustainability Table 1 illustrates this evaluation.

Table 10. Place-making approaches in the context of sustainability

Approach	Three Spheres of Sustainability		
	Social	Environmental	Economic
Livelihoods approach	☑	☑	
Power of 10 approach	☑		☑
Community participation approach	☑	☑	☑
New Urbanism approach	☑		☑
Green planning approach	☑	☑	☑

From the above evaluation, it is evident that place-making is a concept that can be used to change and improve the spaces and places within communities. In the rural planning context, place-making, built fundamentally on various lively and sustainable objectives, can act as a catalyst to affect the planning for sustainable communities.

4. Conclusion

In conclusion, continuous monitoring of the implementation and progress of place-making approaches is imperative. Therefore, transparent management and evaluation of an approach should be maintained to ensure that effectual amendments are made timeously when deemed mandatory. Equally important is that legislation, policies and guidelines that regulate and manages place-making approaches should at all times endeavour to harmonise the needs of the community with the natural layout and resources of the environment, thereby ensuring an effective and sustainable design.

Acknowledgements

This research (or parts thereof) was made possible by the financial contribution of the NRF (National Research Foundation) South Africa.

Any opinion, findings and conclusions or recommendations expressed in this material are those of the author(s) and therefore the NRF does not accept any liability in regard thereto.

References

[1]   H. Girardet. Creating Sustainable Cities. Dartington: Green Books, 1999.

[2]   A. Power. Sustainable communities and sustainable development: a review of sustainable communities. 2004. http://eprints.lse.ac.uk/28313/1/CASEreport23.pdf

[3]   D. Geis, and T. Kutzmark. Developing Sustainable Communities: the future is now. 2006. http://freshstart.ncat.org/articles/future.htm

[4]   D. Jaber .The Language of Sustainability: why words matter. 2009. http://www.greenbiz.com/blog/2009/11/18/language-sustainability-why-words-matter

[5]   United States Environmental Protection Agency. Action Planning and the sustainable Community. 2012. http://www.epa.gov/greenkit/sustain.htm

[6]   W.L. Filho. Dealing with misconceptions of the concept of sustainability. 2000. http://www.esd.leeds.ac.uk/fileadmin/documents/esd/2._Intern ational_Journal_of_Sustainability_in_Higher_Education_2000 _Leal_Filho.pdf

[7]   P.R. Berke. Does Sustainable Development Offer a New Direction for Planning? Challenges for the Twenty-First Century.   2002. http://arroyofilms.com/ftpuser/2nd%20wknd/Berke.pf

[8]   Vanderbilt University Sustainability and Environmental Management Office.. What is Sustainability? 2013. http://www.vanderbilt.edu/sustainvu/who-we-are/what-is-sustainability/

[9]   Peter, C. & Swilling, M. 2012. Sustainable, Resource Efficient Cities   –   Making   it   Happen! http://www.greengrowthknowledge.org/resource/sustainable-resource-efficient-cities-%E2%80%93-making-it-happen Date of access: 22 May 2013.

[10]   Institute for Sustainable Communities.. What is a Sustainable Community?   2014. http://www.iscvt.org/what_we_do/sustainable_community/

[11]   Natural Resources Defense Council. Sustainable Communities. 2012. http://www.nrdc.org/sustainable-communities/

[12]   Teaching and Learning for a Sustainable Future. Sustainable Communities.   2010. http://www.unesco.org/education/tlsf/mods/theme_c/mod17.html

[13]   United Nations (UN). Economic aspects of sustainable development   in   South   Africa.   1997. http://www.un.org/esa/agenda21/natlinfo/countr/safrica/eco.htm

[14]   Project for Public Spaces. What is Placemaking? 2011a. http://www.pps.org/reference/what_is_placemaking/.

[15]   Placemaking Chicago. What is placemaking? 2008. http://www.placemakingchicago.com/about/

[16]   M. Lamit, A. Ghahramanpouri, and S . Sedaghat Nia. A Behavioral Observation of Street Liveliness in Meldrum Walk, Johor Bahru of Malaysia. 2012. http://tuengr.com/V04/003-014.pdf

[17]   J.   Jacobs.   Placemakers.   2011. http://archigar.blogspot.com/2011/02/placemakers-jane-jacobs.html

[18]   J. Grant. Planning the Good Community. 2006. http://www.tandfonline.com/doi/abs/10.1080/1464935080266683#.U0VWzfmSx3k

[19]   Project for Public Spaces. All Placemaking is Creative: how a shared focus on place builds vibrant destinations. 2013. http://www.pps.org/blog/placemaking-as-community-creativity-how-a-shared-focus-on-place-builds-vibrant-destinations/

[20]   R. Veenhoven, and J. Ehrhardt. The cross-national pattern of happiness: test of predictions implied in three theories of happiness. 1995. Social Indicators Research, 34:33-68.

[21]   E.J .Cilliers, W. Timmermans, F. Van den Goorbergh, and J.S.A. Slijkhuis, The Lively Cities (LICI) background document: LICI theory and planning approaches. Part of the LICI project (Lively Cities, made possible by INTERREG IVB North West Europe, European Regional Development Fund, European Territorial Cooperation, 2007-2013. Wageningen University of Applied Sciences, Van Hall Larenstein).

[22]   M. Pacione. Urban Geography: A global perspective. 2005. 2nd ed. New York: Routledge.

[23]   VCEC. 2011. VCEC Regulatory Conference. 2011. http://www.vcec.vic.gov.au/CA256EAF001C7B21/pages/vcec-regulatory-conference-2011#.UuUo6fu6Jdg

[24]   Hobart City Council.. Hobart .Public Spaces and Public Life, a city   with   people   in   mind.   2010. http://www.hobartcity.com.au/Hobart/A_City_with_People_in_Mind

[25]   S. Harrison and P. Dourish, Re-place-ing space: the roles of place and space in collaborative systems. 1996. http://www.dourish.com/publications/1996/cscw96-place.pdf

[26]   H. Soholt, Life, spaces and buildings – turning the traditional planning   process   upside   down.   2004. http://www.walk21.com/papers/Copenhagen%2004%20Soholt%20Life%20spaces%20and%20buildings%20turning%20the%20t.pdf

[27]   The Economist Intelligence Unit. Liveable Cities: challenges and   opportunities   for   policymakers.   2010. https://www.europeanvoice.com/GED/00020000/22400/22491.pdf

[28]   C. Loudier and J.L Dubois. Public spaces: between insecurity and   hospitality.   2001. http://www.ocs.polito.it/biblioteca/verde/uk_PARTIE201_C133.134.pdf

[29]   K.F. Lanham, Planning as placemaking: tensions of scale, culture and identity. Blacksburg, Vancouver: Virginia Polytechnic Institute and State University. 2007.

[30]   Project for Public Spaces. The Atlantic Interviews Fred Kent. 2011b.   http://www.pps.org/blog/the-atlantic-interviews-fred-kent/

[31]   R. A. Williams, Environmental Planning for Sustainable Urban   Development.   2000. http://www.bvsde.paho.org/bvsaidis/cwwa9/will.pdf

[32]   I. Scoones. Livelihoods perspectives and rural development. 2009. Journal of Peasant Studies, 36(1):1-26.

[33]   R. LeGates and , F. Stout. A Short History of Urban Planning. 2013.   www.docslide.com/a-short-history-of-urban-planning/

[34]   Project for Public Spaces. Transforming Cities through Placemaking   and   Public   spaces.   2012. http://www.urbangateway.org/sites/default/ugfiles/Transforming%20Cities%20Through%20Place%20Making_%20PPS_Cynthia%20Nitikin.pdf

[35]   S. Cowan, M. Lakeman, J. Leis, D. Lerch, and J.C. Semenza. The   City   Repair   Project.   2006. www.inthefield.info/city_repair.pdf

[36]   Toronto. Creative city planning framework: a supporting document to the agenda for prosperity: prospectus for a great city. Toronto: AuthentiCity. 2008.

[37]   A. McBride. Community Wisdom + Expert Knowledge = Good   Community   Design.   2013. http://www.pps.org/blog/community-wisdom-expert-knowledge-good-community-design/

[38]   B. Breman, M. Pleijte, S. Ouboter, and A. Buijs, Participatie in waterbeheer.   Een   vak   apart.   2008. http://www.levenmetwater.nl/static/media/files/Praktijkhandleiding_participatie_in_het_waterbeheer.pdf

[39] R. Thornton. What is New Urbanism? - Part 1. , 2010. http://www.examiner.com/article/what-is-new-urbanism-part-1

[40] J. Craven. What is New Urbanism? 2013. http://architecture.about.com/od/communitydesign/g/newurban.htm

[41] Michigan Land Use Institute. 10 Principles of New Urbanism . 2006. http://www.mlui.org/mlui/news-views/articles-from-1995-to-2012.html?archive_id=678#.UzP64fmSx3k

[42] S.M.A. Haq. Urban green spaces: an integrative approach to sustainable environment. 2011. http://www.scirp.org/journal/PaperInformation.aspx?paperID=5881#.Ut_FThD8LIU

[43] C. S. Gomes and E.M. Moretto. A framework of indicators to support urban green area planning: a Brazilian case 2011. study.http://www.iaees.org/publications/journals/piaees/articles/2011-1(1)/A-framework-of-indicators-to-support-urban.pdf

[44] J. Schilling. Towards a Greener Green Space Planning. 2010. http://www.lumes.lu.se/database/alumni/08.10/Thesis/Schilling_Jasper_Thesis_2010.pdf

[45] T. Baycan-Levent and P. Nijkamp, Urban Green Space Policies : a comparative study on performance and success conditions in European cities. 2004. http://dare2.ubvu.vu.nl/bitstream/handle/1871/8932/20040022.pdf?sequence=1

[46] H.D. Kasperidus, I.S. Erjavec, M. Richter, C.S Costa, B. Edlich. B.Guideline for the General Procedure of Developing and Implementing an Urban Green Space Strategy. 2006. http://www.greenkeys-project.net/media/files/greenkeys_strategy_guideline.pdf

26

The Importance of Planning for Green Spaces

Elizelle Juaneé Cilliers

Unit for Environmental Sciences and Management, North-West University, Potchefstroom Campus, South Africa, Potchefstroom

Email address:

juanee.cilliers@nwu.ac.za

Abstract: Green spaces are often perceived as a luxury, especially in rural areas in need of basic services and characterized by housing needs. Recent studies proof the necessity of providing green spaces, captured in terms of the social, environmental, health and economic benefits that such spaces offer to (urban and rural) communities, along with the core linkage to sustainability and enhanced quality of life. Acknowledging the constrains of providing green spaces including issues such as limited municipal budgets, conflicting development priorities, and increasing urbanization placing pressure on space for development, this paper explores the importance of planning for green spaces in terms of the direct and indirect benefits it offers to communities and to the sustainable development approach.

Keywords: Green Spaces, Benefits, Sustainability, Quality of Life

1. Introduction to Green Spaces

The aim of spatial planning is to plan and provide for sustainable living spaces, implying balancing the social needs of the citizens, the development pressure for economic growth and the surrounding environment. Current reality however, suggest of increasing unsustainability linked to diverse and complex reasons such as political, economic and social considerations. However, the prevailing approach to spatial planning is believed to be part of the problem as green spaces are often perceived as a luxury, and not a necessity, especially in rural areas where the value and importance of such spaces are under-prioritised in comparison to providing basic services and meeting housing demands [1]. The concept and importance of green spaces are undervalued in terms of spatial planning approaches. This paper aims to identify the indirect and direct benefits of green spaces in an attempt to enhance the necessity of planning and providing for qualitative green spaces within modern communities.

1.1. Defining Qualitative Green Spaces

Green space planning (and green spaces) refers to land in natural or undeveloped condition that is proximate and easily accessible from residential- and work places. It refers to public and private open spaces in urban and rural areas, primarily covered by vegetation, which are directly (active or passive recreation) or indirectly (positive influence on the urban environment) available to a variety of users and communities. According to [2] green spaces are areas that have contiguous vegetated areas and spaces, such as artificially created city parks, stands with natural vegetation and land areas such as botanical gardens, as well as isolated street trees, street medians and private gardens. Green spaces also include school grounds and sports fields, which can again be divided into formal and informal green spaces [3]. The most common terms for green spaces include "open space", "open areas" and "public space". For the purpose of this paper, the term green space is used. Qualitative green space refers to such green spaces providing a specific function to communities. It recognises the environmentally beneficial role that green spaces may offer, as well as the social, economic and psychological or health benefits [4].

1.2. Planning for Green Spaces

Spatial planning is constantly faced with the challenge to balance "development' and "environmental" pressures. Environmental considerations have recently become an integral part of developmental thinking and decision-making and the green-environment is gaining more and more importance in political, social, and economic terms. There is an expanded scientific understanding that green spaces are substantially beneficial to urban communities [5].

Despite the vision of an integrated, holistic planning process, the current reality suggests that the environment and green spaces are often neglected, and sometimes sacrificed to benefit and enhance development as a result of various factors such as limited municipal budgets, conflicting development priorities, increasing urbanization and the valuation of green spaces. Limited municipal budgets are linked to the perception that green spaces are considered a luxury, a visual attribute of the area and not a necessity to consider in terms of budgeting. It relates to the conflicting development priorities where urban development priorities, such as providing basic needs and services are higher prioritized than the green environment and green spaces with no actual proof of revenue. Increasing urbanization is furthermore placing pressure on space for development, resulting in green spaces being sacrificed. Cities cannot expand housing provisions without sacrificing open space and agricultural land in already populated areas, or on the periphery [6], [7]. The lack of value connected to green spaces are in terms of spatial planning approaches the greatest reason for the under-provision of green spaces in neighbourhood, as communities and local authorities are not aware of the benefits of providing such spaces. Often urban areas and urban developments are valued higher than green spaces; mainly due to the monetary value connected to urban development, reflected in property prices, revenue drawn from development, higher taxes and a better land value and market price, in contradiction to the indirect, immeasurable value of green spaces. Due to these factors, green spaces are often not prioritized in the spatial planning and decision-making process.

This paper highlights the benefits of green spaces in terms of social, economic and environmental benefits, in order to emphasize the importance of such spaces and the necessity to plan and provide green spaces within neighbourhoods.

2. Benefits of Green Spaces

Value is usually determined and quantified from an economic perspective, linked to a financial value. Green spaces, however, are more complex to valuate as it cannot always be related to a quantifiable economic value [8]. Unlike the market for most tangible goods, the market for environmental quality does not yield an observable unit price. However, in order to be able to compete with urban development, the value of green spaces need to be identified and measurable. "The goal is to translate the methods, theories and equations of urban economics and green economics into urban planning approaches which can lead to concrete decision making." [9]. In this way, green spaces will have more weight in the decision-making processes [10], and might be able to survive against the susceptibility to urban pressures [11]. Accordingly some of the social benefits, environmental benefits and economic benefits of green spaces are captured.

2.1. Social Benefits of Green Spaces

Social benefits of green spaces are related to leisure and recreation, the facilitation of social contact and communication, access to and experience of nature, issues influencing human physical and psychological health and well-being and overall sustainability [12], [1]. These social benefits are measured in terms of aesthetic value, qualitative living environments, the positive perception of residents with regard to urban green-space-values, enhanced community cohesion and common interest as a result of green public spaces [13], [14], [15]. Human health and mental health are also part of the social benefits of green-spaces and research in environmental psychology suggested that contact with nature serves psychological restoration [16], [14], [17]. The proximity and accessibility of green spaces in relation to residential areas appears to affect the overall levels of physical activity [18], [19], [20]. Research furthermore proofed the restorative effects of green spaces in terms of stress relief [21], happiness versus aggression [22], [23] and especially the positive social impact on children [24]. Green spaces also contributes to enhanced community cohesion [13], [14], [25], social interaction, lowers levels of fear, less aggressive behaviour, and better neighbour relationships [17].

2.2. Environmental Benefits of Green Spaces

Environmental benefits provided by green spaces include ecosystem services [12] and ecological systems that provide a myriad of services to human societies and in terms of enhanced biodiversity [26]. It relates to storm water management and providing habitats for wild plants and animals [12]. Various studies have been conducted internationally, focusing on street tree costs (tree planting, irrigation and other maintenance) versus calculated benefits (energy savings, reduced atmospheric carbon dioxide, improved air quality, and reduced storm water runoff), to estimate net benefits of green spaces [27], [15]. Green spaces contributes to reducing pollution and enhancing air quality [28], microclimate and heat island effects [29], [30] and noise reduction [28]. Green areas can reduce noise pollution and the visual intrusion from traffic [18]. The greatest environmental benefit of green spaces is the impact on biodiversity and providing refuge to species that are disappearing from urban areas [31], [18].

2.3. Economic Benefits of Green Spaces

The economic benefits of green spaces relates to the economic and financial gain as a direct result of the provided green space, including aspects such as a favourable image for a place, the boost retail sales, increased tourism [32], enhance inward investment in the area [33], and encouraged employment (emphasizing the impact on production values). Economic benefits furthermore relate to the positive impact on property values [10], the value of open spaces and proximity of neighbourhoods to natural areas [34], [35], [36]. Research [35] proofed that proximity to large protected natural areas have a positive influence on housing values. Research conducted [37] on the outdoor environmental quality that contributes to house-buyers preferences were linked to the findings of [34] that concluded that natural parks have the largest statistically significant effect on home sale prices. Proximity to open-space was found to have a statistically significant effect (positive) on a home's sale price [38] and houses that were within one half-block of any type of open space were estimated, on average, to experience the largest positive effect on their sale price [34]. In addition the value of proximity to open space was higher in neighbourhoods that were dense, near the CBD, high-crime or home to many children [39].

2.4. Quantifying the Benefits of Green Spaces

As there is no monetary value connected to green spaces, as it is hard to quantify and measure it in economic terms [40], [41]. Some researchers tried to address this problem by estimating the price of environmental quality using direct elicitation of willingness to pay, travel costs, advertising costs, direct monetary damages, the household production approach, or some combination of the above [42]. The most common qualitative evaluation methods include, but are not limited to the market price method, damage cost avoided, replacement cost or substitute cost method, contingent valuation method, contingent choice method, benefit transfer method, productivity method and the most familiar hedonic pricing method [43]. These methods can in limited ways prove that green spaces have economic value, although it is very case specific and still remains an estimate. Quantifying the value and benefits of green spaces should be explored in an attempt to prioritize green space planning and ensure the realization and implementation of direct and indirect benefits of such spaces within neighbourhoods.

3. Added Value of Green Space Planning

Incorporating green space planning (especially acknowledging the value of green spaces and positive contribution it have on the direct surrounding environment and communities) within current spatial planning approaches will enhance quality of life and contribute to sustainable development objectives, driven by a holistic planning approach.

3.1. Addressing the Sustainable Development Approach

The sustainability concept is increasingly being used to guide planning [48]. Sustainable development, as defined through roughly in literature, always includes the three dimensions of (1) social aspects, (2) the economy and (3) the environment, seeking a state of balance between these dimensions. The current approach to spatial planning and unequal prioritization between pro-development approaches and pro-environmental approaches is probably the greatest reason for not meeting sustainability objectives in urban and rural areas. [49] states that the heart of sustainable development lies in ensuring a better quality of life for everyone, and meeting the four objectives of social progress, effective protection of the environment, prudent use of natural resources and maintenance of high and stable levels of economic growth and employment.

The economy (along with development pressures) and the environment (along with green space protection initiatives) should be planned holistic in order to reach a sustainable state. The approach proposed in this paper is to focus on the social, economic and environmental benefits of green spaces and integrate these benefits as part of the spatial planning process, this to re-establish the balance of sustainable development, in terms of all of the dimensions (social – with the focus to strengthen communities, environment – with the focus to develop spaces that will be attractive and economic – with the focus to enhance the marketability of the area). When all three dimensions are equally valued by local authorities, it can be assumed that it will reflect in the planning and budgeting processes as well. The proposed approach to strengthen the environmental dimension (and to regain the balance of sustainable development) will ensure that the value of green-spaces be more measurable and comparable to development revenues. This implies the identification of direct and indirect values of green spaces and the translation of indirect these benefits of green spaces into monetary values.

3.2. Enhancing Quality of Life in Rural Areas

During the last couple of years Economist tried to quantify the value of green spaces based on various approaches. However, most of these research were conducted in developed countries and proofed the provision of qualitative green spaces could be directly linked to an enhanced quality of life and quality of living environment.

Rural areas, especially, are often neglected in terms of qualitative green spaces, due to other needs being prioritized in these areas. Even though a number of strategies, policies and other implementation programmes are already in place regarding the effective integration and growth of rural communities, rural communities seem to be neglected, enduring great poverty and deprivation [44], along with other social challenges enforced by the strain of poverty, such as limited access to health care [44], [45], enhanced vulnerability [46] and a lack of clean water and qualitative green spaces. Ironically, the contribution of green spaces within rural areas could directly benefit social, environmental and economic challenges that form part of the current reality of these spaces.

The better the living conditions and the equality of life chances, the 'happier' the communities will be [47]. Green spaces, in this sense, can directly influence the sense of place and quality of environment within rural areas, having spin-offs in terms of social benefits, economic benefits and environmental benefits in these areas needed it the most.

In this sense, green-planning research and research regarding the value of green spaces, should be expanded to include the situations in developing countries and rural areas in an attempt to improve the quality of life and quality of environment.

4. Conclusion

The importance of urban green-spaces were known for decades; however, the relationship between urban liveability and green-spaces as incorporated in overall urban green structures has become the focus of international studies especially during the last 10 to 15 years [50].

This paper suggest that the value of green spaces should be firstly identified, in terms of direct and indirect benefits, and secondly quantified and measured in order to be prioritized and comparable to urban developments and other development priorities such as housing provision and commercial developments. Green spaces provide many benefits that are well documented in literature and captured in the paper, as value are subjective to location and community needs, the value of specific green spaces within specific neighbourhoods and areas should be identified, in terms of the local context and characteristics. As there is no one blue-print for delivering sustainable development, there is no blue-print for valuing green spaces, it requires different strategies in different societies.

These identified green space values should then be articulated in monetary terms to compete with development pressures [11]. If the value of green spaces could be expressed in monetary terms, it would consequently have more weight in the development decision-making process [10], as development decisions are often based on comparisons of monetary values, such as cost-benefit analysis. This furthermore stresses the importance to supply public decision-makers (local authorities) with reliable, comparable valuations methods [51].

Acknowledgements

This research (or parts thereof) was made possible by the financial contribution of the NRF (National Research Foundation) South Africa. Any opinion, findings and conclusions or recommendations expressed in this material are those of the author(s) and therefore the NRF does not accept any liability in regard thereto.

References

[1]   E.J. Cilliers and W. Timmermans, "The importance of creative participatory planning in the public place-making process," Environment and Planning B: Planning and Design, vol 41. (EPB 139-098), 2014.

[2]   B. Thaiutsa, L. Puangchit, R. Kjelgren and W. Arunpraparut, "Urban green space, street tree and heritage large tree assessment in Bangkok, Thailand," Forestry and Urban Greening, vol. 7(3), pp. 219-229, 2008.

[3]   M.M. McConnachie and C.M. Shackleton, "Public green space inequality in small towns in South Africa," Habitat International, vol. 34(2), pp. 244–248, 2010.

[4]   C.M. Sutton, on urban open space: a case study of Msunduzi Municipality, South Africa. Canada: Queens University. (Thesis – B.Sc). School of Environmental Studies. 139 p, 2006.

[5]   K.L. Wolf, on public value of nature: economics of Urban trees, parks and open space, Design with Spirit, Washington: Edmond, uEnvironmental Design Research Association, 2004.

[6]   V. McConnell and K. Wiley, "Infill Development: Perspectives and Evidence from Economics and Planning," Resources for the Future RRF, pp. 1-37, 2010.

[7]   SCANPH, Southern California Association of Non-Profit Housing, Density Guide For Affordable Housing Developers, 2003.

[8]   Rics, on the Value of Sustainability: Meeting of the Minds, Asset Strategies, 2004.

[9]   A. Bertaud, The study of urban spatial structures, 2010, http://alain-bertaud.com. (Accessed 4 July 2010).

[10]   J. Luttik, "The value of trees, water and open space as reflected by house prices in the Netherlands," Landscape and Urban Planning, vol. 48, pp. 161-167, 2000.

[11]   A.T. More, T. Stevens, and P.G. Allen, "Valuation of urban parks," Landscape and Urban Planning, vol. 15, pp. 139-152, 1988.

[12]   R. Stiles, "Urban spaces – enhancing the attractiveness and quality of the urban environment," WP3 Joint Strategy. University of Technology, Vienna, December 2006.

[13]   A.E. Kazmierczak and P. James, on the role of urban green spaces in improving social inclusion, Salford: University of Salford, School of Environment and Life Sciences, 2008.

[14]   F.E. Kuo, "The role of arboriculture in a healthy social ecology," Journal of Arboriculture, vol. 29(3), pp. 148-155, 2003.

[15]   E.J. Cilliers, E. Diemont, D.J. Stobbelaar and W. Timmermans, "Sustainable Green Urban Planning: The Green Credit Tool," Journal of Place Management and Development, vol. 3(1), pp. 57-66, 2010.

[16]   A, Van den Berg, T. Hartig, and H. Staats, "Preference for Nature in Urbanized Societies: Stress, Restoration, and the Pursuit of Sustainability," Journal of Social Issues, vol. 63(1), pp. 79-96, 2007.

[17]   S.U. Roger, on health benefits of gardens in hospitals: Plants for People, Texas: Centre for health systems and design, 2003.

[18]   Greenspace Scotland, "Greenspace and quality of life: a critical literature review," Scotland, 2008, Http://www.openspace.eca.ac.uk/pdf/greenspace_and_quality_of_life_literature_review_aug2008.pdf Date of Access: 2 April 2014.

[19]   N. Owen, N. Humpel, E. Leslie, A. Bauman, and J.F. Sallis, "Understanding environmental influences on walking: Review and research agenda," American Journal of Preventive Medicine, vol. 27, pp. 67-76, 2004.

[20]   D.A. Cohen, J.S. Ashwood, M.M. Scott, A. Overton, K.R. Evenson, L.K. Staten, D.Porter, T.L. Mckenzie, and D. Catellier, "Public parks and physical activity among adolescent girls," Pediatrics, vol. 118, pp. E1381-E1389, 2006

[21]   R. Hansmann, S.M. Hug and K. Seeland, K, "Restoration and stress relief through physical activities in forests and parks," Urban Forestry & Urban Greening, vol. 6, pp. 213-225, 2007.

[22]   A. Chiesura, "The role of urban parks for the sustainable city," Landscape and Urban Planning, vol. 68, pp. 129-138, 2004.

[23]   F.E. Kuo and W.C. Sullivan, "Aggression and violence in the inner city - Effects of environment via mental fatigue," Environment and Behavior, vol. 33, pp. 543-571, 2001.

[24] A.F. Taylor, F.E. Kuo and W.C. Sullivan, "Views of nature and self-discipline: Evidence from inner city children," Journal of Environmental Psychology, vol. 22, pp. 49-63, 2002.

[25] E.J. Cilliers, E. Diemont, D.J. Stobbelaar and W. Timmermans, "Sustainable Green Urban Planning: The Workbench Spatial Quality Method," Journal of Place Management and Development, vol. 4(2), pp. 214-224, 2012.

[26] S.S. Cilliers, E.J. Cilliers, C.E. Lubbe, S.J. Siebert, "Ecosystem services of urban green spaces in African countries—perspectives and challenges," Urban Ecosystems, vol. 16(4), pp. 681-702, 2013.

[27] E.G. McPherson, S.E. Maco, J.R. Simpson, P.J. Peper, Q. Xiao, A.M. Van Der Zanden and N. Bell, on Western Washington and Oregon community tree guide: benefits, costs, and strategic planning. Silverton: International Society of Arboriculture, 2002.

[28] P. Bolund and S. Hunhammar, "Ecosystem services in urban areas," Ecological Economics, vol. 29(2), pp.293-301, 1999.

[29] H. Akbari, M. Pomerantz, and H. Taha, "Cool surfaces and shade trees to reduce energy use and improve air quality in urban areas," Solar Energy, vol. 70(3), pp. 295-310, 2001.

[30] E. Alexandri and P. Jones, "Temperature decreases in an urban canyon due to green walls and green roofs in diverse climates," Building and Environment, vol. 43(4), pp. 480-493, 2008.

[31] S. Hodgkison and JM. Hero, "The efficacy of small-scale conservation efforts, as assessed on Australian golf courses." Biological Conservation, vol. 135(4), pp. 576-586, 2007.

[32] H. Woolley, C. Swanwick and N. Dunnet, on nature, role and value of green space in towns and cities: an overview. 2003. www.atypom-link.com/ALEX/doi/abs/10.2148/benv.29.2.94.54467.

[33] Cabe Space, "Paying for parks: Eight models for funding urban green space," London, 2005. www.cabe.org.uk/files/Paying-for-parks-full-report.pdf. (Accessed 20 April 2009).

[34] M. Lutzenhisher and N.A. Netusil, "Effect of Open Spaces on a Home's Sale Price," Contemporary Economic Policy, vol. 19, pp. 291-298, 2001.

[35] S.D. Shultz and D.A King, "The Use of Census Data for Hedonic Price Estimates of OpenSpace Amenities and Land Use," Journal of Real Estate Finance and Economics, vol. 22, pp. 239-252, 2001.

[36] V.K. Smith, C. Poulos and H. Kim, "Treating open space as an urban amenity," Resource and Energy Economics, vol. 24, pp. 107–129, 2002.

[37] C. Jim, Y. Wendy and Y. Chen, "Impacts of urban environmental elements on residential housing prices in Guangzhou (China)," Journal of landscape and urban planning, vol. 78, pp. 422–434, 2006.

[38] B. Bolitzer and N.R. Netusil, "The impact of open spaces on property values in Portland, Oregon," Journal of Environmental Management, vol. 59(3), pp. 185-193, 2000.

[39] S.T. Anderson and S.E. West, "Open space, residential property values and spatial context," Journal of Regional Science and urban economics, vol. 36(6), pp. 773-789, November 2006.

[40] Commissie van Ek, "Amersfoort creatieve stad", 2009. http://www.amersfoortcreatievestad.nl/site/tags/tag/commissie+van+ek/.

[41] A. Herzele and T. Wiedemann, "A monitoring tool for the provision of accessible and attractive urban green spaces," Landscape and Urban Planning, vol. 63(2), pp.109-126, 2002.

[42] D.M. Brasington and D. Hite, "Demand for environmental quality: A spatial hedonic analysis," Regional Science and Urban Economics, vol. 35(1), pp. 57-82, 2005.

[43] Lambert, on economic valuation of wetlands: an important component of wetland management strategies at the river Basin scale, Ramsar Convention, 2003.

[44] Gopaul, M. 2006. The significance of rural areas in South Africa for tourism development through community participation with special reference to Umgababa, a rural area located in the province of KwaZulu-Natal. Pretoria: University of South Africa. (Dissertation – Master of Arts).

[45] Campbell, C., Nair, Y., Maimane, S. & Sibiya, Z. 2008. Supporting people with AIDS and their carers in rural South Africa: Possibilities and challenges. http://eprints.lse.ac.uk/5471/ Date of access: 28 Feb. 2013.

[46] Van der Ploeg, H. Renting, G. Brunori, K. Knickel, J. Mannion, T. Marsden, K. de Roest, E. Sevilla-Guzman, E. and F. Ventura, "Rural Development: From Practices and Policies towards Theory," Sociologis Ruralis, vol. 40(4), pp. 391-408., 2000.

[47] R. Veenhoven and J. Ehrhardt, "The cross-national pattern of happiness: Test of predictions implied in three theories of happiness," Social Indicators Research, vol. 34, pp. 33-68, 1995

[48] [48] K. Krizek and J. Powers, on a planners' guide to sustainable development. PAS 467, Chicago: American Planning Association, 1996.

[49] A. Power, "Sustainable communities and sustainable development a review of sustainable communities", 2004, http://eprints.lse.ac.uk/28313/1/CASEreport23.pdf. Date of access 25 March 2013.

[50] O.H. Caspersen, C.C. Konijnendijk and A.S. Olafsson, "Green space planning and land use: An assessment of urban regional and green structure planning in Greater Copenhagen," Geografisk Tidsskrift, Danish Journal of Geography, vol. 106(2), pp. 7-20, 2006.

[51] E. Defrancesco, P. Rosato and L. Rossetto, on the appraisal approach to valuing environmental resources, Valuing complex natural resource systems: The case of the Lagoon of Venice, Cheltenham, UK: Edward Elgar Publishing, pp 40-57, 2006.

27

The Impact of Deforestation on Soil Conditions in Anambra State of Nigeria

Anyanwu J. C.[1,*], Egbuche C. T.[2], Amaku. G. E.[1], Duruora J. O.[3], Onwuagba, S. M.[1]

[1]Department of Environmental Technology, Federal University of Technology, Owerri, Nigeria
[2]Department of Forestry and Wildlife Techhnology, Federal University of Technology, Owerri, Nigeria
[3]Collage of Education, Nsugbe Anambra State, Nigeria

Email address:
jayjaychimezie@yahoo.com (Anyanwu J. C.)

Abstract: The research was carried out to determine the impact of deforestation on soil conditions in Anambra State. Ten soil samples were collected at random at a depth of 0-35cm below the litter layer from forests and farmlands. The soil samples were collected and analyzed for pH, field capacity, soil moisture, organic carbon, bulk density, soil micro-organism and particle size distribution. The result revealed that soil texture was mostly sandy except in some areas such as Atani, Nzam, Mmiata and Oroma-etiti, where it was generally heavy (clay loam). The result also revealed that the soil samples from the forests have better physical, chemical and biological properties compared to samples from farmlands. The results showed considerable variation for the soil physical, chemical and biological properties across the study area. Soil data were analyzed using Least Significant Difference (LSD). The analysis revealed that the main effect of land use was significant ($p<0.05$) for soil moisture, bulk density, organic carbon, organic matter, pH, viable bacteria number and viable fungal propagule. It was not significant for sand, silt, clay and field capacity. The interaction effect of location and land use on soil properties were significant ($p<0.05$) only for soil moisture, it was not significant for other soil variables. The study recommended, among others, the protection of forests from deforestation so as to maintain good soil conditions in the study area.

Keywords: Soil Texture, Least Significant Difference, Soil Properties, Forests, Farmlands

1. Introduction

The aim of the study is to determine the extent to which deforestation affects soil conditions in Anambra state. Deforestation is the conversion of forested areas to non-forest land use such as arable land, urban use, wasteland or pasture. Anambra State is seriously threatened by deforestation. The 2006 national population and housing survey put the population of Anambra State at 4,182,032 and the population density at 863/Km^2. The quest for more land to meet the needs of the rapidly expanding population in the state as well as the unfavorable economic downturn of many people is socio-economic drivers of deforestation in the state. This increase in population has reduced the fallow periods in the state, consequently little time is allowed for the soil to replenish its nutrients with the result that more forests are cleared for farmlands. The high rate, at which forests are currently converted to agriculture, indicates that the economic return from agriculture is higher than from forests, at least in the short term, and that the land is more valuable deforested than forested [1]. As population increases, the demand for wood continually increases. Trees are harvested for multiple uses ranging from lumber to wood for fuel. The rate of deforestation currently significantly exceeds the rate of forest renewal.

According to [2] when fallow periods are long enough to permit full vegetation regeneration and soil fertility restoration, this cultivation is recognized as ecologically balanced, economically attractive, and culturally integrated. It has been pointed out that certain woody species such as *Dialium guineense, Anthonata macrophylla, Alchornea cordifolia,* dominate the natural fallow system in the humid zone of southeastern Nigeria, where population density is high, the fallow period short, and the soil acid [3]; [4; [5]. Deforestation and cropping have caused a great deal of damage in Anambra State. This uncontrolled deforestation usually accompanied by poor soil management has led to land degradation and ecological imbalance in many parts of the state. Clearing and burning are deforestation methods employed by farmers and hunters in the state. The land clearing and post-clearing soil management methods employed in the state has affected the role of forests as carbon sequestration sites. Large scale deforestation in the state has also occurred as a result of construction works by the local, state and federal governments. Urbanization is another cause of deforestation in the state. The migration of people to urban areas has led to the clearing of forests for residential and other purposes. The citing of industries without recourse to the impact on the environment is also a problem in some parts of the state. Development initiatives such as road and building constructions are supplemented by growing encroachment and illegal logging has further increased the risk of deforestation in the state. The effects of deforestation include loss of soil nutrients, loss of valuable species of economic/medicinal value, siltation of rivers, species extinction, reduced biological diversity, reduced ecosystem stability, reduced plant biomass, and broken food chain. Increased rates of soil erosion have a potential of leading to a rise in river beds and hence increased frequency of flooding, threatening settlements and cultivable land. A major effect of deforestation in the state is increased soil erosion which has displaced people from their native homes, led to destruction of lives and property and collapse of infrastructural facilities in some parts of the state.

2. Materials and Methods

2.1. Study Area

This study was carried out in Anambra State of Southeastern Nigeria with the study sites located in four agro-ecological zones of the state established from the soil and vegetation maps modified from [6]. Anambra is a state in South-Eastern Nigeria. Its boundaries consist of Delta State to the west, Imo State to the south, Enugu State to the east and Kogi State to the north. The indigenous ethnic groups in Anambra State are the Igbo (98% of population) and Igala (2% of the population). Anambra State lies between Longitudes $6^0 35^1 E$ and $7^0 21^1 E$, and Latitudes $5^0 40^1 N$ and $6^0 45^1 N$. The climate is tropical with high annual rainfall ranging from 1,400mm in the north to 2,500mm in the south, with a mean monthly temperature of $27.6^0 C$. Heavy rainfall occurs within the months of April to October while the months of November to February have scanty rainfall, higher temperature and low humidity. Consequently, the natural vegetation in the greater part of the state is tropical dry or deciduous forest, which, in its original form, comprised tall trees with thick undergrowth and numerous climbers.

The four agro-ecological zones of the state were selected based on soil type characterized by vegetation. The soil was used as the basis for choosing the study sites because the soil is a more permanent feature than vegetation which is not stable and can be easily altered over time. The four zones consist of (A) Pale Brown Loamy Alluvial Soils characterized by Fresh Water Swamp Forest, (B) Dip Brown Red Soils Derived from Sandy Deposits characterized by Pennisetum-Dominated Grass Species, (C) Red and Brown Soils Derived from Sandstones and Shales characterized by Lowland Rain Forest, (D) Reddish Brown Gravelly and Pale Clayey Soils Derived from Shales which are characterized by Hyparrhenia-Dominated Grass species. Anambra State lies in the Anambra Basin, the first region where intensive oil exploration was carried out in Nigeria. The Anambra basin has about 6,000m of sedimentary rocks [7]. The sedimentary rocks comprise ancient Cretaceous deltas, somewhat similar to the Niger Delta, with the Nkporo Shale, the Mamu Formation, the Ajali sandstone and the Nsukka formation as the main deposits. The soil types range from alluvial, hydromorphic and ferallitic soils. The alluvial soils are pale brown loamy soils and differ from the hydromorphic soils in being relatively immature and with its horizons not well developed. The alluvial soils are found in the two plain south of Onitsha and Ogbaru, and in the Niger Anambra low plain north of Onitsha. The alluvial soils sustain continuous cropping longer than the hydromorphic and ferallitic soils.

2.2. Data Collection

Relatively stable forests were identified from each of the agro-ecological zones. The forests served as control. Soil samples were collected randomly from each forest and from agricultural lands hundred meters away from the forests. The soil samples were collected for analysis of soil properties and examination of soil micro-organisms. Ten soil samples were collected at random, using soil auger at a depth zone of 0-35cm below the litter layer from each agro-ecological zone. The soil samples were air-dried, gently crushed and made to pass through a 2 mm mesh. Plant residues, gravel and other foreign matter retained on the sieve were discarded [8]). Sieved samples were stored in unused polythene bags and labelled appropriately, before proceeding to the laboratory for analysis of selected soil properties.

2.3. Determination of Parameters

The following parameters were determined: Soil moisture, bulk density, particle size, soil pH, organic carbon, organic matter, field capacity and microbial population. Morphological properties involved characterization of erosion sites. The moisture content of soil was determined by the gravimetric method in which wet soil was oven-dried at 105°C for 24 hours. It was expressed as a percentage of oven-dry soil. Field capacity was measured by sampling the soil in the field after it had been thoroughly wetted to root zone depth by artificial irrigation. Small areas (about 1.50 m square) were selected from different locations. Each was surrounded by a raised bound and enough water was added to wet the upper 15 cm of the enclosed soil, the area being covered with polyethylene sheeting to prevent evaporation. The samples were collected at a depth of 0-35cm and their moisture content determined [9]. The soil core method was used for the determination of bulk density in the field.

Undisturbed soil core was excavated by a core sampler, the bulk density being calculated from the known volume of core and the weight of oven-dry soil. Before each test was performed, soil samples from each site were collected randomly. The samples were later bulked together in order to ensure adequate representation of each site. The bulking procedure for the soil samples was standardized by ensuring that samples collected were of equal volume. Particle size distribution was determined by the hydrometer method [11].

Textural classification was determined by the percentage of sand, silt and clay with the help of the US Department of Agriculture triangular diagram [12]. Soil pH was determined in soil-to-water ratio 1:2.5 with the help of a pH electrode and values were read out from pH meter [13]. Organic carbon was determined using Walkley and Black wet oxidation method. Organic matter was then calculated by a factor of 1.724 (Van Bemmelen's Correction Factor).

2.4. Method of Data Analysis

The data generated from the field studies were analyzed using relevant statistics. Two phases of statistical analysis were employed. First, whether the interaction and/or the main effects of location and land use were significant was analyzed. Second, treatment combinations having significant interactions and main effects were further compared.

Multiple comparisons of means were performed using Least Significant Difference (LSD) test. The Least Significant difference is used to compare means of different treatments that have an equal number of replications. The statistical procedures were performed using GenStat statistical package, GENSTAT Release 7.2 DE, Discovery Edition 3 [14].

3. Results and Discussion

The results of analysis of soil properties obtained from the field are shown below. The analysis helped to establish the impact of deforestation on soil conditions in the study area. Table 1 shows the Main Effect of Land use on soil properties. Table 2 presents the Interaction Effect of Location and Land use on the soil properties. The main effect of land use on soil properties is also presented in Figure 1. The results reveal that soil samples collected from farmlands had lower sand content than samples collected from forests in the study area. The difference in sand contents of farmlands and forests was, however, not significant. The silt and clay contents of soil from farmlands were higher than the silt and clay contents from forests. There were no significant differences between the silt content of farmlands and forests and between the clay contents of farmlands and forests.

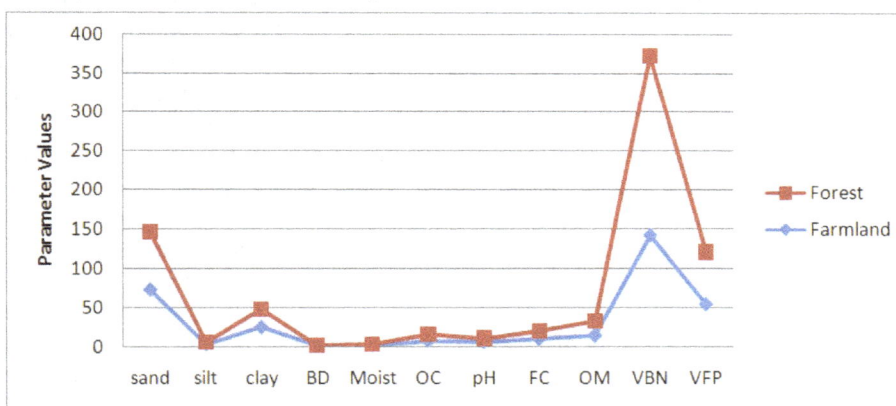

Figure 1. *Soil Properties from Land Use in the Study Area*

Table 1. *Main Effect of Land Use on Soil Properties*

Land Use	Sand %	Silt %	Clay %	Bulk Density g/cm^3	Moisture %	Organic Carbon %	pH	Field Capacity. %	Organic Matter %	Viable Bacteria Number * 10^{-3}g	Viable Fungal Propagule *10^{-3}g
Farmland	72.45	2.95	24.57	1.095	1.122	6.87	5.240	9.67	13.67	142.0	54.0
Forest	73.88	2.71	23.32	0.875	1.455	9.20	5.891	9.68	18.41	231.0	66.7
LSD$_{0.05}$	NS	NS	NS	0.973	0.2015	0.979	NS	NS	2.105	63.0	11.25

Means differ significantly (p < 5%) as established by LSD-test NS = Not Significant.
Source: Fieldwork, 2011

Table 1 further shows that soil samples collected from farmlands had significantly higher bulk density (1.095g/cm^3) than samples collected from forests (0.875g/cm^3). This is understandable because the soils of the farmlands have been exposed to erosion as a result of over cultivation which

results in weaker consistence and thus higher bulk density. The lower bulk density of soil samples obtained from the forests is as a result of their better aggregation and higher consistence. This can be attributed to the fact that trees and other plants in the forest help to bind soil together thereby reducing erosion.

A very important significance of the bulk density of a soil is the amount of surface area a soil has. The higher the bulk density, the lower the surface area of the soil and these results in decreased capacity of the soil to retain water and nutrients. This implies that high bulk density is not desirable in agriculture. One effect of deforestation on soil condition in the study area is the increase in bulk density of the soil resulting in soil compaction, low soil nutrients and erosion. This finding agrees with the observations of [15].

The moisture content of soil from the forests (1.455%) was significantly ($p < 0.05$) higher than that of farmlands (1.122%). The lower moisture content of soils from farmlands is an important effect of deforestation on soil conditions in the study area. The lower the soil moisture, the less able it can sustain vegetation. Deforestation affects the moisture content of the soil by exposing the soil to high solar radiation, which increases the rate of evaporation. Trees help to improve the porosity of the soil thereby retaining moisture. It is also important to note that soil moisture is an important mechanism in controlling the exchange of water and heat energy between the land surface and the atmosphere through evaporation and plant transpiration thus contributing to the development of weather patterns and the incidence of precipitation.

The differences in the organic carbon contents of the soil samples from forests and farmlands were equally significant ($p < 0.05$) with the forests having greater organic carbon content (9.20%) than the farmlands (6.87%). Soil organic carbon is derived from soil organic matter. Organic carbon therefore, is a measure of soil fertility and agricultural productivity potential. The lower organic carbon content of the soil from farmlands is an indication of loss of plant cover occasioned by deforestation and poor agricultural practices. This goes to show that clearing of forests for the purpose of agriculture, in the long run will have a negative effect on soil fertility and productivity. Organic carbon consists of amino acids, organic acids, plant fibers and the biomass of micro-organisms. It drives the energy flow in soils, controls gaseous emissions and dictates the chemical and physical properties of the soil. One implication of this study, therefore, is that deforestation affects soil condition through the loss of soil fertility (i.e. reduced organic carbon content) in the study area. Similar finding has also been reported by [16].

The result further revealed that the pH of soil samples from farmlands was significantly lower (5.240) than for forests (5.891). The destruction of forest cover exposes the soil to erosion and contributes to soil acidification. It must be pointed out that if soil solution is too acidic, plants will not be able to utilize N, P, K and other nutrients they need. Also, plants are more likely to take up toxic metals and die eventually of toxicity. High soil acidity is therefore undesirable for plant growth. Based on the above findings, it can be implied that another effect of deforestation on soil condition in the study area is increased acidity of the soil. In their study, [17] revealed that soils in southeastern Nigeria are very acidic and of low fertility. The differences in field

capacities of soils in forests and farmlands were not statistically significant (Table 1). Soil from farmland had field capacity value of 9.67%, while soil from forest ecosystem had value of 9.68% which is a reflection of their water holding capacities. At field capacity, the water and air contents of the soil are considered to be ideal for crop growth.

Table 1 reveals that organic matter content of soil samples from forests (18.41%) was significantly ($p < 0.05$) higher than the organic matter content of samples collected from farmlands (13.67%). This implies that soil samples from the forests were richer in organic matter when compared to soil samples from the farmlands. This is an indication that soil organic matter declines following deforestation and subsequent cultivation. This finding is in line with that of [18]. It was further observed that the color of most samples was dark, ranging from brown to black. The dark color is as a result of humus. The darker color of the forest soil layers compared to that of the farmlands is an indication that the forest harbors more organic matter in the soil and that loss of forest cover depletes the organic matter content of the soil. This is also an indication that the plough layer is affected by land preparation activities. It was also noted that the deeper the layers, the lighter the color, often grey, and sometimes mottled with yellowish or reddish spots. Soil organic matter directly benefits the soil microbial community and indirectly influences all other organisms, particularly plants. It is rich in nutrients such as nitrogen, phosphorous, sulfur, and micronutrients, and is comprised mostly of carbon. Vegetation generally helps to increase organic matter content of the soil. This is because tree and plant litters decay and are converted to organic matter in the soil thus enriching it. In the course of this study, it was also noticed that in order to amend and restore the soil fertility, most farmers resort to shifting cultivation, thereby causing further decline in forest cover in the state. Measures that can be adopted to increase organic matter in the study area include leaving residues on the soil surface, rotating crops with pasture or perennials, incorporating cover crops into crop rotation and by adding organic residues such as animal manure, litter or sewage sludge. This will no doubt help to reduce extensive agriculture which encourages deforestation.

The result further reveals that the soil microbial load of samples from forests were significantly ($p < 0.05$) higher when compared to that of samples from farmlands. The Viable Bacteria Number (VBN) for farmland was $142.0 * 10^{-3}$g and $231.0 * 10^{-3}$g for forest, while the Viable Fungal Propagule was $54.0 * 10^{-3}$g for farmland and $66.7 * 10^{-3}$g for forest. The preponderance of soil micro-organisms in the forests compared to the farmlands is an indication that the forests are more fertile than the farmlands and that when forests are cleared and converted to farmlands, there is usually loss of soil nutrients and poor soil condition. This is because nutrients tied up in organic matter are not readily available to plants; rather, microbes must first begin the decomposition process and obtain energy from organic carbon. As the organic matter is broken down, nutrients such as nitrogen and phosphorus are released into the soil and are then available for uptake by plants. The implication of this result is that deforestation reduces the soil microorganisms which replenish soil nutrients, thus leading to a decline in the soil nutrients available for plant growth.

Table 2 shows the interaction effect of location and land use on soil properties. The result shows that most of the soil variables were not significantly affected by the combined effect of location and land use. Only soil moisture contents of the farmlands and forests were significantly (p<0.05) affected.

Table 2. Interaction Effect of Location and Land Use on Soil Properties

Location	Land Use	Sand %	Silt %	Clay %	Bulk D. g/cm³	Moisture %	OC %	pH	Field C. %	OM %	VBN *10⁻³g	VFP *10⁻
A	Farmland	55.65	4.30	40.05	0.914	1.632	7.88	6.182	11.15	14.36	194.0	65.6
A	Forest	55.10	3.40	41.50	0.848	2.062	10.64	6.026	11.32	21.88	378.0	99.5
B	Farmland	78.00	2.40	19.55	0.980	1.380	7.08	6.000	8.81	14.52	220.0	71.4
B	Forest	80.25	2.33	17.10	0.894	2.192	9.84	5.946	9.08	19.98	334.0	87.7
C	Farmland	76.85	2.55	20.60	0.967	0.630	6.26	5.042	8.86	13.36	81.0	50.3
C	Forest	81.45	2.45	16.10	0.836	0.788	8.28	4.968	8.84	15.59	77.0	48.0
D	Farmland	79.30	2.55	18.10	0.944	0.844	6.26	4.536	9.88	12.45	73.0	28.9
D	Forest	78.71	2.65	18.60	0.920	0.778	8.02	4.622	9.49	16.19	135.0	31.7
LSD₀.₀₅		NS	NS	NS	NS	0.4030	NS	NS	NS	NS	NS	NS

Means differ significantly (p < 5%) as established by LSD-test. Bulk D. = Bulk Density, OC= organic carbon, FieldC. = Field capacity, OM = Organic Matter, VBN = Viable Bacterial Number, VFP = Viable Fungal Propagule. NS = Not Significant.
Source: Fieldwork, 2011

The above analysis revealed that the main effect of land use (Table 1) was significant (p<0.05) for soil moisture, bulk density, organic carbon, organic matter, pH, viable bacteria number and viable fungal propagule. It was not significant for sand, silt, clay and field capacity. This implies that not all the soil properties were affected by the land use (forest or farmland). In essence, it means that deforestation affects certain soil properties more than others. The interaction effect of location and land use on soil properties (Table 2) were significant (p<0.05) only for soil moisture, which goes to show that deforestation not only affects the moisture content of the soil, but that soil moisture is essential for vegetation regeneration and sustenance. It was not significant for other soil variables, which means that the interaction of location and land use on soil properties did not affect most of the soil properties significantly.

Figure 2. Soil degradation in Oba, Anambra State, Nigeria

4. Conclusion

The study observed that degradation of the soil as a result of deforestation leads to poor soil physical, chemical and biological conditions which are a function of the soil properties. Therefore, protection of forests from deforestation will help to maintain good soil conditions in the study area.

Forest degradation is demographically driven. This is one manifestation of population explosion. There is need to inform and educate the people about links between forest preservation and the need to control population growth. To curb deforestation in Anambra State, policy makers at all levels should begin to see the need for new conservation strategies. The traditional approach of restricting access to isolated forests in areas designated as parks and then employing park guards to protect the forests may not achieve the desired goal, rather modern strategies which should take into consideration the needs of the poor masses living in and near the forests should be adopted as well. The local people should be carried along when planning any conservation strategy. This is because no effective conservation can be achieved without the cooperation of the local people.

The realization of the fact that deforestation can have negative impact leading to poor soil productivity, climate change, erosion, flooding, loss of lives and property, and loss of biodiversity, led to the need for this study in Anambra State. The study suggested ways to mitigate the environmental problems associated with deforestation in the state. It revealed the impact of deforestation on soil conditions in the state. This is in view of the fact that soil degradation is a serious problem that can lead to loss of soil fertility which in turn will affect crop production and productivity with its resultant food insecurity in the state.

References

[1]   Gaston, G., Brown, S., Lorenzini, M., and Singh, K. D. (1998). *State and change in carbon pools in the forests of tropical Africa. Global Change Biology*, Vol. 4, pp. 97-114

[2]   Raintree, J.B., and Warner, K. (1986). Agrofor. Syst. 4, 39-54.

[3]   Obi, J.L., and Tuley, P. (1973). *The Bush Fallow and Ley Farming in the Oil Palm Belt of Southeastern Nigeria.* Land Resour. Div. Misc. Rep. No. 161. ODM., England.

[4]   Okigbo, B.N. (1982). In "Agroforestry in the African Humid Tropics" (L. H. Macdonald, ed.), pp.41-45. United Nations Univ., Tokyo.

[5]   Getahun, A., Wilson, G.F., and Kang, B. T. (1982). In "Agroforestry in the African Humid Tropics" (H. McDonald, eds.), pp.28-36. United Nations Univ., Tokyo.

[6]   Ofomata, G.E.K. (1975). Nigeria in Maps: Eastern States. In G.E.K. Ofomata (Ed). Vegetation Types and Soils. pp.30-45.

Ethiope Publishing House,Benin, Nigeria.

[7]   Olusola, J. O., Ajibola, U. K., and Samuel O. A. (2003). Depositional Environments, Organic Richness, and Petroleum Generating Potential of the Campanian to Maastrichtian Enugu Formation, Anambra Basin,Nigeria.

[8]   Kundu, N.K. and Ghose, M.K. (1997). Studies on the topsoil of an opencast coal mine. Environmental Conservation, 21 (2), 126-132.

[9]   Kundu, N.K. and Ghose, M.K. (1997). Studies on the topsoil of an opencast coal mine. Environmental Conservation, 21 (2), 126-132.

[10] Dakshinamurti, C. and Gupta, R.P. (1968). Practicals in Soil Physics, pp. 1-16. Indian Agricultural Research Institute, New Delhi.

[11] Gee, G.W. and Bauder, J.W. (1986) Particles size analysis. In: Methods of Soil analysis part 1. A. Klute (Eds) Am. Soc. Agron. Madision 101 USA. pp. 38 - 41.

[12] Biswas, T.D. and Mukherjee, S.K. (1994). *Textbook of Soil Science.* Tata McGraw Hill Publishing Co. Ltd., second reprint, New Delhi.

[13] Ghose, A.B., Bajaj, J.C., Hasan, R. and Singh, D. (1983). Soil and Water Testing Methods. Indian Agricultural Research Institute, New Delhi.

[14] Genstat 5 Reference Manual, Clarendon Press, Oxford 1987, 749 pp., ISBN 0-19-852212-6

[15] Ogbodo, E.N., Okorie, P.O. and Utobo, E. B. (2010). Growth and Yield of Lettuce *(Lactuca sativa L.)* at Abakaliki Agro-Ecological Zone of Southeastern Nigeria. World Journal of Agricultural Sciences. Vol 6, No 2, pp. 144-148.

[16] Asadu C. L. A. and Akamigbo F.O.R. (1990). Relative contribution of organic matter and clay fractions to cation exchange capacity of soils in southeastern Nigeria. Samaru. J. Agri. Res. Vol 7, pp 17-23

[17] Asadu C. L. A. and Akamigbo F.O.R. (1990). Relative contribution of organic matter and clay fractions to cation exchange capacity of soils in southeastern Nigeria. Samaru. J. Agri. Res. Vol 7, pp 17-23.

[18] Lemenih, M., Karltun, E., and Tolera, M. (2011). Comparing farmers' perception of soil fertility change with soil properties and crop performance in Beseku, Ethiopia. Land Degraded. Development.

28

A framework for planning green spaces in rural South Africa

Elizelle Juaneé Cilliers

Unit for Environmental Sciences and Management, North-West University, Potchefstroom Campus, South Africa, 2520

Email address:

juanee.cilliers@nwu.ac.za

Abstract: The importance of planning for green spaces is well captured in literature, focusing on the direct and indirect benefits which such spaces provides to various beneficiaries, from local authorities to local communities. However, the planning and implementation of such spaces, especially in a rural context, is complex. The value of spaces is perceived differently by different stakeholders, and this is also true for the rural environments, characterized with unique challenges and needs. This paper captures the value of green spaces and relates it to specific rural considerations, in order to state the value that green spaces can provide to rural areas and communities. The Vaalharts case study is used to explain current realities and best practice options and the paper concludes with a framework for the planning of green spaces in rural South Africa, including all aspects and design elements that should be considered in the planning and provision of green spaces.

Keywords: Green spaces, rural areas, framework, Vaalharts rural area

1. Planning for Green Spaces

1.1. Understanding Green Space

A public green space is defined as every parcel of land classified as a natural surface, judged to be publicly accessible [1]. 'Natural surface' implies predominantly natural area with a sense of quality and the presence of several maintained facilities [2]. Green space includes public and private open spaces in urban and rural areas, primarily covered by vegetation. Public green spaces include parks, forests, golf courses, sports fields and other open nature areas and are seen as the key approach in areas where residential plot sizes are inadequate (as in the case of most rural areas), or the housing stock is dominated by multi-storey buildings [3] in high density urban areas. Qualitative green space, as referred to in this research, include green spaces that provide a specific function to communities, ranging from social, ecological, economic, psychological, health and amenity functions [4], [5].

1.2. Approaches to Planning Green Spaces

There is currently an expanded scientific understanding that green spaces are substantially beneficial to urban and rural communities [6]. An important concept regarding the understanding and conceptualizing of green spaces is their influence on and contribution to shifting the paradigm of spatial segregation of urban landscape functions into complete multifunctional landscapes [2] wherein 'quality' regarding these spaces are linked to the 'value' associated with spaces by recognizing the need of these spaces to reflect the changing social, economic and environmental conditions. Research captured in this journal contributes to the importance of driving the green-agenda and identified various benefits of green spaces in terms of social, economic and environmental benefits, in order to emphasize the importance of such spaces and the necessity to plan and provide green spaces.

1.3. Complexity of Planning for Rural Areas

Rural communities and the development thereof continue to be one of the main priorities within frameworks and constitutions guiding the economic and social development of (especially developing) countries [7]. In South Africa, rural development is an predominant challenge as it is estimated that half of South Africa's population lives in rural areas [8] and that three quarters of the people living below the MLL (minimum living level) live in these rural areas [9].

Planning for rural areas is complex as rural areas have unique challenges and characteristics to take into consideration, such as location challenges (geographic disparities), dispersed rural settlement structures, lack of integrative policies, sociological issues (crime and violence, poor education, lack of facilities), economic issues (declining per capital income, mass unemployment), lack of community participation, level of education, ability to communicate, social issues (provision of basic services, limited access to health care, standard of health, level of poverty), and environmental issues (lack of clean water and lack of qualitative green spaces) [10], [8], [11], [12], [13], [7], [14], [15].

Furthermore, the current planning approach is often project oriented and seldom supports green growth and green space planning [16]. The need for an inclusive approach which considers community participation and stakeholder engagement, whilst considering the wide scale of impact related to green space planning should be recognised, and implemented.

2. Valuing Green Spaces

Green spaces are complex to valuate as the market for environmental quality does not yield an observable unit price. The value of green spaces needs to be identified in order to emphasize the importance of planning for such spaces. The value of green spaces can be divided in two categories, namely indirect benefits and direct benefits, as captured in the following table.

Table 1. Identifying the benefits of green space provision

Benefits		Measurable	Source
Indirect benefits	Environmental benefits	Ecosystem services	[17]
		Enhanced biodiversity	[18]
		Storm water management	[17]
		Habitat provision	[19]
		Reduced carbon dioxide	[20]
		Improved air quality	[21]
		Reduced pollution	[22]
		Microclimate and heat island effect	[23], [24]
		Noise reduction	[22]
		Sustainability	[25]
		Leisure and recreation	[26]
	Social benefits	Social contact, access to experience	[27], [28]
		Physical and psychological health	[29], [30]
		Aesthetic value	[31]
		Quality living space	[32]
		Positive perceptions	[33], [19]
		Community cohesion	[27], [28]
		Levels of physical activity	[21], [34]
		Reduced stress, increase happiness	[35], [36]
		Positive impact on children	[37]
		Lower levels of fear	[29]
		Better neighbourhood relationships	[19]
Direct benefits	Economic benefits	Favourable image of place	[38]
		Boost retail sales and tourism	[38]
		Inward investment in area	[39]
		Encouraged employment	[19]
		Property values	[40], [41]
		Neighbourhood value	[42], [43]
		House buyers preferences	[44], [45]
		Positive influence in crime areas	[46]
		Development impact on children	[46]

These benefits will be explained accordingly.

3. Indirect Benefit of Green Spaces: Social Value

3.1. Local rural Reality and Social Challenges

The issue of safety and security is identified as a precondition for social and economic development in South Africa [14], [9]. Perceptions regarding safety can differ according to several factors found within an environment, including employment status, population, group and area of residences [47]. Therefore it can be inferred that rural communities may be prone to incidents of crime and safety issues as they typically fall under the category of low employment and are situated in areas with low accessibility and environmental quality [7].

Bad perceptions regarding safety and security within a neighbourhood result in other social challenges as communities avoid open spaces and are thereby excluded from recreational possibilities, social interaction, and other benefits that such open spaces provide. With a third of households in rural areas in South Africa avoiding these open spaces, social interaction and growing social cohesion continue to falter [47].

3.2. Green spaces Addressing Social Challenges

Any public space can be contributory to the insecurity of an area if it becomes a place of relegation, neglect, degradation and/or illegible which allures a range of uncivil acts, delinquency and vulnerability [48]. The importance of providing qualitative green spaces within rural areas can enhance social development and address social issues. By creating public green spaces which are fully maintained and cared for, the feeling of being unsafe is eliminated, contributing to the overall value and success of these public green spaces [2].

The priority within rural areas is usually focused on providing basic services, facilities and infrastructure, and the provision of qualitative open spaces is often neglected [49]. Apart from the abovementioned social benefits that green spaces provide to (especially rural) areas, numerous of literature confirms that outdoor play spaces are vital for children's learning and developing stages throughout life [50], [49], [51], [52].

4. Indirect Benefits of Green Spaces: Environmental Value

4.1. Local reality and Environmental Challenges

Green spaces play various roles in creating sustainable urban and rural areas [53]. Green spaces are fundamental areas in human settlements that need intentional planning as it provides the opportunity to enhance sustainability and the appearance of environmental benefits [39].

4.2. Addressing Environmental Challenges in Rural Areas Through Green Spaces

Land use change and environmental quality are closely related, and the nature and location of development can significantly influence both the generation and resolution of environmental problems. Green spaces (linked to ecosystem services) provide various services, such as (1) provisioning services, including food production, medicinal resources, (2) regulating services, including climate and air quality regulation, carbon sequestration and storage, waste-water treatment, erosion prevention, maintenance, of soil fertility, biological control, (3) supporting services, including habitats for species, maintenance of genetic diversity, (4) cultural services, including recreational and mental health, tourism, aesthetic appreciation, sense of place and experience of place [54], [55].

The planning and provision of green spaces should thus consider environmental sustainability and the green space system should be developed with the aim of establishing a network of natural features and compatible land uses that will act as a green network [56].

5. Direct Benefits of Green Spaces: Economic Value

5.1. Local reality and Economic Dimension of Planning

The importance of green spaces were known for decades; however, the relationship between liveability and green spaces as incorporated in overall spatial planning approaches has become the focus of international studies [57]. Often spatial planning decisions are based on pro-developmental approaches, linked to the revenue or economic benefit as a direct result from the development [55]. The understanding that green spaces can contribute to the economic value of an area or development, is not yet fully integrated in spatial planning approaches.

5.2. Green space Provision Addressing Economic Challenges in Rural Areas

Research regarding the economic value of green spaces confirms that most people are willing to pay more for a residential property close to a green space, emphasizing the positive impact on residential property values. Research further proved that green spaces create a favourable image for a place, boost retail sales, attract tourism [38], enhance inward investment in the area [39], and encourage employment (emphasizing the impact on production values).

Rural areas are often neglected in terms of green space provision, and this might be directly linked to the lack of qualitative public spaces, lack of economic opportunities, lack of investment opportunities and lack of other economic spinoffs that were proven to be true when providing green spaces within neighborhoods.

6. Quantifying the Value of Green Spaces

The value of green spaces needs to be identified (as captured in previous sections in terms of social value, economic value and environmental value) in order to emphasize the importance of planning for green spaces. But, the value of green spaces should also be quantified (measurable) in order to have more weight in the decision-making processes [40], and survive against the susceptibility to urban pressures [50].

The green space values, as stated, are divided in indirect benefits and direct benefits. Indirect benefits are hard to quantify in monetary terms and include social aspects and environmental aspects. Direct benefits refer to the direct economic benefit as a result of the provision of green spaces within an area, and are more easily translated into monetary terms and financial gains.

Literature identifies four main approaches to determine the economic value green-spaces in terms of: (1) the economic approach, (2) the development approach, (3) the ethical or moral approach and (4) the utilitarian approach [4], [58], [40]. [59], [60], [61]. The scope of this paper is not to review these different methods but to identify the different values of green spaces and propose that these values be quantified as part of formal spatial planning processes.

In the process of quantifying values of green spaces, the beneficiaries of the green values should be identified, as green spaces have different value for different beneficiaries. In this sense, green spaces provide different values to communities, in comparison to the values it provide to local authorities. These beneficiaries and related values are captured in the following table.

Table 2. Beneficiaries and values of green spaces

	Local authority benefit	Local community benefit
Environmental — Social	More citizen participation	Accessibility to green spaces
	Voluntary contribution	Cultural benefits
	Social capital	Better living spaces
	Aesthetic spaces	Health benefits
	Maintenance of green space	Availability of green spaces
	Higher biodiversity	Aesthetic area
	Sustained eco-systems	Recreational spaces
	Sustainable spaces	Quality of life
	Emission removal	Better working spaces
Economic	Green proximity tax	Inward investment
	Energy savings	Higher property prices
	Higher market values	
	Storm water mitigation	

7. Implementation: Vaalharts Case Study

Vaalharts is a rural area located in the greater Polokwane municipality district, in the North West and Northern Cape provinces of South Africa. The Vaalharts area claims the second largest irrigation scheme in the Southern Hemisphere namely the Vaalharts Irrigation Scheme [62]. Accordingly the environmental, social and economic issues present in this area will be captured.

7.2. Current Reality and Environmental Issues in Vaalharts

The environment is characterized by natural green areas, agricultural areas, rural residential settlements and nature reserves. There are mostly natural indigenous plants and agricultural vegetation present in the area. The area is highly accessible and residents are dependent on the environment for provisioning, regulating, supporting and cultural services. The abundance of water (due to the Vaalharts Irrigation Scheme) enhances the potential and provision of sustainable green spaces, and the Vaalharts area has the potential to support the needed ecosystem services.

However, green spaces in this area are mostly left unplanned with no vision, usage or maintenance plans. Green spaces are isolated from each other and most open spaces are homogeneous (similar in appearance as well as function). A lack of finances for the planning of green spaces is part of the current reality, along with a lack of knowledge on environmental benefits amongst communities and authorities.'

7.3. Social issues identified in Vaalharts

The social composition of residents is characterised by rural communities with limited education, training or opportunities to social and economic development. A conclusive need-assessment was conducted [62] based on data obtained from a sample of 31 willing individuals and a stratified sample of 958 randomly drawn participants in order to identify social issues and needs [7]. The research identified social challenges in terms of lack of basic government services, lack of educational related services, agricultural challenges, lack of community facilities, health and welfare challenges, safety and security issues and lack of emergency services [62].

Community members identified core actions to address these social problems [62], as (1) provision of recreation facilities, (2) enhancing sports, arts and culture and youth activities in the communities and (3) providing help and support for people with disabilities, the elderly and people who are terminally ill.

7.4. Economic Issues Identified in Vaalharts

Characterized as a rural area, there are limited economic opportunities in the Vaalharts area. The need-assessment identified the lack of employment and lack of adequate facilities as core economic issues to be addressed [62]. The Vaalharts Irrigation Scheme is probably the sole initiator of economic potential in the area. The area is in need of an integrated development approach to guide future planning. The consideration of the economic value that green spaces can provide to this area need to form part of such an integrated approach.

7.5. Green space Provision Addressing Identified Issues

Based on the literature investigation presented in this research, green spaces can contribute and address (some) of the identified issues of the Vaalharts area.

The planning and provision of qualitative green spaces in this area should be directly linked to the Vaalharts Irrigation Scheme and access to water, enabling the provision of sustainable green spaces and ecosystem services. An integrated plan, driving the green agenda, should link current isolated spaces and provide a future vision and management plan, for different typologies of green spaces. Authorities and communities needs to be informed about the benefits of green spaces in order to ensure the success of such spaces. Providing qualitative green spaces ensures adequate community engagement possibilities and the strengthening of social cohesion. It provides recreational opportunities that will reflect in health and welfare benefits. Qualitative green spaces can be directly beneficial for child-development as it relates to the planning of child-friendly spaces. The provision of qualitative green spaces will enhance the identity of the area and increase inward investment in the area. This will result in further economic spinoffs. Accordingly an integrated framework for the planning and development of green spaces in rural areas will be presented, based on the theoretical investigations, international best practice approaches and findings in terms of the local reality and rural challenges.

8. Conclusions: Integrated Framework for Planning Green Spaces

An integrated framework to guide local planning approaches, focusing on environmental benefits, social benefits and economic benefits of green spaces should be inclusive of the following aspects:

8.1. Identifying the value of green spaces

The value of space is subjective and should be identified within the local context, considering the unique challenges and opportunities present in the specific area. The following aspects need to be considered as point of departure when planning green spaces:

Table 3. Aspects to consider as point of departure

Aspects to consider	Description
Typology of green spaces	Identify different types of green spaces for possible implementation
African considerations	Consider unique challenges and cultural needs
Policies and legislation	Refer to guiding legal requirements
Participatory planning	Include the community in the formal planning process
Place-making elements	Consider elements that will enhance identity of place and contribute to value of the space
Layout and design	Consider natural restrictions and opportunities present in the area

8.2. Identifying methods and beneficiaries of green spaces

The different methods to value green spaces, along with the identification of the beneficiaries of the planed green space should be determined accordingly.

The chosen method to value green spaces should fit the local context and environment. The economic approach and development approach should be used in areas where residential values are known and related to market values. This is, for example, not possible in rural areas where subsidy housing, backyard rentals and informal housing provision are predominant. The ethical approach or utilitarian approach would thus be a better method to utilize in rural areas, subject to availability of data and resources.

In the same sense, beneficiaries should be identified for each space and area. The added value as a result of the planning and provision of green spaces should be known to the different beneficiaries in order to enhance social capital and the buy-in from these stakeholders.

8.3. Quantifying the Value of Green Spaces

The value of green spaces should, as final output, be quantified in order to emphasize the value of these spaces and have more weight in the decision-making process. The following issues and design elements need to be considered in the valuing of current green spaces and planning of future qualitative green spaces. A multi-criteria analysis can be used to quantify these issues.

1. Indirect benefits:
   a. Social values
      - Identity of place
      - Attractions provided
      - Flexibility
      - Seasonal opportunities
      - Accessibility
      - Visibility
      - Lighting
      - Landscaping
      - Signage
      - Access control
      - Proximity to nodes
      - Safety considerations
      - Maintenance
      - Recreation possibilities
      - Social contact
      - Physical and mental health
      - Aesthetic value
      - Quality living space
      - Positive perceptions
      - Community cohesion
      - Levels of physical activity
      - Reduced stress levels
      - Increased happiness
      - Social cohesion
   b. Environmental values
      - Ecosystem services
      - Biodiversity
      - Storm water management
      - Habitat provision
      - Environmental quality
      - Reduced pollution
      - Microclimate regulation
      - Noise reduction
      - Enhanced sustainability
      - Linked spaces
      - Integrative approach
   c. Child-friendly spaces
      - Safety
      - Natural setting
      - Access
      - Sociability
      - Integration
      - Located near schools
      - Walkable distances
      - Transportation options
      - Adequate surfaces and edges
      - Public furniture
      - Amenities,
      - Safe and visible entrances
      - Appropriate signage
      - Development opportunities
      - Participatory planning
      - Multi-disciplinary approach
2. Direct benefits
   d. Economic values
      - Image of place
      - Increased retail sales
      - Increased tourism
      - Inward investment in area
      - Encouraged employment
      - Higher property values
      - Increased neighborhood value
      - Positive influence in crime areas

Acknowledgements

This research (or parts thereof) was made possible by the financial contribution of the NRF (National Research Foundation) South Africa. Any opinion, findings and conclusions or recommendations expressed in this material are those of the author(s) and therefore the NRF does not accept any liability in regard thereto.

References

[1] O. Barbosa, J.A. Tratalos, P.R. Armsworth, R.G. Davies, R.A. Fuller, P. Johnson and K.J. Gaston, "Who benefits from access to green space? A case study from Sheffield, UK", Landscape and Urban Planning, vol. 83, pp. 187-195, 2007.

[2] C.M. Shackleton and A. Blair, "Perceptions and use of public green space is influenced by its relative abundance in two small towns in South Africa", Landscape and urban Planning, vol. 113, pp. 104-112, 2013.

[3] M.M. McConnachie and C.M. Shackleton, "Public green space inequality in small towns in South Africa," Habitat International, vol. 34(2), pp. 244–248, 2010.

[4] C.M. Sutton, on urban open space: a case study of Msunduzi Municipality, South Africa. Canada: Queens University. (Thesis – B.Sc). School of Environmental Studies. 139 p, 2006.

[5] E. Lange, S. Hehl-Lange and M.J. Brewer, "Scenario-visualization for the assessment of perceived green space qualities at the urban–rural fringe", Journal of Environmental Management, vol. 89, pp. 245-256, 2007.

[6] K.L. Wolf, on public value of nature: economics of Urban trees, parks and open space, Design with Spirit, Washington: Edmond, uEnvironmental Design Research Association, 2004.

[7] N. De Jong, Addressing social issues in rural communities by planning for lively places and green spaces, Dissertation submitted to the North-West University, Potchefstroom, 2013.

[8] Campbell, C., Nair, Y., Maimane, S. & Sibiya, Z. 2008. Supporting people with AIDS and their carers in rural South Africa: Possibilities and challenges. http://eprints.lse.ac.uk/5471/ Date of access: 28 Feb. 2013.

[9] Department of Rural Development and Land Reform, Rural Development Framework, Pretoria, South Africa, 1997.

[10] Gopaul, M. 2006. The significance of rural areas in South Africa for tourism development through community participation with special reference to Umgababa, a rural area located in the province of KwaZulu-Natal. Pretoria: University of South Africa. (Dissertation – Master of Arts).

[11] Van der Ploeg, H. Renting, G. Brunori, K. Knickel, J. Mannion, T. Marsden, K. de Roest, E. Sevilla-Guzman, E. and F. Ventura, "Rural Development: From Practices and Policies towards Theory," Sociologis Ruralis, vol. 40(4), pp. 391-408., 2000.

[12] L. Lategan, A study of the current South African housing environment with specific reference to possible alternative approaches to improve living conditions, Masters dissertation submitted to the North-West University, South Africa, 2012.

[13] South Africa, Department of Housing, White Paper on a New Housing Policy and Strategy for South Africa, 1994/

[14] Department of Rural Development and Land Reform, Integrated Sustainable Rural Development Strategy (ISRDS), Pretoria, South Africa, 2000.

[15] T. Cannon, J. Twigg and J. Rowell, Social vulnerability, sustainable livelihoods and disasters, Chatham, UK: Natural Resources Institute, University of Greenwhich, 2005.

[16] D.C. Okeke, An analysis of spatial development paradigm for enhancing regional integration within national and its supporting spatial systems in Africa, Doctoral Degree submitted to the North-West University, Potchefstroom, 2014.

[17] R. Stiles, "Urban spaces – enhancing the attractiveness and quality of the urban environment," WP3 Joint Strategy. University of Technology, Vienna, December 2006.

[18] S.S. Cilliers, E.J. Cilliers, C.E. Lubbe, S.J. Siebert, "Ecosystem services of urban green spaces in African countries—perspectives and challenges," Urban Ecosystems, vol. 16(4), pp. 681-702, 2013.

[19] E.J. Cilliers, E. Diemont, D.J. Stobbelaar and W. Timmermans, "Sustainable Green Urban Planning: The Workbench Spatial Quality Method," Journal of Place Management and Development, vol. 4(2), pp. 214-224, 2012.

[20] E.G. McPherson, S.E. Maco, J.R. Simpson, P.J. Peper, Q. Xiao, A.M. Van Der Zanden and N. Bell, on Western Washington and Oregon community tree guide: benefits, costs, and strategic planning. Silverton: International Society of Arboriculture, 2002.

[21] Greenspace Scotland, "Greenspace and quality of life: a critical literature review," Scotland, 2008, Http://www.openspace.eca.ac.uk/pdf/greenspace_and_quality_of_life_literature_review_aug2008.pdf Date of Access: 2 April 2014.

[22] P. Bolund and S. Hunhammar, "Ecosystem services in urban areas," Ecological Economics, vol. 29(2), pp.293-301, 1999.

[23] H. Akbari, M. Pomerantz, and H. Taha, "Cool surfaces and shade trees to reduce energy use and improve air quality in urban areas," Solar Energy, vol. 70(3), pp. 295-310, 2001.

[24] E. Alexandri and P. Jones, "Temperature decreases in an urban canyon due to green walls and green roofs in diverse climates," Building and Environment, vol. 43(4), pp. 480-493, 2008.

[25] S. Hodgkison and JM. Hero, "The efficacy of small-scale conservation efforts, as assessed on Australian golf courses." Biological Conservation, vol. 135(4), pp. 576-586, 2007.

[26] E.J. Cilliers and W. Timmermans, "The importance of creative participatory planning in the public place-making process," Environment and Planning B: Planning and Design, vol 41. (EPB 139-098), 2014.

[27] A.E. Kazmierczak and P. James, on the role of urban green spaces in improving social inclusion, Salford: University of Salford, School of Environment and Life Sciences, 2008.

[28] F.E. Kuo, "The role of arboriculture in a healthy social ecology," Journal of Arboriculture, vol. 29(3), pp. 148-155, 2003.

[29] A, Van den Berg, T. Hartig, and H. Staats, "Preference for Nature in Urbanized Societies: Stress, Restoration, and the Pursuit of Sustainability," Journal of Social Issues, vol. 63(1), pp. 79-96, 2007.

[30] S.U. Roger, on health benefits of gardens in hospitals: Plants for People, Texas: Centre for health systems and design, 2003.

[31] B. Thaiutsa, L. Puangchit, R. Kjelgren and W. Arunpraparut, "Urban green space, street tree and heritage large tree assessment in Bangkok, Thailand," Forestry and Urban Greening, vol. 7(3), pp. 219-229, 2008.

[32] E.J. Cilliers, E. Diemont, D.J. Stobbelaar and W. Timmermans, "Sustainable Green Urban Planning: The Green Credit Tool," Journal of Place Management and Development, vol. 3(1), pp. 57-66, 2010.

[33] A. Chiesura, "The role of urban parks for the sustainable city," Landscape and Urban Planning, vol. 68, pp. 129-138, 2004.

[34] D.A. Cohen, J.S. Ashwood, M.M. Scott, A. Overton, K.R. Evenson, L.K. Staten, D.Porter, T.L. Mckenzie, and D. Catellier, "Public parks and physical activity among adolescent girls," Pediatrics, vol. 118, pp. E1381-E1389, 2006

[35] N. Owen, N. Humpel, E. Leslie, A. Bauman, and J.F. Sallis, "Understanding environmental influences on walking: Review and research agenda," American Journal of Preventive Medicine, vol. 27, pp. 67-76, 2004.

[36] R. Hansmann, S.M. Hug and K. Seeland, K, "Restoration and stress relief through physical activities in forests and parks," Urban Forestry & Urban Greening, vol. 6, pp. 213-225, 2007.

[37] A.F. Taylor, F.E. Kuo and W.C. Sullivan, "Views of nature and self-discipline: Evidence from inner city children," Journal of Environmental Psychology, vol. 22, pp. 49-63, 2002.

[38] H. Woolley, C. Swanwick and N. Dunnet, on nature, role and value of green space in towns and cities: an overview. 2003. www.atypom-link.com/ALEX/doi/abs/10.2148/benv.29.2.94. 54467.

[39] Cabe Space, "Paying for parks: Eight models for funding urban green space," London, 2005. www.cabe.org.uk/files/Paying-for-parks-full-report.pdf. (Accessed 20 April 2009).

[40] J. Luttik, "The value of trees, water and open space as reflected by house prices in the Netherlands," Landscape and Urban Planning, vol. 48, pp. 161-167, 2000.

[41] M. Lutzenhisher and N.A. Netusil, "Effect of Open Spaces on a Home's Sale Price," Contemporary Economic Policy, vol. 19, pp. 291-298, 2001.

[42] V.K. Smith, C. Poulos and H. Kim, "Treating open space as an urban amenity," Resource and Energy Economics, vol. 24, pp. 107–129, 2002.

[43] S.D. Shultz and D.A King, "The Use of Census Data for Hedonic Price Estimates of OpenSpace Amenities and Land Use," Journal of Real Estate Finance and Economics, vol. 22, pp. 239-252, 2001.

[44] C. Jim, Y. Wendy and Y. Chen, "Impacts of urban environmental elements on residential housing prices in Guangzhou (China)," Journal of landscape and urban planning, vol. 78, pp. 422–434, 2006.

[45] B. Bolitzer and N.R. Netusil, "The impact of open spaces on property values in Portland, Oregon," Journal of Environmental Management, vol. 59(3), pp. 185-193, 2000.

[46] S.T. Anderson and S.E. West, "Open space, residential property values and spatial context," Journal of Regional Science and urban economics, vol. 36(6), pp. 773-789, November 2006.

[47] Statistics South Africa, Victims of crime survey, Pretoria, South Africa, 2011.

[48] C. Loudier and J.L. Dubois, Public spaces: Between insecurity and hospitality, 2001.

[49] A. Shackell, Design for Play: A guide to creating successful play spaces, England, 2008.

[50] A.T. More, T. Stevens, and P.G. Allen, "Valuation of urban parks," Landscape and Urban Planning, vol. 15, pp. 139-152, 1988.

[51] J. Zomervrucht, Gradually grow to cycle: Experiences with a child-friendly public space, Netherlands, 2005.

[52] C.C. Marcus and C. Francis, People Places: Design Guidelines for Urban Open Spaces, 2nd ed. New York: Wiley, 1998.

[53] J. Byrne and N. Sipe, Green and Open Space Planning for Urban Consolidation – A Review of the Literature and Best Practice, Brisbane: Griffith University, 2010.

[54] TEEB, The Economics of Ecosystems & Biodiversity: Ecosystem Services in Urban Management, 2011.

[55] L. Cilliers, Evaluating the spatial and environmental benefits of green space: An international and local comparison on rural areas, Dissertation submitted to the North-West University, Potchefstroom, 2014.

[56] H. Rohr, Water Sensitive Planning: An integrated approach towards sustainable urban water system planning in South Africa, Masters Dissertation submitted to the North-West University, Potchefstroom, 2012.

[57] O.H. Caspersen, C.C. Konijnendijk and A.S. Olafsson, "Green space planning and land use: An assessment of urban regional and green structure planning in Greater Copenhagen," Geografisk Tidsskrift, Danish Journal of Geography, vol. 106(2), pp. 7-20, 2006.

[58] C.J. Fausold and R. Lilieholm, "The economic value of open space: A review and synthesis", Environmental Management, vol. 23(3), 1999.

[59] M. Meadows, "The ecological resource base: biodiversity and conservation", the geography of South Africa in a changing world. Cape Town: Oxford University Press South Africa, 1999.

[60] S.J. Schmidt, "The evolving relationship between open space preservation and local planning practice", Journal of Planning History, vol. 7(2), pp. 91-112, 2008.

[61] C. Thompson, "Urban open space in the 21st century", Landscape and Urban Planning, vol. 2(7), pp. 59-72, 2002.

[62] H. Coetzee, Research Report 2: Needs assessment conducted in the Vaalharts Region, North West and Northern Cape Provinces, South Africa (A North West living labs baseline projects), Potchefstroom: Research Logistics cc, 2011.

29

African Approaches to Spatial and Green Planning

Okeke D. C.

Research Unit for Environmental Sciences, Department of Urban and Regional Planning, North-West University, Potchefstroom Campus, Potchefstroom, South Africa

Email address:

dcokeke2000@yahoo.co.uk, donald.okeke@unn.edu.ng

Abstract: As spatial planning evolved two notions of green planning emerged: traditional and sustainability notions. The former identifies with the practice in African countries when traditional urbanism is the vogue and popular design tradition in planning managed eco-centric settlements. In this context, spatial and green planning fused and drew impetus from the spirituality and traditional institutions of African societies. The sustainability notion of green planning is a recent phenomenon that is common with developed countries although it is assuming global dimension. It came with systemic changes which redefined the instrumentality of spatial planning. In effect, spatial and green planning literarily demerged and the later found expression in green growth otherwise sustainable development. This paper recalls the legacy of green planning in traditional urbanism and the lessons it holds for sustainable urbanism in contemporary societies.

Keywords: Spatial Planning, Green Planning, Green Growth, African Cities, Negritude

1. Introduction

Before spatial planning was conceptualized as a scientific body of knowledge there were sets of activities which performed the role of planning. These set of activities informed traditional urbanism. They cradled many civilizations in the pre-industrial and pre-capitalist period. Then, Africa made its contribution to spatial planning and this was epitomized in the physical expression of Empires and Kingdoms for the kingdom building societies. The non-kingdom building societies had their peculiar settlement pattern. In both instance, especially in the later, spatial planning was eco-centric and this was driven by the worldview of traditional societies, their survivalist outlook and political structure. Traditional urbanism reflects the spirituality of the people and for the animist societies, like the Ibos in south-east Nigeria, their settlement pattern was directed at transcendental ends. Living with nature was pronounced and this was maintained through code systems of symbols. Minimal interference with nature led development activities.

This outlook changed with the inception of modern town planning following colonization. Since the colonial experience, Africa has leaned on borrowed planning culture, which diffused or, as a matter of fact, was imposed from the global north. The new planning culture was driven by economic rather than cultural and traditionalvaluesystems. The need for change followed imperial interest and consummated with the inception of colonial towns, which are known to serve as trade outlets and conduit for resource marketing. Green planning component of spatial planning lost favor with the ensuing modern urbanism that is built on technological innovations. Over the years, deepening environmental crisis compelled global reaction for remedial measure and this was sought through sustainable development, the application of which is circumstantial in Africa.

The notion of sustainable development inspires green growth. In neo-liberal sense, green growth is not far, in practical terms, from environmental economics built on spatial determinism. However, viewed from environmental perspective, it redefines traditional spatial planning and renews attention on socio-cultural factors and normative value system that affect the use of environment. This revision to traditional standards is being orchestrated as though it is a new concept, safe for its neo-liberal economic agenda, as it relates to Africa. It is argued that the principle of sustainable development is indeed not a new phenomenon [1]. However, in its new shell, given its imperial background, it has problems with recognizing the authority of traditional institutions. Unfortunately, Africa had to contend with this scenario from a dependent status in the world system.

Otherwise, the neo-liberal notion of green growth prevails.

Africa is subject to the incipient participatory process in planning in a more compelling manner than the originators of the concept in other to contend with sustainable urbanism. The participatory process, which implodes sprawl, defines planning paradigm for project development in Africa. In developed countries the planning paradigm engages urban design to contemplate different models of compact city concept as panacea for sustainability. Incidentally, the attributes of the compact city, which includes workability, high density development, limited size, etc are not too far from the standards of settlement pattern in traditional African societies.

Basically, sustainability is a growth concept notwithstanding its green planning paradigm. As a matter of fact green planning lost its environmental essence in the process of linking green planning with growth in neo-liberal perspective. This paper adopts the annals approach to review the African experience. Hindsight is engaged to track spatial planning of earlier epoch and the evolution of green planning in the context of changing shades of urbanism during the pre-modern, modern and post-modernist planning periods in Africa. This body of knowledge is linked with modern planning experience to explore the application of urban growth boundaries as instrument for green planning in Africa.

2. Spatial Planning in Africa Under Popular Design Tradition

The African original contribution to spatial planning was done under the popular design tradition. Then the conception of spatial planning as a body of knowledge is a remote scientific knowledge. The nature of spatial planning in Africa is therefore better understood from the point of appreciating its end product - the African city.

Long before Europeans appeared on the scene, according to [2], Africans 'have created cultures and civilizations, evolved systems of government and systems of thought, and pursued the inner life of the spirit with a passion that has produced some of the finest art known to man'. With the abundance of valuable natural resources in place and their mindset focused, the thrust of their development policy was led by traditions of culture, which were translated into cities in concrete terms through the instrumentality of liberal arts. This rich heritage of African ancestry was rediscovered by intellectual explorers and reserved in a cultural movement called *negritude* to rescue a main section of humanity from unhappy misunderstanding [2].

Traditional urbanism prevailed in three out of the seven discernable stages in the history of African civilization and state building that nurtured African cities. These stages include; the period prior to 10th century when traditional African Kingdoms flourished; between 10th century and 15th century, the mercantilist period marked by the Trans-Sahara Trade; and between 15th century and mid 19th century, the slave trade period. The other stages are; between mid 19th

century and 1960, the colonial period; between 1960 and 1970, the independence decade; between 1970 and year 2000, the neo-liberal development concept period; and from 2000 until now the New perspectives on Africa Development (NEPAD) and African Union period.

The planning cultures built on culture and spirituality, which created these towns lost impetus and are unable to revive since colonial experience. According to [3], the development of many African settlements got frozen at a quasi-urban developmental stage. The trend of events witnessed urban planning driven by economic growth imperatives determined with western values. Today, many African cities are considered to be in crisis, as measured by the 'formal' institutional order of late capitalist modernity based on individualism as the basis of social reproduction; on citizenship in a representative democracy administered by a constitutional state as the basis of political relations; and on utilitarian rationality in a system of generational commodity production and market exchange (mediated by state redistribution) as the basis of economic relations [4].

3. Urbanism in Africa

Traditional urbanism manifested regional variations in Africa. In Anglophone African countries there were three aspects of interface between town boundaries and community or settlement form: urban pattern, distribution and change. Urban pattern in all aspect of walled cities was characterized with heterogeneous inhabitants arranged in large grains of high density homogeneous quarters segregated along the lines of professional groups, peer groups and kinship ties, etc. Specialist quarters featured prominently especially in Sudanese cities and Kanem Bornu cities in Hausa land. The contrary was the case in the forest area cities of Yoruba land where cities did not feature specialist areas: all craft work was carried on in houses and people practicing the same craft were not grouped together in any way; although each quarter of the town was fairly homogeneous.

Urbanization was generally introverted and urban distribution was concentrated with high dwelling density, mainly for defense reasons and symbiotic living. Nucleated urban form, not extending beyond five kilometers with peripheral greenbelt hemmed in within the town wall, was manifest due to gentrification processes. This caused attention to focus at the center where the hub of political, commercial and religious activities is located or concentrated. Residence at the periphery was uncomfortable due to security reasons. This characterization of the urban form of walled cities in the middle ages was universal irrespective of the nature of city boundaries. The same urban typology can be gleaned in the design of most European cities as well as in central parts of many older cities in North America, Australia and New Zealand [5].

In Francophone African countries the pre-colonial city (imperial city), was socially segregated. This was always the case whenever those in power wanted to plan urban space. According to [6]:...the imperial cities, the capitals of theocratic states, the holy cities and the trading posts along the coast were all segregated, but the degree of *hierarchisation* was dependent on local conditions. The essential elements of the pre-colonial city were the palaces, the mosques and the trading posts.....Urban planning in the imperial city in areas plied by large trucks involved the creation of special spaces, and the use of major highways and elevated sites, all of which addressed the political, military, religious and economic concerns caused by the social conditions.

Writing in Francophone African context, [7] in a review of the literature on the development of the urban phenomenon in the region confirmed that different types of cities have co-existed since the rise of the great empires of the 8th century. These are imperial cities, trading posts, and colonial cities. This collaborates the various account by Arab merchants and Portuguese explorers concerning tales of flourishing African cities some of them founded on the successive sessions of Ghana (8th century -11th century), Mali (12th century-14th century) and Songhai (15th century-16th century) empires [8]. There were tales of desert area Sudanese towns of Tekrur, Audoghast, Oulata (Walata), Timbuktu, Gao and Agadez; the southern belt of towns which included Segou, Djenne, Ouagadougou, Oyo, Katsina and Kano; the east Africa city-states of Mogadishu, Mombasa, Zanziba, Kilwa; and the 'Kraals' of Zulu Kingdoms in southern Africa. These imperial cities grew up when international trade between the Mediterranean, Europe and the Far East was expanding rapidly [9].

With the inception of modern urbanism in mid-19th century African city heritage paled into insignificance. African cities remained organic as Africa served within the period (perhaps until now) as source region for global economy and ostensibly for the growth of western civilization. This was, and is still, being achieved through unfavorable trade relations - the type that led to colonization. The unfortunate situation truncated city development in divergent ways in sub-Saharan Africa. Strategically, the integrated cosmology of traditional Africa was replaced with single-minded utilitarian objectives which produced utilitarian designs for cities in Africa. The design options bulldozed away cultural symbols, behavior, and beliefs that determined the system of base of traditional African cities. Cities in Africa became hybrids, an inevitable product of intervening culture and policy formulation hegemony spurn abroad.

Overall, from the mid 19th century, cities in Africa were no longer 'African cities' both in character and in function, because the institutional framework on which they existed altered significantly. 'African cities' became cities in the Diaspora in their homeland as the world system bear on contemporary African development. As a matter of fact, it seems, since the attempt in the colonial period to import liberal capitalism and make it blend with the social nature of African society, cities in Africa tend to drift awkwardly along unfamiliar courses charted by globalization or more precisely, neo-liberalism as global economic orthodoxy.

In mid-1990s sustainable urbanism gained attention as an advocacy of the UN-Habitat. Sustainability, which is the theoretical base of sustainable urbanism, has been severally explained. The principles of sustainability concept is not as confusing as its practice and more so in the African context. There are wide ranging definitions of sustainability. They all talk about development that does not jeopardize the ability of future generations to meet their own needs. Urban productivity and regional integration as well as concerns for unsustainable level of resource consumption in cities, especially those characterized by urban sprawl, led sustainability to appear in urban planning theory. Urban sustainability therefore emerged as a planning concept. The objectives of urban sustainability as suggested by both UN's Agenda 21 and Habitat Agenda [10] include: 'a compact urban form; the preservation of open space and sensitive ecosystems; reduced automobile use; reduced waste and pollution; the creation of liveable and community-oriented human environments; decent, affordable, and appropriately located housing; improved social equity and opportunities for the least advantaged; and the development of a restorative local economy'.

With this background [11] identified six operational principles of sustainable development thus: harmony with nature, livable built environment, place-based economy, that is, a local economy should strive to operate within natural system limits, equity, polluters pay, and responsible regionalism. With these provisions it is fairly clear what sustainability is out to achieve, but reactions to these principles is where the problem lies. This reflects in the instruments and strategies applied in the administration of sustainable development. Reactions in developed countries indicate recourse to spatial models of urban sustainability. Some notable examples include the urban compaction model in Britain and New Zealand, urban consolidation or urban intensification model in Australia, growth limit and the rise of new urbanism in USA, regional urban containment model in Britain, amongst others. In essence renewed focus on quality urban design proclaimed sustainable urbanism.

The situation in Africa did not reflect the same focus on quality urban design. Rather, there is an increasing tendency of addressing cross-cutting issues and not core issues in planning. Environmental management and decentralization policies now effectively usurp urban policies. Hence, attention drifts from spatio-physical aspects of urban form, expressed in the urbanity of cities, to urban quality issues that dwell on degradation in socio-economic and environmental terms. This explains in part the frail relation between modernist and post-modernist planning. More significantly, it explains to a great extent the paradigm shift in planning, which has had a chequered history epitomized in the 1980s.

The trend of a paradigm shift in planning has regional peculiarities in Africa. For Francophone Africa it moved from physical planning to action planning leading to planning tools such as urban reference plans, urban audit plans, urban contract plans and urban grid plans, all for purposes of implementing urban projects. Next to action planning is strategic development planning, with its strengths and weaknesses, and finally there is a move to localizing the Millennium Development Goals (MDGs). For Anglophone Africa the successive approaches are comprehensive master planning, action planning, structure planning, strategic planning, community planning, sustainable cities programme (SCP), city development strategies (CDS), and an infrastructure-led development approach. Reference [12] claimed that a more comprehensive urban-rural inter-linkages perspective or regional planning approach to planning is being advanced and promoted courtesy of UN-Habitat contributions. He further claimed that many countries have imbibed the new perspective, although this is not very visible in literature. Moreover, a lot depend on the nature of linkage in anticipation because linkages could be backward and unproductive. Such linkages are based on survivalist objectives which UN tend to advocate for the so-called 'poor environments' found in Africa.

As a result of these trends, thematic treatment of planning is the vogue, hence the emergence of sectoral planning. The subject matter of planning transits to poverty issues, thus pro-poor planning coupled with the consideration of informality serve as core characteristic features of planning initiatives. In the circumstances, independent nations tend to find their own synthesis depending on their local conditions, however, seldom with regard to regional (territorial) integration. Most of the planning initiatives are driven by the prevailing neo-liberal planning theory although the resilience of formal planning theory is noticed and it is remarkably acknowledged by UN-Habitat. The economic basis of growth which neo-liberal planning theory seeks further redefines the sustainability notion of green planning. In the circumstance, green planning gradually assumes the status of an economic concept,subject to the economic goal of sustainability.

4. Spatial Planning in Africa Under Professional Design Tradition

There are basically two dimensions of spatial planning in Africa and they are; urban design which defines the urbanity of cities and regional integration otherwise spatial distribution of development projects which defines the space economy. Both dimensions work complimentarily in pre-modern planning as a unified activity under local authority. In modern planning their determinant factors tended to polarize under globalization. The former is guided by planning rationality while the later is guided by market force. Planning rationality aligns with urban planning and design which deals with land use management and identifies with the traditional notion of green planning. In this case spatial equilibrium is the focal point therefore spatial planning is synonymous with urban planning. On the other hand, spatial planning which defines the space economy theoretically addresses the economic and spatio-physical bases of development. In practical terms, since the modernist planning period, spatial planning seems not to address spatio-physical bases of development. Its attention increasingly focus on the

economic basis of development, hence it is concerned more with spatial determinism in economics otherwise environmental economics. It therefore connects with the sustainability notion of green planning. These seemingly divergent tendencies influence current thoughts about spatial planning in Africa.

In a general reassessment of urban planning in African cities [12] master planning prevailed in the 1930's-1960's, followed by disjointed incrementalism in 1970's, structural planning in 1980's and lastly action planning since 1990's. With the inception of colonial urbanization in mid 19th century until the independence decade in the 1960s the instruments for urban design in Africa, precisely Anglophone Africa, were imported as direct product of professional design tradition in urban planning in Europe; especially British town planning laws, the Town and Country Planning Act of 1947. A different approach somehow applied in Francophone Africa where colonial authorities fabricated planning instruments insitu given circumstantial conditions. According to [13] both approaches focused on legislative provisions: …for the planning of regions, district and local areas, development control, subdivisions and consolidations, acquisition and disposal of land with the objective of conserving and improving the physical environment and in particular promoting health, safety, order, amenity, convenience and general welfare, as well as efficiency and economy in the process of development and improvement of communications; authorization of the making of regional plans, master plans and local plans, whether urban or rural; the protection of urban and rural amenities and the preservation of buildings and trees and generally to regulate the appearance of the townscape and landscape; the acquisition of land; the control over development including use, of land and buildings; regulation of subdivision and the consolidation of pieces of land; and matters incidental to or connected with the foregoing.

Planning legislation was backed-up with planning standards, housing standards, building codes, land acquisition acts, and scores of by-laws relating to model building, health, hawkers/vendors, shop licensing, liquor, and premise. Other back-up provisions include General Development Order/Interim Development Order, labor relations, public health act, local government act, urban council, district council, and rural councils act, local government finances act, local government service act, mines and mineral act, etc. [13].Admittedly this represents a fairly comprehensive list of legal provisions but remarkably it de-emphasized provisions for green conservation and culture. The focus was on developing the city as engines of growth and not as crucibles of development which is culture specific.

Master plans prepared within the first half of 20th century especially by French architects under the influence of Le Corbusier are ubiquitous. Within the same period Japan experimented directly with imposed master planning and Western urban forms in what were then its own colonies of Taiwan, Korea, China and Manchuria [14].India and Latin American cities involvement is outstanding.

The national governments inherited the colonial planning instruments. By 1970s the master planning instrument had undergone severe criticisms. There are two shades of criticism; those against master planning instrument and those against modernist planning which engage master planning instrument. The criticism against master planning dates back to the mid-20th century in Europe and in mid-1960s and early 1970s it filtered to Africa. In Europe critics questioned the validity of MPA in a pluralistic society. In Africa critics claim that it is not indigenous hence non-compliant with local institutions and that it is restrictive and exclusionary. The planning standards are also held to be restrictive, imperial, inflexible, non-standardized, and unrealistic. Similar criticism applies to development control, which in addition to the already mentioned shortcomings is regarded as negative and reactive. In some instances there are reservations about the adequacy of the statutory provisions. For example land use acts tend to create land administration problems; mines and minerals act usurps development control; General Development Order conflicts with by-laws; and there is high ignorance and corrupt practices in the application of by-laws.

The nationalist governments applied national development plans which adopted the notion of spatial planning based on the spatial distribution of development projects. A variety of these national development plans focused mainly on regional economic development. These development plans have spatial content such as those of Nigeria in the 1970's to 1990's before the rolling plan periods. They contest for position with predating UN-Habitat alternatives. Most of the UN-Habitat models (IUDIP, PEDP, etc) were introduced as follow-up to the launch of neo-liberal theories of development in Africa in the 1980's. Meanwhile traditional master planning continues in several contexts [17].

In early 1980s democratization of the planning machinery ensued and attention focused on community participation. In Francophone Africa the period 1960-1990 marked the municipal administration phase when decentralization slowed down after an initial fast take off at the end of Second World War. Decentralization process revived in 1990-2000 period, which marked the urban management phase when planning consultancy peaked off, and 2000 until now, seen as local development phase, marked the passage of rhetoric on decentralization to actually putting in place local development mechanisms involving various categories of stakeholders.

It is therefore not uncommon that urban institutional and regulatory frameworks that are used in several Anglophone African countries and indeed preferred by central or state level governments are multi-purpose bodies or organs, variously called "Urban Development Corporations", "Urban Development Boards", "Urban Development Authorities", "Planning Authorities", etc. – which had since national political independence been used with significant positive effects and outcomes [12].These organizations are held to effect the envisaged integrated city development, ensuring that planning and development decisions are made in the overall public interest. It is however recognized that public interest may not necessarily always be the same as the interest of all.

Indeed, the urban planning process began to involve, admit of and be affected by a wider variety of participant actors including from the government (central, state/regional and local levels), community and neighborhood associations as well as other civil society stakeholders and interest groups. Considering the weak status of civil society in Africa inclusive participatory planning is more or less induced as a fall out of international development discourse. The discourse built support for the new planning methodology with "subsidiarity principle" a concept which advocates that "decisions should be made and services provided at the lowest level that is cost-effective without creating too many over-spill effects" [12].

In Anglophone Africa the new process, which is experiencing difficult to pick up, tends to significantly slow down planning decisions, administrative and delivery processes and participation is compelled by lack of political will and logistics to be limited to consultation. In Francophone Africa collaborative planning approaches actually started to evolve in the 1960s.The new approaches are technocratic and normative, consulting the beneficiaries is nonetheless mandatory [15].However, the welfare state took responsibility for planning and relegated people's participation to the background.

Current trend indicate the review and revision of most regulations with the aim of creating enabling rather than restrictive environment. A lot of reviews have already taken place especially with planning laws but pragmatic application is still lacking as in Nigeria with the new Urban and Regional Planning Decree 88 of 1992.Most reviews take the form of capacity building and reorientation that build on existing theoretical foundations, however they seek decentralization and democratization of planning decisions. The prime motive is to accommodate the informal sector in urban planning and to make planning instruments implementation-oriented. This is considered an imperial agenda, considering the remote causal factors of the informal sector which are connected with the installation and sustenance of extroverted space economy in Africa.

Urban planning practice in Africa is yet to step into the arena of urban sustainability that is built on design-oriented approach to urban resource management. This feat is not being addressed with the incidence of 'architectural approach to planning' which provides contemporary critics ([16], [17], [18]) foundation for their continual criticism against master planning instrument. Rather than urban sustainability attention is focused on environmental sustainability which indicates preference for environmental action plan. Therefore plans are encouraged to be either sector specific or project specific or specific to levels of intervention that is national, regional or urban level. These approaches are given multiple variants of adjectival qualifications that are sometimes confusing. In fact some of the new ideas via-off the professional mandate of spatial planning practice. A typical example is the Environmental Planning and Management

(EPM) approach of UN-Habitat sustainable cities program. What transpired in reality in EPM is nowhere near the green planning that is associated with sustainable development. It was more or less pro-poor planning efforts, which were appraised by the UN to be unsuccessful because it did not resonate with local institutions [13]. .

At least in theory, international criticism of master planning generated some dynamics in the planning system of some African countries. In South Africa, with a relatively developed planning system in the region, fortified with scores of policies, guidelines and legislation aptly summarized by [19], measures consistent with current planning thoughts were adopted. The overall objective was to alienate master planning. Master plans which prevailed in pre-independence period [20] transited to strategic plans in the form of 'guide plans' and later 'structure plans' to manage the overall growth of areas [21].After 1994 it got to strategic spatial planning. Eventually from 2000, spatial frameworks were required as an element of statutory integrated development plans (IDPs): strategic plans intended to guide the work of municipalities [22].What transpired in practice in early 2000 and apparently until now was heavily criticized to be below expectation ([23], [24], [25], [18], [19].

As it is consistently the case, detailed end-state planning continued to direct land use change in spite of spatial development planning that was theoretically in vogue. Reference [22] reaffirmed this position in the analysis they provided for Ekurhuleni metropolitan municipality, South Africa. Internationally, in the 1980s, cities in China and East and South-east Asia hitherto without institutionalized planning system adopted master planning amidst the contemplative scenario for new innovations in urban planning. Remarkably Singapore and Hongkong within the Asian bloc have long standing and successful experience with master planning. The new entrant China was formally rehabilitated with the City Planning Act of 1989, which set up a comprehensive urban planning system based on the production of master plans to guide the growth of China's burgeoning new cities [26][27].Naturally they adopted the new master planning approach which emphasized concern for implementation.

Summarily there are about five discernable features that mark urban planning in Africa. They are, the resilience of informality in the transition from traditional (pre-colonial) to modern (colonial) urban planning, the evolution of legislation in urban planning, poor relationship between rhetorical and practical meaning of urban planning concept, the controversy of master planning, and the commitment of external assistance agencies to sponsor new approaches to urban planning or what could be termed 'neo-liberal urban planning' approach in Africa. The ambivalence of the African society towards these new approaches to spatial planning cannot be ignored. However, the new approaches bear overlapping influence that makes it difficult to discern the direction of urban planning in Africa. The tendency is for individual nations to find their own synthesis for urban planning depending on their local conditions but with little regard for

regional integration. Except perhaps for South Africa, common impression indicate the incidence of declining performance of urban planning and the erosion of its relevance in the scheme of national development. In the whole scenario it seems a fundamental misconception of urban planning exists where urban planning and project planning tend to be perceived interchangeably and used synonymously. This reflects in the mixed-bag of development of contemporary African cities, a phenomenon that is responsible for the inherent process of sprawling urban growth, urban crime, poor environmental quality, declining productivity, and dysfunctional infrastructural and activity systems, etc.

5. Contribution: Urban Growth Boundary Instrument for Green Planning in Africa

The sustainability notion of green planning is a redundant concept in Africa given the poor performance of the SCP/EPM initiative and several other neo-liberal planning initiatives directed at sustainable development. It is not clear how this notion performs in developed countries where quality urban design holds sway for urban sustainability. Most of the principles of urban design concerning green planning draw from the traditional notion. This is why green planning still relates to forest reserves, recreation areas, greenbelts, national parks, zoological gardens, etc. The phenomenon of smart growth, Greenfield and Brownfield development comes into play as spatial planning serves as conservation instrument.

Growth boundaries play a determinant role in traditional urbanism, both for developed and developing countries. In Africa, growth boundaries were conceptualized as town walls. The town wall took many forms and it is used for many purposes especially delimiting Brownfield development. This instrument in urban design played a formidable role in the traditional notion of green planning. Today, the same town wall concept is re-invented as Urban Growth Boundaries (UGBs) and applied as best practice mechanism for urban growth management. Reminiscences of this ancient practice hold lessons for green growth in contemporary planning.

The traditional form of UGB was 'Green Belt'. In developed countries, Britain pioneered this practice in mid-20th century as a reaction to finding planning solution to spatial distortions in economic land use responsible for urban productivity decline. Simply the rationale for 'Green Belt' in developed countries is spatial growth-related. Britain adopted the 'Green Belt' and UGB concepts that provide limits for urban expansion. As a matter of fact England is regarded as home of 'Green Belts' and UGB. London is surrounded by a boundary and a 900 square mile Green Belt. Copenhagen is also surrounded by a boundary and 'green wedges' of open space.

Majority of traditional African cities, ranging from the city-states of the southern Sudan empires to the cities of forest area empires of West Africa, adopted town walls. Town walls of great length (Kano, 22km; Ibadan, 16km; Old Oyo, 25km) identified these city states around 15th century to 18th century.

The boundary wall for towns was mud structures and for villages boundary wall was matting or corn stalks. Until the 18th century the boundaries of cities of the coast of East Africa were poorly constructed because the inhabitants relied on cooperation with the hinterland, defense was not a critical issue. For inland East Africa the towns of traditional Chwezi kingdom were surrounded by defensive ditches sometimes cut right into the bedrock, and within the towns the chief's house and cattle Kraal were also surrounded by ditches [2].In Central Africa as in Ibo-land of West Africa the subconscious formality of invisible boundaries recognized by the inhabitants was a common feature in the axial plan of cities. Some cities of the Sudanese empire such as Kumbi were surrounded with the moat and some others especially in Zululand under Shaka in southern Africa was surrounded with barren 'buffer zone' deliberately created to work against penetration by migrants.

Town boundaries commonly defined with town wall should not be mistaken for internal walls referred to as urban boundaries that are manifest in contemporary cities. Both features are conceptually and mutually different, although a relationship is being observed between the two concepts. While internal walls or urban boundaries explain divisions or boundaries within the urban fabric, town walls or town boundaries define the limits within which urban activities are confined. In other words boundary walls circumscribe the perimeter boundary or limits of traditional settlements. This limit was not provided entirely then by constructed mud walls. In most cases the mud walls were used to provide protection for the exposed flanks of the settlement which was not covered by natural barriers.

However, irrespective of the features that make up these boundaries the town boundaries were intended to provide defense against threats of predating animals and human enemies. Also it gave identity to its inhabitants and protects the authority of sovereign city states. Beyond physical protection provided by the sense of enclosure, the walls provide psychological reassurance. Hence town walls were more than bricks and mortar; they were in effect boundaries between two worlds responsible for 'them and us' syndrome associated with social relations in traditional cities. This is responsible for the cause-effect relationship that tend to exist between town walls (external boundaries) and urban boundaries (internal boundaries) in which case the presence of town walls cause the disintegration of internal boundaries.

The rationale for town walls in Africa relates to territorial definition, defense mechanism and limits of urban activities. Transition from walled cities occurred with the inception of modern urbanism when automobile development assumed prominence as dominant morphological factor that influence urban form. Cities could now spread up to twenty or thirty kilometers outwards depending on available technology. The growth factor which informs the sustainability notion of green planning lacks the spatial dimension which the rationale for UGB instrument shares with town walls. Sustainability and UGB are rarely related in literature but their relationship could be conveniently implied.

Irrespective of sustainability, the UGB concept is currently being applied in America, Europe, Middle East, North and South Africa. ASEAN countries are favorably disposed to borderless cities. East and West African countries seem to be non-aligned. The UGB concept was introduced in South Africa in the 1970's by the Natal Town and Regional Planning Commission of the Province of Natal (now known as KwaZulu-Natal) in the regional guide plans for Durban and Pietermaritzburg. The concept was at that stage termed an Urban Fence. The urban fence strategy was incorporated in the Integrated Development Plan that is required for all local authorities in South Africa. This plan would as one of its components include a Spatial Development Framework plan which would normally, certainly for the larger metropolitan areas, indicate an Urban Edge beyond which urban type development would be severely limited or restricted (Metropolitan Durban - Draft Guide Plan, Natal Town and Regional Planning Reports Volume 28, 1974.).However,[28] in his study on why South Africa continues to build unsustainable cities indicated that UGB concept cannot be applied in South Africa. The North African experience is barely recorded.

6. Conclusion

The primary challenge facing Africa is to stabilize its urban system and in the process introvert the space economy to localize productivity. Therefore spatial planning needs to be guided by spatial integration targeted at redressing distortions in the urban region, which are responsible for urban productivity decline. Paradigm shift is inevitable to mobilize operations in this direction, but not in the sense of neo-liberal planning. Neo-liberalism itself will be subject to change to an alternative development ideology that is compliant with the objectives of African renaissance. Neo-mercantilism contends for this position, considering the epistemology of African civilization. It will serve as thinking instrument for paradigm shift in planning, which will favor territorial planning. This is top in the agenda for further research. The current drift towards neo-liberal participatory planning which is project oriented is antithetical to green planning in Africa. It seldom supports green growth and cannot be used to establish green infrastructure network - an activity that exists at the realm of (re)modeling the urban form. According to Amundsen, [29] green infrastructure is both a process and product, presumably in the same manner with formal spatial planning for Africa.

Africa cannot afford to alienate spatial planning in favor of increased emphasis on public participation and consensus in planning whereby the wishes of individual, small groups and the popularity of politicians shape urban destiny. So far there has been an over-reaction on the part of urban planners leading to excessive participation and the neglect of physical planning. Africa must learn from quality urban design practiced by developed countries. The basic requirement now is the revision of current trend towards framework planning. This contribution postulates the use of urban growth boundaries (UGBs) to capacitate master planning as a strategic move towards managing Greenfield and Brownfield development in Africa. The overall effect of reworking the space economy is the ultimate goal. Policy reforms are therefore imperative to re-instate *inter alia* the traditional notion of green planning in the global concept of sustainable urbanism. There is no gainsaying that the sustainability notion of green planning is arguably an illusion in neo-liberal urbanism in Africa.

References

[1]   C. B. Schoeman,Urban planning and the interface with environmental management and transportation planning.Series H: Inaugural Address: NR. 2010, pp. 1-40.

[2]   B. Davison, and The Editors of Time-Life BooksAfrica KingdomsTime-Life Books, New York. 1966, p.37

[3]   C. Rakodi, 2 Global forces, urban change, and urban management in Africa. (In Carole, R. ed The urban challenge in Africa: Growth and management of its large cities. United Nations University Press. TokyoNew YorkParis. 1997

[4]   P. Jenkins, African cities: competing claims on urban land.(In Locatelli, F. & Nugent, P.edsAfrican cities; competing claims and urban spaces.Brill. Leiden.: 2009, pp81-107.

[5]   J. Arbury,From urban sprawl to compact city – an analysis of urban growth management in Auckland.(Unpublished Thesis Auckland University, Auckland) (http://portal.jarbury.net/thesis.pdf)2005Date of access: 14th May 2011.

[6]   K. Attahi; D. Hinin-Moustapha, &K. Appessikaa,Revisiting urban planning in the Sub-Saharan francophone Africa. (Being a regional study prepared for revisiting urban planning: global report on human settlements). 2009 Available from http://www.unhabitat.org/grhs/2009.

[7]   C. Coquery-Vidrovitch,Processus d'urbanization en Afrique.Tome I &II Harmattan. Paris. 1988

[8]   C. Coquery-Vidrovitch,The Process of Urbanization in Africa (From theorigins to the beginning of independence)African Studies Review 34,1991,pp.73.

[9]   S. Denyer,African traditional architecture: an historical and geographical perspective.Africana Publishing Company. New York. 1978, pp. 31.

[10] S. Wheeler,Planning for metropolitan sustainability.Journal of Planning Education and Research. 20.2000, pp. 134.

[11] P. Berke, &M. Manta-Conroy,Are we planning for sustainable development? An evaluation of 30 comprehensive plans.Journal of the American Planning Association. 66(1): pp. 21-33. 2000.

[12] D. Okpala, Regional overview of the status of urban planning and planning practice in Anglophone (Sub-Saharan) Africa.(Being a regional study prepared for revisiting urban planning: global report on human settlements.) 2009 Available from http://www.unhabitat.org/grhs/2009.

[13] United Nations Centerfor Human Settlement (UNCHS)Reassessment of urban planning and development regulations in African cities.Nairobi, Kenya: UNCHS.1999.

[14] V. Watson, The planned city sweeps the poor away…:urban planning and 21st century urbanization.Progress in planning. 72:153-193. 2009.

[15] UN-Habitat. Sustainable urbanization: revisiting the role of urban planning, global report on human settlement. Nairobi: UN-Habitat. 2009.

[16] A. Mabin, &D. Smit,Reconstructing South Africa's cities? The making of urban planning 1900-2000.Planning Perspectives.12:193-223. 1997.

[17] A. Todes,Urban spatial policy(In Pillay, U., Tomlinson, R. and Toit, J. du, (eds),Democracy and delivery: urban policy in South Africa.Cape Town: HSRC Press. 2006.

[18] V. Watson,The usefulness of normative planning theories in the context of Sub-Saharan Africa.Planning Theory. 1(1):27–52. 2002.

[19] E. J. Cilliers,Creating Sustainable Urban Form: The Urban Development Boundary as a Planning Tool for Sustainable Urban Form: Implications for the Gauteng City Region.South Africa:Vdm Verlag. 2010.

[20] D. Dewar, &R. Uytenbogaardt,South African cities: A manifesto for change.Cape Town:University of Cape Town. 1991.

[21] A.Mabin,Urban crisis, growth management and the history of metropolitan planning in South Africa.(Paper presented at the South African Institute of Town and Regional Planners conference of growth management and cities in crisis.Cape Town. 1994.

[22] A, Todes, A, Karam, N, Klug, N. Malaza,Beyond master planning? New approaches to spatial planning in Ekurhuleni, South Africa.Habitat International, 34:pp. 414-419. 2010.

[23] P. Harrison, A. Todes,&V. Watson,Planning and transformation: lessons from the South African experience.London:Routledge. 2008.

[24] A. Todes,Rethinking spatial planning.Town and Regional Planning. 53:10-14. 2008.

[25] I. Turok,Persistent polarization post-Apartheid? Progress towards urban integration in Cape Town, urban change and policy research group discussion paper 1.Glasgow:University of Glasgow. 2000.

[26] J. Friedmann,Globalization and the emerging culture of planning.Progress in Planning. 64(3): pp. 183–234. 2005.

[27] J. Friedmann,China's urban transition.Minneapolis, MN: University of Minnesota Press, Minneapolis. 2005.

[28] M.D. Schoonraad, Some reasons why we built unsustainable cities in South Africa. (In Strategies for a sustainable built environment). Department of Town and Regional Planning, University of Pretoria, Pretoria. August 2000.

[29] O. M. Amundsen, W. Allen, &K.Hoellen,Green infrastructure planning: recent advances and applications.PAS Memo May/June. American Planning Association.2009.

30

Socio-Economic Importance and Livelihood Utilization of Bamboo *(Bambusa vulgris)* in Imo State Southeast Nigeria

Nwaihu E. C.[1], Egbuche C. T.[1], Onuoha G. N.[2], Ibe A. E.[1], Umeojiakor A. O.[1], Chukwu A. O.[3]

[1]Department of Forestry and Wildlife Technology, Federal University of Technology, Owerri, Imo State Nigeria
[2]Department of chemistry, Federal University of Technology, Owerri, Imo State Nigeria
[3]Department of Agricultural Economics and Extension, Federal University of Technology, Owerri, Imo State Nigeria

Email address:
ctoochi@yahoo.co.uk (Egbuche C. T.)

Abstract: The study was conducted in the three Local Government Areas of Mbaise; namely Aboh, Ahiazu and Ezinihitte to assess the socio-economic importance of Bamboo (*Bambusa vulgaris*) in Mbaise. Respondents for the study were selected from the three Local Government Areas, and each of Local Government produced three Communities to give a total of Nine Communities to give a total of one hundred and eighty (180) respondents. The entire selection was by random and purposive sampling technique. The objectives of the study were to; determine the various uses of bamboo and to ascertain the contribution of bamboo in the present dispensation to the socio-economic well being of the people in the study area. The data were collected using questionnaire, oral interview schedule and field visits. Data collected were analyzed using descriptive statistics such as percentages, frequency distribution table and return on investment (ROI). The result showed that majority (78%) of the respondents were male, 58% were between the age of 41-50 years, 76% were married, 38% had FSLC education. The result further revealed that 130 of the respondents use bamboo for staking of yams, 164 for erosion control, 168 for watershed while 68 indicated using it for building construction. Equally, 68 stated using it to confer aesthetic beauty on structures, 148 uses it for thatching/fencing and barn construction while an insignificant number (26) use it to construct platform, for goats. The result further revealed that 73% of the respondents are owners as well as dealers, 29% are middlemen dealers (Major) while 7% are middlemen dealers (Minor). Analysis done on return on investment indicated that dealers on bamboo made 92k, 82k and 76k profit for every naira invested in the business of supply for building, staking of yams and fencing barn construction and thatching respectively. From the foregoing, bamboo contributes in no small measure to employment and income generation in the study area.

Keywords: Socio-Economic, *Bambusa vulgris* Importance, Livelihood, Imo State Nigeria

1. Introduction

Bamboo (*Bambusa*) is a drought resistant perennial woody grass belonging to the group angiosperm and the order monocotyledon. They are in the sub-family of *bambusoideae* and family *poacae* Forest composition was assessed in the last century in terms of the commercial value of timber.

Seldom was other forest component considered to have economic value [1]. In the 1900's, when vast area of tropical forests were exploited of timber for local use, bamboos and other non-timber products such as mushrooms snails, wildlife, vegetables of both food and medicinal plant like *Gnetum Africana, Vernonia amygdalina, Dennetia tripetaila, Gercina cola, Afriamomu meleguata, Xylopia ethiopic* etc, were considered as having no economic value and as such ignored.

Bamboo is the fastest growing timber like substance on earth, growing skyward as fast as 121cm in 24hours period and can also reach maximal growth rate exceeding one meter per hour for short period of time [2]. A Culm can reach its full maturity in a matter of two to three months which makes it one of the forests growing, highest yielding renewable natural Resources [3]. Bamboo reaches maturity in three to five years of growth compared with timber products that take a longer of period 30-50 years to mature [4]. Bamboo is widely distributed in South East, South-South and middle belt of Nigeria based on ecological disposition of these rain forest regions [5]. Here in Nigeria, Bamboo is generally called Indian Bamboo and originated from South East Asia [6].

Bamboo has made significant contributions to rural and urban livelihood apart from its ecosystems role. The single most important item of forest product used extensively in the tropical rural communities is bamboo. It plays important role in the environment, including hydrologic cycle, soil conservation and preservation of biodiversity. Bamboos are important resources (renewable resources) of economic welfare and rural development. The socio economic role of Bamboo in particular and forestry in general boils down to poverty alleviation, generation of employment, income and improved standard of living in general. The utilization of bamboo and forest resources by the rural people posses a great potential to the solution of rural poverty, unemployment and backwardness. The forest and its resources (Bamboo) together with the benefits they provide in form of income. Watershed and environmental protection enable the rural people to secure a stable livelihood. Rural people have great skills and are capable of transforming Bamboo and other forest products into simple products which can attract higher prices both within rural and urban markets [7].

In the rural areas bamboo is extensively used in making fencing, planted around ponds as water shed to reduce evaporation, carbon sequestration, making of wooden gong, used as staking materials for yams, decking of storey buildings, supporting lodging banana and plantain, building thatched houses. It is widely used by raffia palm wine tappers as ladder for climbing in some localities, used as improvised television pole. The leaves are sources of organic manure while when dried, it is used in rural households as fuel wood [8]. Sale of Bamboo generates income used to improve standard of living of rural people which significantly reduces rural poverty. Poverty has turned out to be a siege on humanity reducing the dignity and peace of man. Poverty in Nigeria is severe and widespread. It is highest in the rural areas [9].

Over the years, as a result of the increasing level of poverty, the Nigeria rural sector has been regarded as the backward sector of the economy. A vast majority of the inhabitant's rural communities are suffering from adverse environment, unemployment, poverty and disease. Rural poverty is undesirable. For millions of Nigerians rural inhabitants, life is neither satisfying nor decent. Their incomes are low, diets are inadequate, and often uncertain [10]. Hardship renders many of them redundant; malnutrition and hunger threaten their existence and survival [11]. In view of all these problems, it is important that forest exploitation/Bamboo exploitation should be done wisely to ensure its sustainability. It is however unfortunate that Bamboo resources utilization in support of rural livelihood has been characterized by abuse and misuse [12]. Wise use of this resource is closely related with food security, energy needs poverty alleviation and environmental protection. Unfortunately, most Bamboos harvested are at a rate which exceeds natural growth, so current utilization is anything, but sustainable [13], [14], Over-exploitation associated with growing human populations, destruction of tropical forests and new demand for industrial uses, especially by the pulp

and paper industry, has resulted in wide-scale decimation of Bamboo stocks, from vast forests bamboo in South and Southeast Asia at the beginning of 20th century. The greatest losses are borne by the poor, especially the rural poor, as once abundant and cheap material that provided sustenance, shelter, and income has become scarce and expensive. Truly, the present crisis in the availability of bamboo is testament to its remarkable utility [15] and [16].

Consistent increase in use pressure on Bamboo vulgaris, if unchecked could threaten the existence of the species. In this regard sources of pressure on its population need to be identified, its economic impact on the livelihood of the people and measures to mitigate its extinction. Bamboo is a resource that is closely related with food security if sold, source of energy, poverty alleviation and environmental protection. Little attention was paid to it in the past, because it was seen as not having substantial benefits in terms of generating revenue. This assumption has changed because people now realize that with proper management techniques, bamboo could be sustaining to the economy. Bamboo as a non-timber forest material can replace or use as an improvised alternative for furniture, building material, paper, ladder etc. it is a strong and reliable material for construction of scaffold by architects. The scaffold is used in building [17].

Bamboo is regarded as poor mans timber and plays a vital role in the socio-economic livelihood of the rural population [18]. Since time immemorial in human society, it contributes to the subsistence need of over a billion people worldwide [19]. Traditionally, it is used as fuel wood, rural housing (thatched house), Shelter belt, fencing, chair and various other purposes [20]. In modern days, it is used as industrial raw material for pulp and paper. Ready market exists for bamboo in the construction industry. Farmers in the yam cultivating areas of Nigeria, have high demand for Bamboo for staking vines of yams. Along the riverine areas of Nigeria, houses are built entirely with Bamboo with only raffia palm used for roffing [21]. It is source of income to the rural villages and farmers. Bamboos are plant of global interest because of their distinctive life form, their ecological importance and the wide range of uses and values to humans [22]. Above all, it is used in, making handicrafts, props for banana and plantain.

Through new techniques, Bamboo has been combined with modern materials like reinforced concrete including housing, bridges and observation towers. These architects have made a deliberate attempt to increase social acceptance of bamboo and promote its adoption as an inexpensive and environmentally friendly building materials among both the rich and the poor [23]. Bamboo are a significant structural component of many forest ecosystems and play a major role in ecosystem dynamics through their distinctive cycle of mass flowering and subsequent die – off [24]. Bamboo is fire resistant and because it can regenerate quickly, it is possible then to imagine that the rotten leaves and steams add humus to the soil and could be principal sources of mulch for maintaining and improving soil fertility [25]. In the urban and rural areas, restaurants are usually constructed with bamboo to enhance their beauty in some five star hotels like Sheraton; NICON Hilton Abuja, Nigeria, Bamboos are used as ornamentals. Bamboo is an exquisite component of landscape design. For the human environment, Bamboo provides shade, acts as windbreaks, provides a aesthetic value, has anti-erosion property that sticks the soil together along fragile river banks of deforested areas and in places vulnerable to erosion menace. The litter (Leaves) make fodder for animals. Bamboo is mystical. It is planted as a symbol of strength, flexibility, tenacity, endurance and compromise, and has for centuries been integral to religious ceremonies arts music and daily life. In the music industry it acts as flute or xylophone which creates some of the most beautiful sounds in music.

2. Methodology

The study was carried out in Mbaise, Imo State Southeast Nigeria. Mbaise is a large area of three Large Local Government areas namely: Aboh Ahiazu and Ezinihitte Local Government Areas. It is located in the rain forest area and is densely populated with a population estimate of 1 million people [28]. It is bounded in the North by Ehime Mbano, in the South by Isiala Ngwa in Abia State, in the East by Obowo and in the West by Owerri North and Ngor Okpuala LGA's. the people are great farmers producing farm / agricultural produce like yams, cassava, maize, palm oil, plantain, vegetables, Livestock, poultry, fish etc. the people have rich cultural heritage. Mbaise is bounded by latitude $5^0 25$ and $5^0 45$ N and longitudes $6^0 58$ and $7^0 10$ E [29]. Two distinct seasons are identified – rainy (wet) and dry season. Most of the mean annual rainfall of about 2152 mm occurs during the wet season, April to October, and is associated with moisture – laden maritime southwest rain bearing winds from the Atlantic Ocean. The temperature ranges between 23^0c to 26^0c which creates an annual relative humidity of75% with humidity reaching 90% in the rainy season. There is a period of dry spell called, "August break" which last for one week. The dry season experiences two months Harmattan from December to late February. The hottest months are between January and [30]. The people of Mbaise are highly educated which makes education the highest industry of the people. Respondents for the study were chosen from the three Local Government Areas. Random and purposive sampling techniques were used to select respondents. Three (3) communities were randomly selected from each local government area (LGA) to get Nine (9) communities. From the nine communities, twenty (20) respondents were selected on purposive basis from a list submitted by the community leader from each LGA. The sample size is one hundred and eight (180) respondents for the study. The data for the study were collected using questionnaire, oral interview schedule and field visits. The data were analyzed using percentages, return on investment frequency distribution table.

3. Results and Discussion

3.1. Socio-Economic Characteristics of Respondents

Table shows that 78% of the respondents are males while 22% are females. The same table shows that 8.99% of the respondents are within 21-30 years, 26.67% are 31-40 years of age, 57.78% are 51 years of age and above while 6.67 falls within 41-50 years. Analysis shows that 24% of the respondents are single while 76% are married. Table 1 equally shows that 37.78% have educational qualification of First School Leaving certificate. 35.56% O'Level, (WAEC/GCE/NABTEB), 8.89% OND, 2.22% have Higher National Diploma (HND), 13.33% B.Sc and other degree while 2.2% have no formal education.

Table 1. Distribution of Respondents according to socio-Economic characteristics

Sex	Frequency	Percentage (%)
Male	140	78
Female	40	22
Total	180	100
Age		
21-30	16	8.88
31-40	48	26.67
41-50	12	6.67
51 and above	104	57.78
Total	180	100.0
Marital status		
Single	44	24
Married	136	76
Total	180	100
Qualification		
FSLC	68	37.78
O'Level	64	35.56
OND	16	8.89
HND	4	2.22
B.Sc	24	13.33
No formal Education	4	2.22
Total	180	100

Source: field data, 2014

3.2. Various Uses of Bamboo (Bambusa vulgaris)

Table 2. Distribution of Respondents according to the various uses they made of Bamboo.

Bamboo uses	Number of users
Staking of yam	130
Erosion control	164
Watershed	168
Architectural building/construction	108
Aesthetic decoration	68
Fuel wood	68
Local furniture making	140
Construction of platform/pallet for goats	26
Thatching/fencing	148
Barn construction	168

Sources: field data 2014

The various uses of Bamboo (*Bambusa vulgaris*) were shown in table 2. From the analysis, 130, 164, 168 and 68 indicated using Bamboo for staking of yam, erosion control, watershed, and fuel wood respectively. 108 pointed out using it for building construction while 140 indicated using it for local furniture making. Equally, 68, 148 and 168 indicated using it as aesthetics, thatching and barn construction. An insignificant number (26) indicated using it for construction of pallet/platform for goats. Apart from these bamboo helps to maintain ecosystem balance such as clean air through carbon sequestration, as well as maintenance of biodiversity and attractive landscape.

3.3. Multiple Responses: Dealers on Bamboo

Majority of the dealers (132) or 73% are owner dealers. They sell to middlemen who buy and supply to end users for building construction (decking of houses/storey building), especially in urban areas where they are extensively used for architectural construction. The middlemen are into full time business as they invade the rural area and buy these materials in large quantities and supply to builders. Also 36 or 20% of dealers buy from Bamboo owners and sell to yam farmers for staking of yam. 12% or 7% of the respondents supply to people who use it for thatching residential fencing and barn construction.

Table 3. *Distribution of respondents dealers According to uses of bamboo*

Attribute	Frequency	Percentage (%)
Owner dealers who sell to middlemen who purchase and resell to builders	132	73
Middlemen who sell to end users for thatching, fencing and barn construction	12	7
Middle men who sell to yam farmers for yam staking	36	20
Total	180	100

Source: field data, 2014.

Return on investment on bamboo accruable to dealers: It can be seen from investment analysis that investment on bamboo has a high return as much timber forest products can yield. This obviously revealed that dealers on bamboo made 92k and 82k profit from every Naira investment in the business of supply for building and staking of yams. The amount from supply for building is higher because it is an all season business while yam staking is season barred. Return on fencing, Barn construction and thatching of 76k is low because these activities are equally seasonal. From the foregoing, Bamboo Contributes in no small measures to employment and income generation to producers and dealers (Middlemen) in both rural and urban areas.

Table 4. *Table return on Bamboo Investment*

Attribute	ROI for Dealers	ROI f or Middlemen
Owner dealer	90.51	70.09
Major Middlemen	81.53	52.60
Minor Middlemen	76.24	90.38

Source: Field survey 2014

4. Conclusion

Based on the result of the findings, Bamboo plant and its products have made significant contribution to the socio-economic well being of Mbaise people. Bamboo is the fastest growing wood like substance on the planet. There is growing consensus that non-Timber forest products are not only crucial to ecosystem but also invaluable in Mbaise and environ. Bamboo is extensively used in the study area for yam staking, fencing, architectural building, construction of yam barns etc. As a result of the increasing demand, more lands should be made available for Bamboo Planting. This can be done in marginal soils that may not support growing of arable crops.

References

[1] Kigomo, B. (2007): guideline for growing Bamboo, Kenya Forestry Research Institute, KEERI pp 45.

[2] Abd. Latif, M (1993) Effects of age and height of three Bamboo species on their machining properties. Journal of tropical forest science 5 (4): 528-525,000.

[3] UN (1972): Sharing experiences of incentive measure for conservation on biological diversity, third meeting Beunos Aires.

[4] Lee, A.W.C, B. Xuesong and N.P Perry (1994) selected physical and mechanical properties of giant timber Bamboo grown in South Carolina. Forest production journal 44 (9): 40-46.

[5] RMRDC (2004); Bamboo occurrence and utilization in Nigeria Raw Materials Research and Development Council Publication, 1994.

[6] Balakrishman Nair, N (1990) forward. In Bamboos current Research (I.V. Ramanuja Rao; R. Gnanabaran and C.B. Sastry, Eds). Proceeding of the International Bamboo Workshop held in Cochin, India from Nov. 14-18, 1988. Pp viii.

[7] Ibe, A. E (2013): sustainable forest Resources management for Rural Livelihood and food security in Imo State, Nigeria. In: research for Development (R4D) Responses to Food security and Poverty reduction in Africa. A festschrift: Chigozie Cyril Asiabaka Braima D. Jame (Eds) pp 239-252.

[8] Canagarajah, S., Ngwafor J., and Thomas S. (1997), Evolution of Poverty and Welfare in Nigeria; 1985-1992, World Bank Policy Research working Papers 1715 Washington D.C. pp 23.

[9] Canagarajah, S., Ngwafor J., and Thomas S. (1997), Evolution of Poverty and Welfare in Nigeria; 1985-1992, World Bank Policy Research working Papers 1715 Washington D.C. pp 23.

[10] Ikojo, H. A. (2001). The Role of Forestry in Ameliorating Environmental Problems in Rural Communities in Nigeria. In: Popoola, L., Abu, J. E and Oni P. I. (Eds) Proceedings of the 27th Annual Conference of Forestry Association of Nigeria held at Abuja, FTC between 17th-21st September, 2001 pp 116-128

[11] Ibe, A. E (2013): sustainable forest Resources management for Rural Livelihood and food security in Imo State, Nigeria. In: research for Development (R4D) Responses to Food security and Poverty reduction in Africa. A festschrift: Chigozie Cyril Asiabaka Braima D. Jame (Eds) pp 239-252.

[12] Ibe, A. E (2013): sustainable forest Resources management for Rural Livelihood and food security in Imo State, Nigeria. In: research for Development (R4D) Responses to Food security and Poverty reduction in Africa. A festschrift: Chigozie Cyril Asiabaka Braima D. Jame (Eds) pp 239-252.

[13] IFAR/INBAR (1991). Research needs for Bamboo and rattan in the year 2000. Tropical tree crops Program. International Fund for Agricultural Research/International Network for bamboo and Rattan Singapore. (IFAR / INBAR, 199, Tawari, 1992),

[14] Tewari, D.N. (1997): A monograph on Bamboo. International book distribution Dehra dun (India).

[15] Sastry C.B. and Webb, D. (1990) Preface in Bamboos current research (iV. Ramanuja Rao, R Gnanabaran and C.B Sastory, pds). Proceeding of the international Bamboo workshop held in Cochin. India Nov. 14-18, 1988.. pp ix-x

[16] Kwiyamba, S. (2005): Bamboo trade and Poverty alleviation in Ileje district pp Media Ltd pp 3

[17] Okafor, J.C. Omoridon and P.S Amaza (1994) Non-Timber forestry Products: Tropical Forestry Action Programme Study Report pp 185.

[18] Prasad, R. (1990) bamboo *(dendiocalanusstrictus)* resources of the outer himalayasand siveliks of western utter Pradesh; a conservation plea for habitat restoration, In bamboo current research (l, V, Ramanuja Rao. R Ganabaran and C. B Sastry, (els). Proceedings of the international, Bamboo Workshop held in Cochin India Nov, 14- 18 1988 pp 34 33

[19] Raizda, J.S. (1986): Preservation treatment of green Bamboos under low Pneumatic pressure treatment pp 21.

[20] Sharma, S.N. and M.I Medra (1980). Varistin of specific gravity and tangential shrinkage in the wall thickness ofbamboo and its possible influence on trend of the shrinkage moisture content characteristics. Ind for Bull 259.

[21] Clark, L (1997): Bamboos: the centre piece of the grass family. In Chepman, G.P (Eals): the Bamboos, Linnean Society symposium series Number 19, Academic Press London, UK pp 501-512.

[22] Guitirrez, J. (2000): The development of non-traditional Bamboo technologies in Costa Rica" pp 2-6.

[23] Keeley, C.N. and Bond, I.D (1999): Agro forestry in the content of Land cleaning and conservation. Academy of Science.

[24] INBAR (2000): the Tropical tree crops and their uses. International Network Research.

[25] Limon W (1992): the Microscopic examination of woody material. Waston and microscope record number 30.

[26] National Population Commission (NPC) (2006). Provisional census Figure 2006 Nigeria census

[27] Acholonu Alex, D.W. (2008). Water quality studies of Nworie River in Owerri, Nigeria' Mississippi Academy of science. Retrieved 2014-9-11.

[28] Monanu S. and Inyang, F (1975) Climatic regimes. In" Nigeria maps G.E.K Ofomata (Ed) 29-229. Ethiope Publishers House Benin.

PERMISSIONS

LIST OF CONTRIBUTORS

Wondimu Bekele
Oromia Agricultural Research Institute, Mechara Agricultural Research Center, West Hararghe Zone, Mechara, Ethiopia

Ketema Belete and Tamado Tana
College of Agriculture and Environmental Science, Department of Plant Science, Haramaya University, Dire Dawa, Ethiopia

E. U. Onweremadu, E. I. Uzor and L. C. Agim
Department of Soil Science and Technology, Federal University of Technology, Owerri Nigeria

Egbuche C. T.
Department of, Forestry and Wildlife Technology, Federal University of Technology, Owerri, Nigeria

D. J. Njoku
Department of Environmental Technology, Federal University of Technology, Owerri, Nigeria

A. C. Udebuani
Department of Biotechnology, Federal University of Technology, Owerri, Nigeria

Berhane Sibhatu, Gebremeskel Gebrekorkos and Kasaye Abera
Department of Agronomy, Ethiopian Institute of Agricultural Research, Mehoni Agricultural Research Center, Maichew, Ethiopia

Hayelom Berhe
Land and Water Research Process, Ethiopian Institute of Agricultural Research, Mehoni Agricultural Research Center, Maichew, Ethiopia

E. U. Onweremadu
Department of Soil Science and Technology Federal University of Technology, Owerri, Nigeria

A. C. Udebuani
Department of Biotechnology, Federal University of Technology, Owerri, Nigeria

Egbuche C. T.
Department of Forestry and Wildlife Technology, Federal University of Technology, Owerri, Nigeria

Ndukwu B. N.
Department of Agricultural Extension, Federal University of Technology, Owerri, Nigeria

Egbuche C. T., Nwaihu E. C. and Umeojiakor A. O.
Department of Forestry and Wildlife Technology, School of Agriculture and Agricultural Technology, Federal University of Technology Owerri, Nigeria

Zhang Jia'en
College of Agriculture, South China University of Agriculture Guangzhou, China

Okechukwu Ukaga
College of Food, Agriculture and Natural Resources, University of Minnesota Extension USA

Simona Rainis
Crita S.c.a.r.l. - Research Center for Technological Innovation in Agriculture, Via Pozzuolo, Udine, Italy

Ennio Pittino and Giordano Chiopris
Ersa - Regional Agency for Rural Development of Friuli Venezia Giulia, Via Del Montesanto, Gorizia, Italy

Umeojiakor A. O., Egbuche C. T., Ubaekwe R. E. and Nwaihu E. C.
Department of Forestry and Wildlife Technology, Federal University of Technology, Owerri, Nigeria

Nkwopara U. N.
Department of Soil Science and Technology, Federal University of Technology, Owerri, Nigeria
College of Resources and Environment, Key Laboratory of Subtropical Agricultural Resources and Environment, MOA, Huazhong Agricultural University, Wuhan, China

Emenyonu-Chris C. M., Ihem E. E., Ndukwu B. N., Onweremadu E. U. and Ahukaemere C. M.
Department of Soil Science and Technology, Federal University of Technology, Owerri, Nigeria

Egbuche C. T.
Department of Forestry and Wildlife, Federal University of Technology, Owerri, Nigeria

Hu H.
College of Resources and Environment, Key Laboratory of Subtropical Agricultural Resources and Environment, MOA, Huazhong Agricultural University, Wuhan, China

S. N. Obasi and U. P. Iwuanyanwu
Department of Agricultural Technology, Imo State Polytechnic, Umuagwo Imo State, Nigeria

E. U. Onweremadu
Soil Science Department, Federal University of Technology, Owerri, Imo State Nigeria

Egbuche C. T.
Department of Forestry and Wildlife Technology, Federal University of Technology, Owerri, Imo State Nigeria

Barkat Ali Kalwar, Mehmood Ahmed Kalwar and Madan Lal
Department of Nutrition and Animal Product Technology, Faculty of AHV, Science, SAU, Tandojam-Sindh

Hakim Ali Sahito and Zaibun Nisa Memon
Department of Zoology, Faculty of Natural Sciences, SALU- Khairpur- Sindh

Habtamu Assaye
College of Agriculture and Environmental Sciences, Bahir Dar University, Bahir Dar, Ethiopia

Zerihun Asrat
School of Forestry, Hawassa University, Hawassa, Ethiopia

Ihejirika G. O., Nwufo M. I., Ibeawuchi I. I., Obilo O. P., Ofor M. O., Ogbedeh K. O.,
Okoli N. A., Mbuka C. O., Agu G. N. and Ojiako F. O.

Department of Crop Science Technology, Federal University of Technology, Owerri, Nigeria

Akalazu J. N.
Department of Plant Science and Biotechnology, IMO State University, Owerri, Nigeria

Emenike H. I.
Cooperative Information Network, OBAFEMI AWOLOWO University, Ile-Ife, Osun State, Nigeria

Egbuche C. T.
Department of Forestry and Wildlife Technology, Federal University of Technology Owerri, Imo State Nigeria

College of Forest Ecology, South China Agricultural University Guangzhou, China

Nwaihu E. C., Umeojiakor A. O. and A. E. Ibe
Department of Forestry and Wildlife Technology, Federal University of Technology Owerri, Imo State Nigeria

Su Zhiyoa
College of Forest Ecology, South China Agricultural University Guangzhou, China

Anyanwu J. C.
Department of Environmental Technology, Federal University of Technology Owerri, Nigeria

Onweremadu E. U.
Department of Soil Science Technology, Federal University of Technology Owerri, Nigeria
Tahmina Siddika, Ripon Kumar Adhikary, Md. Hasan-Uj-Jaman, Shoumo Khondoker,

Nazia Tabassum and Md. Farid Uz Zaman
Department of Fisheries & Marine Bioscience, Jessore University of Science & Technology, Jessore, Bangladesh

Ihem E. E., Osuji G. E., Onweremadu E. U., Uzoho B. U., Nkwopara U. N., Ahukemere C. M.,
Onwudike S. O., Ndukwu B. N., Osisi A. S. and Okoli N. H.
Department of Soil Science and Technology, Federal University of Technology, Owerri, Imo State, Nigeria

G. M. Abdur Rahman and Champa Bati Dutta
Economics Discipline, Khulna University, Khulna, Bangladesh

Md. Jamal Faruque
Bangladesh Agricultural Development Corporation (BADC), Khulna, Bangladesh

Mehedi Hashan Sohel
Department of Soil Science, EXIM Bank Agricultural University Bangladesh, Chapainawabgonj, Bangladesh

Abu Sayed
Department of Agricultural Engineering, EXIM Bank Agricultural University Bangladesh, Chapainawabgonj, Bangladesh

Faruwa Francis Akinyele, Egbuche C.T., Umeojiakor A. O. and Ulocha O. B.
Department of Forestry and Wildlife Technology, Federal University of Technology, Owerri, Imo State, Nigeria

Luan Cilliers
Unit for Environmental Sciences and Management, North-West University, Potchefstroom Campus, South Africa

Ibeawuchi I. I. and Ihejirika G. O.
Department of Crop Science and Technology, School of Agriculture and Agricultural Technology, Federal University of Technology, Owerri, Imo State, Nigeria

Egbuche C. T.
Department of Forestry and Wildlife, School of Agriculture and Agricultural Technology, Federal University of Technology, Owerri, Imo State, Nigeria

Jaja E. T.
Department of Applied and Environmental Biology, Rivers State University of Science and Technology, Nkpolu, Port Harcourt, Rivers State, Nigeria

Nicolene de Jong
Unit for Environmental Sciences and Management, North-West University, Potchefstroom, South Africa

Zhan Goosen
Unit for Environmental Sciences and Management, North West University, Potchefstroom, South Africa

Nwaihu E. C., Egbuche C. T., Ibe A. E. and Umeojiakor A. O.
Department of Forestry and Wildlife Technology, Federal University of Technology Owerri, Imo State Nigeria

Onuoha G. N.
Department of Chemistry, Federal University of Technology Owerri, Imo State Nigeria

Chukwu, A. O.
Department of Agric Economics, Extension and Rural Development, Imo State University Owerri, Nigeria

Hildegard E. Rohr
Faculty of Natural Sciences, North-West University, Potchefstroom, South Africa

Sanmarie Schlebusch
Unit for Environmental Sciences and Management, North-West University, Potchefstroom Campus, South Africa

Elizelle Juaneé Cilliers
Unit for Environmental Sciences and Management, North-West University, Potchefstroom Campus, South Africa, Potchefstroom

Anyanwu J. C., Amaku. G. E. and Onwuagba, S. M.
Department of Environmental Technology, Federal University of Technology, Owerri, Nigeria

Egbuche C. T.
Department of Forestry and Wildlife Techhnology, Federal University of Technology, Owerri, Nigeria

Duruora J. O.
Collage of Education, Nsugbe Anambra State, Nigeria

Elizelle Juaneé Cilliers
Unit for Environmental Sciences and Management, North-West University, Potchefstroom Campus, South Africa, 2520

Okeke D. C.
Research Unit for Environmental Sciences, Department of Urban and Regional Planning, North-West University, Potchefstroom Campus, Potchefstroom, South Africa

Nwaihu E. C., Egbuche C. T., Ibe A. E. and Umeojiakor A. O.
Department of Forestry and Wildlife Technology, Federal University of Technology, Owerri, Imo State Nigeria

Onuoha G. N.
Department of chemistry, Federal University of Technology, Owerri, Imo State Nigeria

Chukwu A. O.
Department of Agricultural Economics and Extension, Federal University of Technology, Owerri, Imo State Nigeria

Index